Supply Chain Disruptions: Theory and Practice of Managing Risk

Haresh Gurnani · Anuj Mehrotra
Saibal Ray
Editors

Supply Chain Disruptions: Theory and Practice of Managing Risk

Springer

Prof. Haresh Gurnani
Department of Management
University of Miami
Coral Gables, FL
USA
e-mail: haresh@miami.edu

Prof. Anuj Mehrotra
University of Miami
Coral Gables, FL
USA
e-mail: anuj@miami.edu

Prof. Saibal Ray
Desaulets Faculty of Management
McGill University
Sherbrooke St (W) 1001
Montreal, QC H3A 1G5
Canada
e-mail: saibal.ray@mcgill.ca

ISBN 978-0-85729-777-8 e-ISBN 978-0-85729-778-5
DOI 10.1007/978-0-85729-778-5
Springer London Dordrecht Heidelberg New York

British Library Cataloguing in Publication Data
A catalogue record for this book is available from the British Library

Library of Congress Control Number: 2011938009

Cover design: eStudio Calamar S.L.

Printed on acid-free paper

Springer is part of Springer Science+Business Media (www.springer.com)

Preface

One of the most critical issues facing supply chain managers in todays globalized and highly-uncertain business environment is how to proactively deal with disruptions that might affect the complicated supply networks characterizing modern enterprises. This book presents state-of the-art perspective addressing this particular issue. The distinctive features of this book are:

(i) it demonstrates that effective management of supply disruptions necessitates both strategic and tactical measures—the former involving optimal design of supply networks, and the latter involving approaches like inventory, financial and demand management;
(ii) it shows that managers ought to use all available levers at their disposal throughout the supply network—like sourcing and pricing strategies, providing financial subsidies, encouraging information sharing and incentive alignment between supply chain partners—in order to tackle supply disruptions; and
(iii) it brings together up-to-date, methodologically-rigorous research from academicians with the latest operational risk management practices used in industry to demonstrate how academic researchers and practitioners can learn from each other.

Consequently, this book is not only suitable for students and professors who are interested in pursuing research or teaching courses in the rapidly growing area of supply chain risk management, but also acts as a ready reference for practitioners who are interested in understanding the theoretical underpinnings of effective supply disruption management techniques.

We would like to thank all the authors who have contributed to this book: Zumbul Atan, Goker Aydin, Volodymyr Babich, Natashia Boland, Atanu Chaudhari, Awi Federgreun, Kevin Hendricks, Wally Hopp, Seyed Iravani, Jussi Keppo, Walid Klibi, Gary Lynch, Alain Martel, Zigeng Liu, Romesh Saigal, Martin Savelsbergh, Amanda Schmitt, Kashi Singh, Vinod Singhal, Larry Snyder, Brian Tomlin, Adam Wadecki, Owen Wu, Nan Yang, and Fuqiang Zhang. This book comprises of 12 chapters that highlight the use of different approaches to managing disruption risk. In what follows, we summarize the key features from each chapter.

In Chap. 1, Hendricks and Singhal empirically identify four developments that have resulted in a dramatic increase in the attention surrounding supply chain disruptions in recent times. They summarize the financial consequences of disruptions and offer insights into the factors causing disruptions and through use of examples, they discuss some strategies and practices in managing the disruptions. They highlight the trade-off between efficiency of supply chains and the associated high risk of disruptions. Indeed this chapter sets the stage for the book by highlighting why effective supply chain risk management is an important issue for todays enterprises. In Chap. 2, Hopp, Iravani and Liu propose a general framework to effectively mitigate the impact of supply chain disruptions, thereby managing the associated risks. Their framework outlines prevention strategies such as systematically classifying potential disruptions and concentrating on reducing the risk of high-impact disruptions; response strategies to detect and develop swift measures to counter the threats due to disruptions; protection strategies to contain the impact of the disruptions and suggest development of a recovery plan to lessen the impact after a disruption through recovery strategies.

In Chaps. 3–5, the authors illustrate a protection strategy through effective management of inventory and procurement policies. In Chap. 3, Schmitt and Tomlin study the use of diversification to manage supply disruptions. Diversification refers to use of multiple supply sources on an ongoing basis, which provides as natural hedge should any one source becomes unavailable. They also discuss emergency backup sourcing which is as an example of a recovery strategy after a disruption has occurred. In Chap. 4, Federgruen and Yang more specifically discuss the use of diversification for procurement. They discuss the issues of identifying the number and specific suppliers from a set of potential suppliers. They also highlight the risks associated on the demand side and address the issues of how ones inventory strategy should be set in presence of simultaneous supply and demand risks and whether trade-offs between reliability and cost differentials among the suppliers can be effectively captured. In Chap. 5, Atan and Snyder discuss the optimal management of inventory systems requiring higher inventory levels beyond those that would be required in a disruption-free environment and suggest that an inventory based approach is a preferred strategy if disruptions tend to be frequent but short in duration, versus other strategies such as supply diversification which are more useful if disruptions are rare but catastrophic in nature.

Chapters 6 and 7 deal with use of financial instruments as levers for mitigating supply risk. In Chap. 6, Wadecki, Babich, and Wu discuss how manufacturers increase the reliability of suppliers by offering subsidies to reduce the risk of supply disruptions due to supplier bankruptcies. They examine the optimal subsidy decisions of manufacturers and include the competition among manufacturers and their choice of dedicated or shared suppliers. They conclude that both the manufacturer and the consumers gain when monopolistic manufacturers use a shared supplier and that when manufacturers use dedicated suppliers, the overall decreased subsidies for the suppliers make them less reliable and negatively impact the consumers who suffer from manufacturer competition. In Chap. 7, Babich et al. conclude that alternative nancing sources (internal nancing and trade credit loans) are substitutable and that the rm is inclined to use more suppliers if the internal nancing is not available. They also address the question of whether rms operating in developing economies should contract with more suppliers than rms operating in developed economies.

In Chap. 8, Zhang shows how information sharing and contractual mechanisms that align incentives among channel partners can be effective in managing supply risk. He does so using a framework that captures the increased focus on the delivery performance of the suppliers as a result of growth in outsourcing/offshoring. Since increased demand on delivery performance may potentially result in higher costs to the supplier to maintain higher capacity or inventory, the buyer needs to carefully design incentive schemes to induce the right action from the supplier in a setting where the buyer faces uncertainties about the supplier's cost structure when negotiating the supply contract. The issue of the buyer's supply (procurement) contract design problem under both asymmetric cost information and delivery performance consideration is addressed. Some simple, but suboptimal, mechanisms for the buyer that only specify a target delivery performance and do not require the supplier's cost information as an input are proposed. These simple mechanisms yield nearlyoptimal outcome for the buyer in a variety of settings.

The importance of robust design of supply networks so that they perform well even after a disruption by making additional investments in existing infrastructure has been widely noted in the literature. Chapters 9 and 10 discuss this stream in detail. In Chap. 9, Martel and Klibi highlight that the complexity of supply chain networks and their reengineering gives rise to major projects which must be carefully planned and managed. These projects must follow a comprehensive analysis and design methodology taking into account all the problem facets, and they must be supported with appropriate computer-aided analysis and modeling tools. They propose a comprehensive SCN reengineering methodology to illustrate their approach on the location-transportation problem under uncertainty. In Chap. 10, Boland and Savelsbergh present a range of models to support and automate various aspects of coal chain planning for the complex logistics of PortWaratah Coal Services (PWCS), located in Newcastle, NSW, Australia operating the world's largest coal export facility, sharing its service among around 30 mines owned by about 15 different coal mining companies in the Hunter Valley. They also discuss the challenges and opportunities to handle disruptions in an operation of this magnitude. In Chap. 11, Chaudhuri and Singh describe two case studies—one from the aerospace industry in which a risk assessment methodology was proactively developed as a part of new product program and one from the pharmaceutical industry in which the need for risk assessment was realized due to yield losses of the product, after it was launched. These case studies highlight the scope for using detailed step-by-step analysis for supplier risk assessment and control.

Through a number of real-life examples in Chap. 12, Lynch illustrates the importance of identifying, measuring, mitigating, financing, validating, and monitoring risks to reduce the negative impact of disruptions. While the first chapter of the book establishes why supply chain risk management is important, the last chapter provides a roadmap of how to implement such a program in practice, thus providing a thorough coverage of the domain.

We thank the following individuals for providing helpful reviews: Milind Dawande, Mehmet Gumus, Xinxin Hu, Xiao Huang, Sammi Tang, Navneet Vidyarthi, Ling Wang, Yusen Xia, Lei Xei, Tallys Yunes, and Dan Zhang. We would also like to thank Spinger Verlag (London) for their willingness to work with us on this project, especially, Anthony Doyle, Beverley Ford, and Claire Protherough. Finally, we would like to thank Daniel Andrés Díaz Pachón for his help in formatting the book.

University of Miami — Haresh Gurnani

Anuj Mehrotra

McGill University — Saibal Ray

Contents

1 Supply Chain Disruptions and Corporate Performance 1
Kevin B. Hendricks and Vinod R. Singhal

2 Mitigating the Impact of Disruptions in Supply Chains 21
Wallace J. Hopp, Seyed M. R. Iravani and Zigeng Liu

3 Sourcing Strategies to Manage Supply Disruptions 51
Amanda J. Schmitt and Brian Tomlin

4 Supply Chain Management Under Simultaneous Supply and Demand Risks 73
Awi Federgruen and Nan Yang

5 Inventory Strategies to Manage Supply Disruptions........... 115
Zümbül Atan and Lawrence V. Snyder

6 Manufacturer Competition and Subsidies to Suppliers......... 141
Adam A. Wadecki, Volodymyr Babich and Owen Q. Wu

7 Supply Contracting Under Information Asymmetry and Delivery Performance Consideration 165
Fuqiang Zhang

8 Risk, Financing and the Optimal Number of Suppliers 195
Volodymyr Babich, Göker Aydın, Pierre-Yves Brunet, Jussi Keppo and Romesh Saigal

9 A Reengineering Methodology for Supply Chain Networks Operating Under Disruptions 241
Alain Martel and Walid Klibi

10 Optimizing the Hunter Valley Coal Chain 275
Natashia L. Boland and Martin W. P. Savelsbergh

11 Risk Assessment of Supply Chain During New Product Development: Applications in Discrete and Process Manufacturing Industries 303
Atanu Chaudhuri and Kashi N. Singh

12 Supply Chain Risk Management 319
Gary S. Lynch

Chapter 1
Supply Chain Disruptions and Corporate Performance

Kevin B. Hendricks and Vinod R. Singhal

1.1 Introduction

Managers are becoming increasingly aware that their companys reputation, earnings consistency, and ability to deliver better shareholder returns are increasingly dependent on how well they manage supply chain disruptions. Although firms have always faced the risk of supply chain disruptions, the attention it receives has increased dramatically in recent years. This is likely driven by at least four developments. First, supply chains have become more complex due to globalization, outsourcing, single sourcing, and the focus on removing slack from supply chains. While many of these strategies have improved performance, these strategies have also made supply chains more prone to disruptions.

Second, the focus on supply chain disruptions has increased following a number of costly and highly-publicized supply chain disruptions. National and local media are filled with news reports on the increase in supply chain disruptions, and the fact that many companies are unable to cope with these disruptions. Some recent examples, include the disruptions due to Mattels recall of 21 million toys due to safety issues [6]; Boeings unexpected delay in introducing its much anticipated 787 Dreamliner because of difficulties in coordinating global suppliers [21]; and recall of contaminated meat, pet foods, and pharmaceuticals products [9, 16].

Third, academicians and practitioners are discussing the impact of supply chain disruptions on performance as well as highlighting the need to adopt practices that can prevent disruptions [5, 11, 13, 17, 19, 24]. A survey by FM Global of more than

K. B. Hendricks
School of Business & Economics, Wilfrid Laurier University,
Waterloo Ontario N2L-3C5, Canada
e-mail: khendricks@wlu.ca

V. R. Singhal (✉)
College of Management, Georgia Institute of Technology,
Atlanta, GA 30332, USA
e-mail: vinod.singhal@mgt.gatech.edu

H. Gurnani et al. (eds.), *Supply Chain Disruptions: Theory and Practice of Managing Risk*, DOI: 10.1007/978-0-85729-778-5_1,

600 financial executives finds that supply chain risks pose the most significant threat to profitability [28]. A survey by Accenture of 151 supply chain executives finds that 73% indicate that their firms experienced supply chain disruptions in the past 5 years [10]. Various studies identify drivers of supply chain risk, and develop frameworks and strategies for managing, and mitigating supply chain risk [3, 7, 8, 11, 18, 20, 26, 30, 31].

Finally, the passage of the Sarbanes–Oxley Act of 2002 makes senior executives more responsible for forecasts of performance and protection of shareholder value. This has heightened the need to identify and manage various risks, including supply chain disruptions.

This chapter addresses three issues that are critical in managing supply chain disruptions. First, it summarizes evidence from recent empirical research on the financial consequences of supply chain disruptions [13–15]. One of the reasons why many companies are not adequately prepared for responding to supply chain disruptions is that they do not have a good understanding of the magnitude and persistence of the negative consequences of disruptions on financial performance. While anecdotes make for splashy headlines, they do not provide the objective evidence that many senior executives are looking for to better understand the financial consequences of supply chain disruptions to make decisions about the initiatives and investments they should undertake to manage disruptions. The financial consequences are examined by documenting the impact of supply chain disruptions on shareholder returns, share price volatility, and profitability. Second, it offers insights into the factors that can increase the chances of disruptions to guide managers as they assess the chances of disruptions. Third, it highlights some of the strategies and practices in managing disruptions using examples from Wal-Mart, Mattel, and Boeing.

The evidence and discussion presented in this chapter is important for a number of reasons. As mentioned above, it fills a gap in the literature regarding the financial consequences of demand-supply mismatches. Supply chain disruptions are a form of demand-supply mismatches. Although the conventional belief is that supply-demand mismatches will have negative financial consequences, there is very little rigorous empirical evidence on the magnitude and severity of the financial consequences.

Efficiency, reliability, and responsiveness of supply chains are key drivers of a firms profitability. References [17, 24] suggest that much of the supply chain management efforts in the recent past have focused on increasing the efficiency (lowering costs) of supply chain operations, and less on increasing the robustness and reliability of supply chains. This could partly be because unlike efficiency, it is much harder to place a value on robustness and reliability. Disruptions are an indication that a firms supply chain is not reliable and robust. By associating disruptions with financial outcomes, we provide an estimate on the value of reliable and robust supply chain performance.

This chapter also adds to the recent research that has begun to quantify the impact of supply chain management strategies and practices on operating performance. One stream of research has focused on developing mathematical models of supply chain issues to understand how alternate ways of managing supply chains affect capital costs, operating costs, inventories, and service levels (see for example, [1, 4, 5, 22,

29]. Another stream of research has attempted to empirically establish the relationship between supply chain practices and performance. The approach used is to develop conceptual and theoretical frameworks of the drivers of supply chain performance, identify supply chain practices, use surveys to measure the intensity with which these practices are implemented, and link these to performance changes reported by survey respondents [12, 23, 25, 27]. Although significant research has been done on the relationship between supply chain performance and financial performance, most of the existing evidence is based on hypothetical or self-reported data. Hence, it is not clear how well the evidence correlates to actual performance.

The next section describes the sample, performance metrics and methodology for estimating the financial impacts. Section 1.3 presents results on the impact of supply chain disruptions on shareholder value, share price volatility, and profitability. Section 1.4 discusses the various drivers of supply chain disruptions. Section 1.5 discusses what firms can do to reduce the frequency of disruptions and mitigate the negative consequences of disruptions. The final section summarizes the chapter.

1.2 Sample, Performance Metrics and Methodology

The evidence presented in this report is based on an analysis of more than 800 supply chain disruptions that were publicly announced during 1989–2001. These announcements appeared in the Wall street journal and/or the Dow Jones news service, and were about publicly traded companies that experienced production or shipping delays. Some examples of such announcements are:

- *Sony sees shortage of playstation* 2 s for holiday season, *Wall street journal, September* 28, 2000. The article indicated that because of component shortages, Sony has cut in half the number of PlayStation two machines it can manufacture for delivery.
- *"Motorola 4th quarter wireless sales growth lower than order growth", The Dow Jones news service, November* 18, 1999. In this case Motorola announced that its inability to meet demand was due to the shortage of certain types of components and that the supply of these components is not expected to match demand sometime till 2000.
- *Boeing pushing for record production, finds parts shortages, delivery delay, The wall street journal, June* 26, 1997. The article discusses reasons for the parts shortages, the severity of the problems, and the possible implications.
- *Apple Computer Inc. Cuts 4th period Forecast Citing Parts Shortages, Product Delays, The wall street journal, September* 15, 1995. Apple announced that earnings would drop because of chronic and persistent part shortages of key components and delays in increasing production of new products.

The performance effects of the above-mentioned instances of supply chain disruptions are estimated by examining performance over a 3-year time period starting

1 year before the disruption announcement date and ending 2 years after the disruption announcement date. Two stock-market-based metrics are used in the analysis:

- Shareholder returns are measured by stock returns that include changes in stock prices as well as any dividends declared.
- Share price volatility.

The effect of disruptions on profitability is examined using the following measures:

- Operating income (sales minus cost of goods sold minus selling and general administration).
- Return on sales (operating income divided by sales).
- Return on assets (operating income divided by total assets).

To control for industry and economy affects that can influence changes in the above performance measures, the performance of the disruption experiencing firms is compared against benchmarks of firms that are in the same industry with similar size and performance characteristics.

1.3 The Effect of Supply Chain Disruptions on Corporate Performance

1.3.1 The Effect of Supply Chain Disruptions on Shareholder Value

Figure 1.1 depicts the shareholder value effects on the day supply chain disruptions are publicly announced. The effects that can be attributed to disruptions is estimated by comparing the stock returns of disruptions experiencing firms against four different benchmarks that serve to control for normal market and industry influences on stock returns.

The evidence indicates that supply chain disruptions are viewed very negatively by the market. On average shareholders of disruption experiencing firms lose:

- 7.18% relative to the benchmark that consists of the portfolio of all firms that have similar prior-performance, size, and market-to-book ratio of equity to the disruption experiencing firm (portfolio matched benchmark).
- 7.17% relative to the firm that has similar prior-performance and market-to-book ratio of equity, and is closest in size to the disruption experiencing firm (size matched benchmark).
- 6.81% relative to the firm that has similar size and market-to-book ratio of equity, and is closest in terms of prior-performance to the disruption experiencing firm (performance matched benchmark).

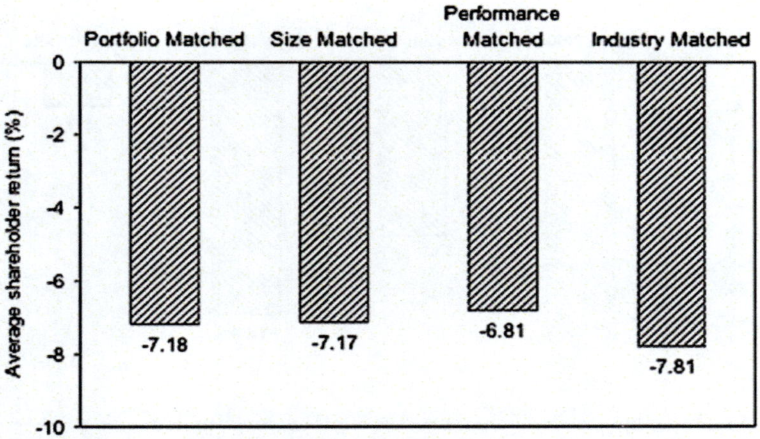

Fig. 1.1 The average shareholder return on the day information about disruptions is publicly announced. Portfolio, size, performance, and industry matched are different set of benchmarks used to estimate the relative stock price performance of the firms that experience disruptions

- 7.81% relative to the firm that has similar size, prior performance, and market-to-book ratio of equity, and is closest in terms of the industry to the disruption experiencing firm (industry matched benchmark).

When one examines the relative stock price performance during the periods before and after the disruption announcement, the shareholder value effects are much worse than those depicted in Fig. 1.1. Figure 1.2 depicts the stock price performance starting 1 year before and ending 2 years after the disruption announcement date. The stock price performance is measured relative to the portfolio of all firms that have similar prior-performance, size, and market-to-book ratio of equity to the disruption experiencing firm (i.e. portfolio matched).

During the year before the disruption announcement, stocks of disruption experiencing firms underperformed their benchmark portfolio by nearly 14%. Even after the announcement of disruptions, firms continue to experience worsening stock price performance. In the year after the disruption announcement firms on average lose another 10.45% relative to their benchmark portfolios. Although the negative trend continues in the second year after disruption, the magnitude of underperformance of 1.77% is not as high as that during the year before and the first year after the disruption announcement. More importantly, the results show that firms do not recover during this period from the negative stock price performance that they experienced in the prior two years, indicating that the loss associated with disruptions is not a short-term effect.

Figure 1.3 depicts the extent of shareholder value loss associated with disruptions over the three-year period. Depending on the benchmark used the average level of underperformance on shareholder returns ranges from 33 to 40%. One way to judge the economic significance of this level of underperformance is the fact that on average

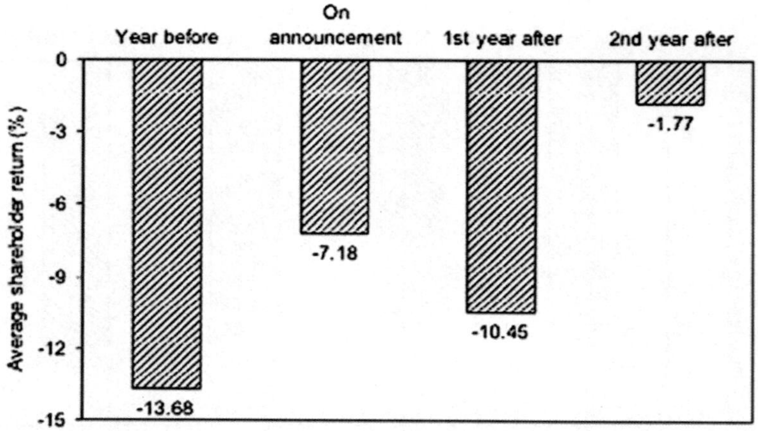

Fig. 1.2 The average shareholder returns during the year before the disruption announcement, on announcement, and each of the two years after the disruption announcement. The shareholder returns are estimated relative to the portfolio of all firms that have similar prior-performance, size, and market-to-book ratio of equity to the disruption experiencing firm

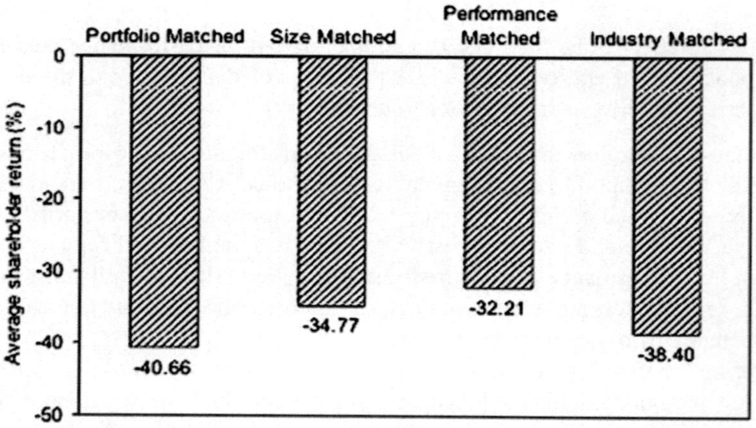

Fig. 1.3 The average shareholder returns relative to various benchmarks measured over a three-year period that begins a year before the disruption announcement and ends two years after the disruption announcement. Portfolio, size, performance, and industry matched are different set of benchmarks used to estimate the relative stock price performance of the firms that experience disruptions

stocks have gained 12% annually in the last two decades. Even if a firm experiences one major supply chain disruption every 10 years, the annual return would be close to 8–9%, which is a significant difference when one takes into account the effect compounding over long periods. Clearly, it pays to avoid supply chain disruptions. These results also underscore the importance of why senior executives must be aware of and actively involved in monitoring and managing the performance of their firms supply chain.

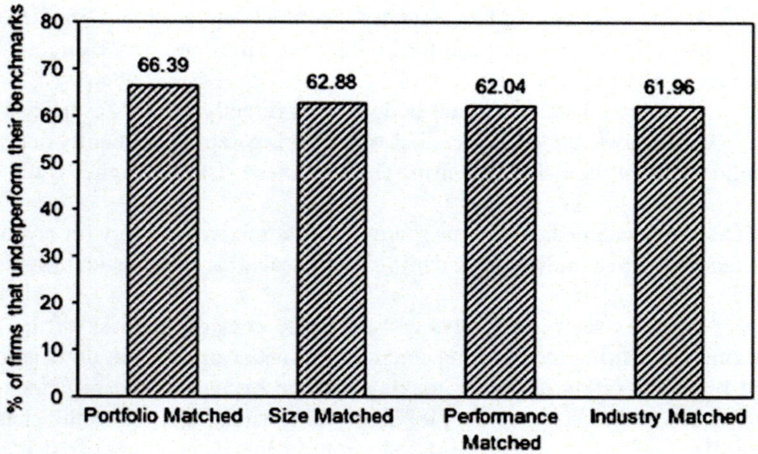

Fig. 1.4 The percent of disruption experiencing firms that underperform their benchmarks over a three-year period that begins a year before the disruption announcement and ends two years after the disruption announcement. Portfolio, size, performance, and industry matched are different set of benchmarks used to estimate the relative stock price performance of the firms that experience disruptions

The average level of share price underperformance documented in Fig. 1.3 is not driven by a few outliers or special cases. Figure 1.4 shows that anywhere from 62 to 68% of the firms that experience disruption underperform their respective benchmarks over a three-year period, which is a statistically-significant level of underperformance.

In summary, Figs. 1.1 through 1.4 indicate the following:

- Supply chain disruptions result in significant short-term and long-term shareholder value losses. Thirtythree to fourty percent stock price underperformance over 3 years is both economically and statistically significant.
- Firms that experience disruptions do not recover quickly from the stock price underperformance. Disruptions have a long-term devastating effect on shareholder value.

1.3.2 The Effect of Supply Chain Disruptions on Share Price Volatility

Supply chain disruptions can create uncertainty about a firms future prospects and can raise concerns about its management capability as disruptions indicate management inability to manage and control crucial business processes. Disruptions may also lead to questions and concerns about a firms business strategy. Disruptions could therefore increase the overall risk of the firm. Understanding how disruptions can affect the risk of the firm is important for a number of reasons:

- Risk is a critical factor used by investors to value a firms securities. Risk influences the return that investors demand for holding securities and hence directly affects the pricing of securities.
- The discount rate used in capital budgeting is directly related to the risk of the firm. Furthermore, the cost of capital when raising capital via equity and/or debt is influenced by the risk of the firm. The higher the risk, the higher is the cost of capital.
- Increased risk can make the firms shares a less attractive currency for acquisitions as potential targets may be less willing to do deals that depends on volatile share prices.
- Rating agencies such as Moodys and S&P 500 consider the risk of the firm in determining a firms credit rating. Increase in risks can result in downgrading of debt by credit rating agencies, making it more expensive and difficult to raise capital. It can also increase the probability of financial distress as the chances of the firm not being able to cover its fixed commitments increase as the risk increases.
- Risk changes can create conflicts between the various stakeholders. An increase in share price volatility transfers wealth from bondholders to shareholders, a potential source of conflict that may require management time and attention. Risk-averse employees may demand higher compensation to work for a firm that has high risk. Suppliers and customers may also be wary of dealing with the firm that has high risk and may demand some form of assurances and guarantees before doing business with the firm, thereby raising the cost of doing business for the firm.

To estimate the effect of disruptions on risk, this study compared the share price volatility before and after the disruption announcement date. Share price volatility is measured by the standard deviations of stock returns, which are estimated annually for 4 years, starting 2 years before through 2 years after the disruption announcement. To control the other factors that could affect volatility, percent changes in the standard deviation of stock returns of the disruption experiencing firms are compared against that of a matched control sample.

Figure 1.5 gives share price volatility (standard deviation of stock returns) using daily stock returns for the firms that experienced supply chain disruptions. The figure indicates that the share price volatility is monotonically increasing starting 2 year before the disruption announcement and ending 2 years after the disruption. For example, the standard deviation of stock returns in the second years before the disruption announcement was 4.13% and since then has steadily increased to 5.05% in the second year after the disruption announcement. The evidence supports the view that disruptions increase the share price volatility, and hence the risk of the firm.

One can get a better idea of the extent of share price volatility changes by comparing the change in the share price volatility of disruption experiencing firms against the change in share price volatility experienced by a control sample. Figure 1.6 reports these results. The results indicate that after adjusting for other factors that could affect share price volatility there is still a significant increase in volatility that can be attributed to the disruption. Much of this increase happens after the disruption announcement. For example, the share price volatility increases by 13.5% in the year

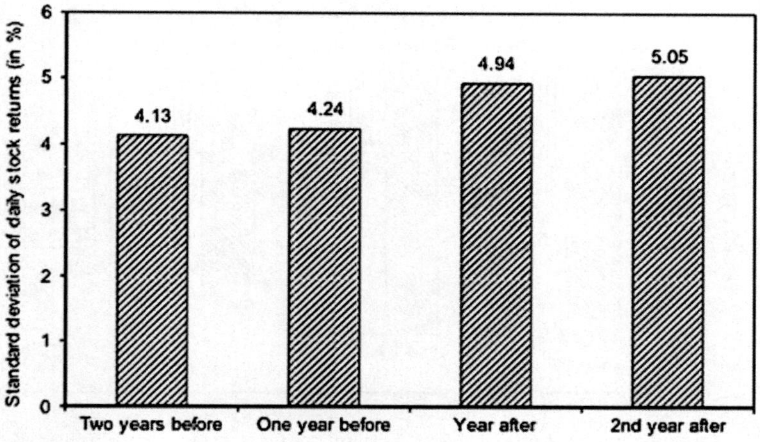

Fig. 1.5 Estimated standard deviation of stock returns over a four-year time period for the sample of firms that experienced disruptions

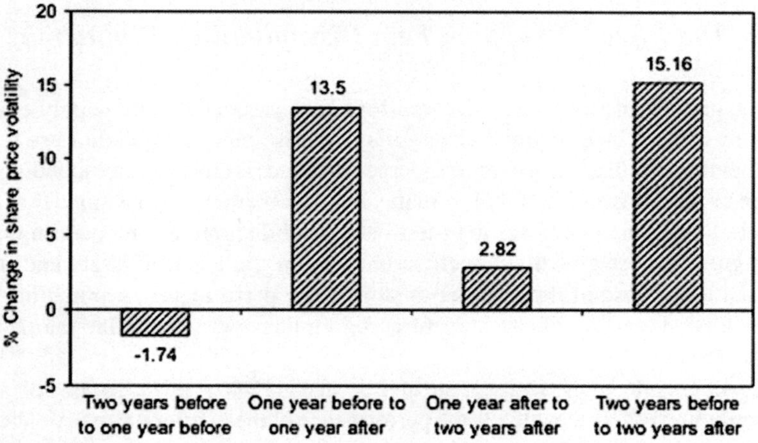

Fig. 1.6 Estimated percent changes in standard deviation of stock returns over a four year time period. The reported percent changes are the difference between the percent changes of the disruption experiencing firms and its control firms

after the disruption when compared to the volatility one year before the disruption announcement. Furthermore, the share price volatility remains at this high level for at least the next year or two. Overall, disruptions increase the risk of the firm.

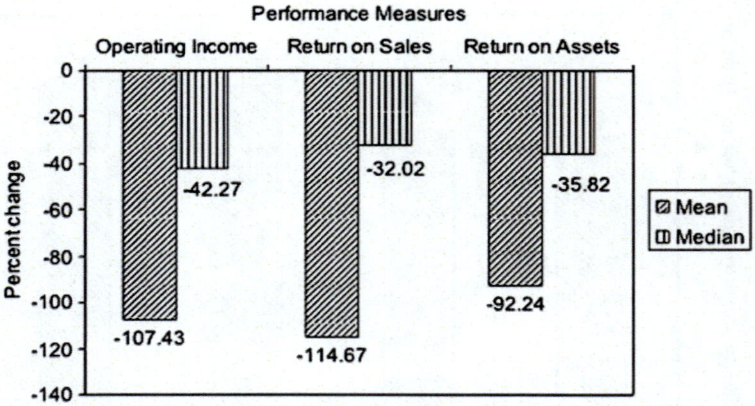

Fig. 1.7 Control-adjusted changes in profitability-related measures from supply chain disruptions. Performance effects are estimated starting one year before and ending two years after the disruption announcement

1.3.3 The Effect of Supply Chain Disruptions on Profitability

The magnitude of stock price underperformance associated with supply chain disruptions and the lack of any recovery may surprise many and could raise the issue whether the significant stock price underperformance is due to a corresponding reduction in profitability or it is simply a matter of stock market overreaction. This issue is explored by documenting the long-term effects of disruptions on operating income, sales growth, cost growth, as well as changes in the level of assets and inventories. As in the case of the analysis of stock price performance, profitability effects are estimated starting one year before and ending two years after the disruption announcement.

The key results of this analysis are highlighted in Figs. 1.7, 1.8. To control industry, economy, and others that affects the performance of the disruption-experienced firms is compared to controls using the three different control samples. Since the three control samples give similar results, the results from the control sample where most of the sample firms are matched are reported. Since accounting data are more prone to extreme values or outliers, the average values reported are those obtained after trimming 1% on each tail. The median changes, which are less influenced by outliers, are also reported.

The results indicate that supply chain disruptions have a devastating effect on profitability. Figure 1.7 shows that firms which experience disruptions on average experience a 107% decrease in operating income, 114% decrease in return on sales, and 92% decrease in return on assets. Outliers are not driving the negative mean changes in operating income-based measures. The median of the percent changes in operating income, return on sales, and return on assets are -42, -32, and -35%, respectively.

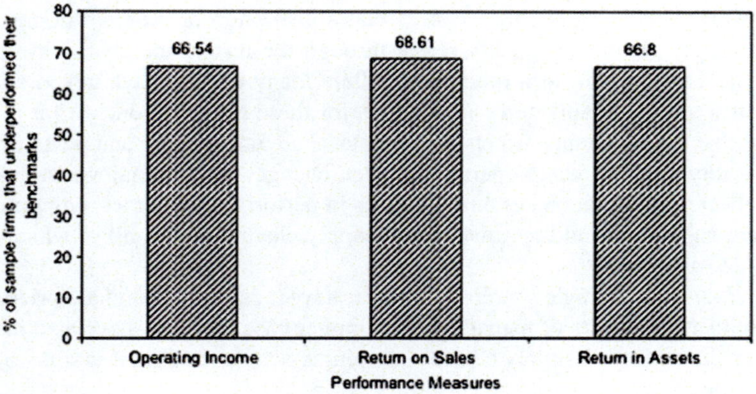

Fig. 1.8 The percent of disruption experiencing firms that underperform their benchmarks. Performance effects are estimated starting one year before and ending two years after the disruption announcement

The proportion of firms experiencing negative performance (see Fig. 1.8) indicates that disruptions are bad news across the board. For example, nearly 67–69% of the sample firms experienced a negative change in operating income.

1.4 Drivers of Supply Chain Disruptions

The analysis of the effect of supply chain disruptions on financial performance is valuable because it provides firms with a perspective on the economic effect of poor supply chain performance. The evidence clearly indicates that ignoring the possibility of supply chain disruptions can have devastating economic consequences. As one reflects on this evidence, a natural question is what are the primary drivers of supply chain disruptions? Given the recent heightened awareness of the risk of supply chain disruptions many experts have offered insights into the factors that can increase the chances of disruptions. Some of these factors are discussed next with the intention that these factors can serve as guideline for managers as they assess the chances of disruptions in their supply chains. The chances of experiencing disruptions are higher now and in the future than in the past because of some recent trends and practices in managing supply chains:

- *Increased complexity*: the complexity of supply chains has increased due to global sourcing, managing large number of supply chain partners, the need to co-ordinate across many tiers of supply chains, and dealing with long lead times. This increased complexity makes it harder to match demand and supply, thereby increasing the risk of disruptions. The risk is further compounded when various supply chain partners focus on local optimization, when there is lack of collaboration among supply chain partners, and when there is lack of flexibility in the supply chain.
- *Outsourcing and partnerships*: increased reliance on outsourcing and partnering has heightened interdependencies among different nodes of the global supply

networks and increased the chances that a disruption or problem in one link of the supply chain can quickly ripple through the rest of the chain, bringing the whole supply chain to a quick halt. While many experts have talked about the virtues of outsourcing and partnerships, for these to truly work well it is important that supply chain partners collaborate, share information and plans, and have visibility in each others operations. Such changes require major investments in connected information systems, changes in performance metrics, commitment to share gains, and building trust among supply chain partners, all of which are not easy to achieve.

- *Single sourcing*: single sourcing strategies have reduced the purchase price and the administrative costs of managing the supplier base, but may have also increased the vulnerability of supply chains if the single-source supplier is unable to deliver on time.
- *Limited buffers*: focus on reducing inventory and excess capacity and squeezing slack in supply chains has more tightly coupled the various links leaving little room for errors. Just-in-time delivery and zero inventory are commonly cited goals but without careful consideration of the fact that these strategies can make the supply chain brittle.
- *Focus on efficiency*: supply chains have focused too much on improving efficiency (reducing costs). Firms are responding to the cost squeeze at the expense of increasing the risk of disruptions. Most firms do not seem to consider the inverse relationship between efficiency and risk. Strategies for improving efficiency can increase the risk of disruptions.
- *Over-concentration of operations*: in their drive to take advantage of economies of scale, volume discounts, and lower transaction cost, firms have over-concentrated their operations at a particular location, or with their suppliers or customers. Over-concentration reduces the flexibility of the supply chain to react to changes in the environment and leads to a fragile supply chain that is susceptible to disruptions.
- *Poor planning and execution*: poor planning and execution capabilities result in more incidents of demand-supply mismatches. Plans are often too aggregate, lack details, and are based on inaccurate inventory and capacity information. Lack of good information systems hinders the ability of the organization to be aware of what is happening. Lack of forward looking metrics affects the ability of firms to anticipate future problems and be pro-active in dealing with these problems. Firms also have limited visibility into what is happening in upstream and downstream supply chain partners. Most firms have limited abilities and capabilities to identify and manage supply chain exceptions. This is further compounded by the lack of synchronization and feedback between supply chain planning and supply chain execution.

1.5 What Can Firms do to Mitigate the Chances of Disruptions?

There are no doubts that many of the above-mentioned practices and trends have led to improvements in supply chain performance and profitability. Nonetheless, they may have also contributed to supply chains becoming more susceptible and vulnerable to disruptions. The challenge therefore is to devise approaches that can deal more effectively with disruptions, while not sacrificing efficiency. Some of these approaches are briefly outlined below:

- *Improving the accuracy of demand forecasts*: one of the primary reasons for demand-supply mismatches is inaccurate forecasts. Bringing some quantitative rigor to forecasting can certainly help improve the accuracy and reliability of forecasts. Firms should consider not only the expected demand forecast but also the demand forecast error (variance) in developing plans. This would give planners an idea of what kind of deviation may happen from the mean value. Firms should also recognize that long-term forecasts are inherently less accurate than short-term forecasts as well as the fact that disaggregate forecasts are less accurate than aggregate forecasts. These considerations will enable planners to look more carefully at the forecasts they receive from sales and marketing. Forecasts often go bad when firms do not dynamically adjust forecasts, and fail to consider events outside their own organizations that could have a material effect on forecasts. Furthermore, firms often make forecasts assuming static lead times, transit time, capacity, and transportation and distribution routes. These assumptions must constantly be questioned to make adjustments as and when needed. Long planning time horizons that are frozen also makes it harder to develop accurate forecasts.
- *Integrate and synchronize planning and execution*: firms have become sophisticated in their planning activities. But plans are often insulated from execution reality. In many cases plans are tossed over the wall for execution. Managers responsible for execution make adjustments to these plans to reflect current operating conditions. Such adjustments can grow over time but are seldom communicated to the planners, resulting in lack of integration between development and execution of plans. By better coordinating and integrating planning and execution many of the problems with supply-demand mismatches can be avoided.
- *Reduce the mean and variance of lead time*: forecasting inaccuracy and disconnect between planning and execution can be particularly devastating when lead times are long and highly variable. Reducing the mean and variance of lead time can help reduce the level of uncertainties in the supply chain. Some of the following practices can help reduce the mean and variance of lead times:
  - Remove non-value added steps and activities.
  - Improve the reliability and robustness of manufacturing, administrative, and logistics processes.
  - Pay close attention to critical processes, resources, and material.

  - Incorporate dynamic lead-time considerations in planning and quoting delivery times.

- *Collaborate and cooperate with supply chain partners*: although the concepts of collaboration and cooperation among supply chain partners have been around for a long time, achieving this has not been easy. The evidence presented in this study provides an economic rationale why supply chain partners must engage in these practices. The precursor for collaboration and cooperation is developing trust among supply chain partners, agreeing upfront on how to share the benefits, and showing a willingness to change from the old mindset. Once these elements are in place, supply chain partners must do joint decision-making and problem solving, as well as sharing information about strategies, plans, and performance with each other. These activities can go a long way in reducing information distortion and lack of synchronization that currently plague supply chains and contribute to disruptions.
- *Invest in visibility*: to reduce the probability of disruptions, firms must be fully aware of what is happening in their supply chain. This includes internal operations, customers, suppliers, and location of inventory, capacity, and critical assets. The following may be needed to develop visibility:

  - Identify and select leading indicators of supply chain performance (suppliers, internal operations, and customers).
  - Collect and analyze data on these indicator.
  - Set benchmark levels for these indicators.
  - Monitor these indicators against the benchmark.
  - Communicate deviations from expected performance to managers at the appropriate levels on a real-time basis.
  - Develop and implement processes for dealing with deviations.

- *Build flexibility in the supply chain*: firms must make careful and deliberate decisions to build flexibility at appropriate points in their supply chains to enhance responsiveness. There are multiple dimensions of flexibility and what will be appropriate for a firm depends on its operating environment.

  - *Building flexibility on the product design side*: standardization, modularity, and use of common parts and platforms can offer the capability to react to sudden shift in demand and disruptions in delivery in parts.
  - *Building sourcing flexibility*: this can be achieved by using flexible contracts as well as use of spot markets to purchase parts and supplies. Spot markets can be used to both acquire parts to meet unexpected increase in demands as well as dispose of excess inventory if demand is below expectation.
  - *Building manufacturing flexibility*: this can be accomplished by acquiring flexible capacity that can be used to switch quickly among different products as the demand dictates. Firms should also consider segmenting their capacity into base and reactive capacity, where the base capacity is committed earlier to products whose demand can be accurately forecasted and reactive capacity is

committed later for products where forecasting is inherently complex. Such would be the case for products with short product life cycles as well as products with volatile demand. Late differentiation of products can also be used as a strategy to increase manufacturing flexibility.

- *Postponement strategy*: postponement or delayed differentiation is a strategy that delays product differentiation at a point closer to the time when there is demand for the product. This involves designing and manufacturing standard or generic products that can be quickly and inexpensively configured and customized once actual customer demand is known. By postponing differentiation of products, the chances of producing products that the market may ultimately not want are minimized, thereby reducing the chances of demand-supply mismatches. Key success factors for implementing this strategy include:
  - Cross-functional teams that represent the design and manufacturing functions.
  - Product and process reengineering to increase standardization.
  - Modularity.
  - Common parts and platforms.
  - Collaboration with customers and suppliers.
  - Performance measures and objectives that resolve conflicts and ensures accountability.
- *Invest in technology*: investment in appropriate technology can go a long way in reducing the chances of disruptions. Web based technologies are now available that can link databases across supply chain partners to provide visibility of inventory, capacity, status of equipment, and orders across the extended supply chains. Supply chain event management systems have the ability to track critical events and when these events do not unfold as expected send out alerts and messages to notify appropriate managers to take corrective actions. This enables the firm to identify supply chain problems earlier rather than later and operate in a proactive rather than reactive mode. RFID technology has the promise to improve the accuracy of inventory counts as well as provide real-time information on the status of orders and shipments in transit and what is being purchased by customers. Such access to real-time information alleviates information distortions and provides true demand and supply signals, all of which can reduce the chances of demand-supply mismatches.

Although there are a number of strategies that firms can use to mitigate the chances of disruptions, which of these would be appropriate for a particular firm depends on the firms operating environment. To identify what strategies to adopt, firms need a systematic process for risk management that is carefully and regularly applied. The process should be championed at the highest executive level as this is critical for bringing about awareness of the importance of managing disruption risk. A broad plan for developing and implementing such a process could be as follows:

1. *Assemble a cross-functional team of risk experts*: in most organizations, risk management is housed at the corporate level in insurance, legal and audit services. But supply chain disruption risks require a different type of arrangement.

The knowledge of supply chain risks lies in marketing, operations, procurement, logistics, and information technology. Thus, the cross-functional team must include members from these areas as they have dealt in the past with disruptions and have sufficient experience to identify and quantify risks. To provide credibility and visibility to the team, top management must support and champion the teams activities and efforts by making a case for the importance of risk considerations and driving changes that mitigate risks.

2. *Characterize the major sources of risk*: the cross-functional team must regularly scan the internal and external environment to identify the vulnerable points of their supply chain. This involves identifying the primary sources of risk, estimating the probability of each risk happening, estimating the financial impact of the risk, the amount it would cost to recover from the risk, and the amount of time it would take to recover from the risk. Precise estimates on these issues may not be easy to get and therefore as a first step it would be appropriate to gather some qualitative data such as high- or low-frequency of occurring, high- or low-financial impact, and easy or hard to recover.
3. *Assess and prioritize risks*: once the primary sources of risk have been identified and agreed upon, the next step is to assess and prioritize the risks that should be of serious concern to the firm. Top management and the board should be made aware of the high-risk issues. Various alternatives should be considered to mitigate the high-risk factors. Such alternatives include developing contingency plans to deal with the risk it should surface, options for spreading risks through insurance, forward contracts, flexible contracts, and making organizational changes in how the supply chain is designed and operated so that these risks are mitigated in the future.
4. *Monitor risk and take actions as needed*: once the primary risks issues have been identified and contingency plans have been developed, firms should set a system to monitor risks. Leading indicators need to be tracked, control limits need to be set to determine out of control conditions, two-way communication with suppliers and customers must be done on a continuous basis, and visibility systems must be in place. When risks surface the appropriate contingency plans are activated and the effectiveness of these plans in mitigating the risk is continuously monitored.
5. *Improve the risk management process*: firms must continuously strive to improve their risk management processes. As and when risk is dealt with, effort must be made to document the outcomes of the risk mitigation plans and highlight what worked and what did not work. These lessons should be shared across the organizations and used to improve the risk management process. Benchmarking a firms process against other firms that have well functioning risk management process can identify best practices and help make a firms process more robust and effective.

Firms must develop capabilities to deal with supply chain risks. Developing these capabilities requires leadership, commitment of resources, and detailed and meticulous planning. Building robust capabilities for dealing with supply chain risks involves the following steps:

1. *Analyze what could potentially go wrong*: this may require brainstorming, thinking about the unthinkable, observing disruptions that your company and other companies have experienced, and involving experts in creating scenarios of what could go wrong.
2. *Identify and analyze possible alternatives to deal with different types of risks*: this may require benchmarking of best practices with other companies, scenario analysis, and idea generation. Various alternatives should be considered to mitigate the high-risk factors. Such alternatives include developing contingency plans to deal with the risk should it surface, options for sharing and transferring risks through insurance, forward contracts, flexible contracts, and making changes in how the supply chain is designed and operated so that these risks are mitigated in the future.
3. *Develop plans to deal with disruptions*: this involves outlining what needs to be done to deal with disruptions, when it will be done, how it will be done, and who will do it. The plan needs to assign responsibility and authority to employees to carry out the plans. Without such plans, employees are left clueless about what to do, which actually creates more chaos and magnifies the negative consequences of disruptions.
4. *Monitor the situation*: companies should develop a system to monitor risks. Leading indicators need to be tracked, control limits need to be set to determine out of control conditions, two-way communication with suppliers and customers must be done on a continuous basis, and visibility systems must be in place.
5. *Execute the plan*: when disruptions occur, the appropriate plans are activated and the effectiveness of these plans in mitigating the negative impact is continuously monitored and adjustments need to be made on real-time basis.
6. *Improve the risk management process*: firms must continuously strive to improve their risk management processes. As and when risk is dealt with, efforts must be made to document the outcomes of the risk mitigation plans and highlight what worked and what did not work. These lessons should be shared across the organizations and used to improve the risk management process. Benchmarking a firms process against other firms that have well functioning risk management process can identify best practices and help make a firms process more robust and effective.

1.6 Summary

The evidence presented in this chapter makes a compelling case that ignoring the risk of supply chain disruptions can have serious negative economic consequences. Based on a sample of more than 800 supply chain disruption announcements, the evidence indicates that firms that suffer supply chain disruptions experience 33–40% lower stock returns relative to their benchmarks, 13.5% increase in share price volatility, 107% drop in operating income, 7% lower sales growth, and 11% increase in costs. By any standard these are very significant economic losses. More importantly, firms do not quickly recover from these losses. The evidence indicates that firms continue to operate for at least two years at a lower performance level after experiencing disruptions. Given the significant economic losses, firms cannot afford such disruptions even if they occur infrequently.

The evidence presented in this study underscores why supply chain management issues deserve close attention by senior executives and board members. Heightened scrutiny of corporate governance makes executives more directly responsible for earnings forecasts and prediction. To the extent that supply chain disruptions can devastate corporate performance, senior executives must be fully aware of the performance of their supply chains.

As discussed, overemphasis on efficiency and removing slack from the system can make supply chains vulnerable, unreliable, and non-responsive. While efficient and lean supply chains are desirable objectives, they should not come at the expense of reliability and responsiveness. There is a trade-off between efficiency of supply chains and risk of disruptions within supply chains.

It is quite common to find practitioner and academics talk about changes in supply chain management practices and investments in terms of their effect on efficiency and cost savings. Risk issues are often ignored because they cannot be easily quantified. Yet the evidence presented in this report strongly suggests that investing in supply chain reliability and responsiveness is equally important, if not more as investing in cost reduction. Such investments should be viewed as insurance against avoiding shareholder value destruction while disruptions happen. Given the evidence presented in this report, senior management must ask the question of whether they can afford not to proactively prevent and manage supply chain disruptions risk.

References

1. Aviv Y (2001) The effect of collaborative forecasting on supply chain performance. Manag Sci 47:1326–1343
2. Barnes-Schuster D, Bassok Y, Anupindi R (2002) Coordination and flexibility in supply contracts with options. Manuf Serv Oper Manag 4:171–207
3. Billington C, Johnson B, Triantis A (2002) A real options perspective on supply chain management in high technology. J Appl Corp Finance 15(2):32–43
4. Cachon G, Fisher M (2000) Supply chain inventory management and the value of shared information.. Manag Sci 46:1032–1048
5. Cachon G, Lariviere MA (2001) Contracting to assure supply: How to share demand forecasts in a supply chain.. Manag Sci 47:629–646
6. Casey N, Pasztor A (2007) Safety agency, mattel clash over disclosures. Wall Str J, 4 September, A1
7. Chopra S, Sodhi MS (2004) Managing risk to avoid supply-chain breakdown. Sloan Manag Rev 46(1):53–61
8. Craighead CW, Blackhurst J, Rungtusanatham MJ, Handfield RB (2007) The severity of supply chain disruptions: design characteristics and mitigation capabilities. Decis Sci 38:131–156
9. Fairclough G (2008) How a Heparin maker in China tackles risks—plant now supplying US must make frequent checks to monitor difficult market. Wall Str J, 10 March, B1.

10. Ferrer J, Karlberg J, Hintlian J (2007) Integration: the key to global success. Supply Chain Manag Rev, March 1–4
11. Fisher M (1997) What is the right supply chain for your product?. Harvard Bus Rev 75(2): 105–116
12. Frohlich MT, Westbrook R (2001) Arcs of Integration: an international study of supply chain strategies. J Oper Manag 19:185–200
13. Hendricks KB, Singhal VR (2003) The effect of supply chain glitches on shareholder value. J Oper Manag 21:501–522
14. Hendricks KB, Singhal VR (2005) An Empirical analysis of the effect of supply chain disruptions on long-run stock price performance and risk of the firm. Prod and Oper Manag 14:35–52
15. Hendricks KB, Singhal VR (2005) Association between supply chain glitches and operating performance. Manag Sci 51:695–711
16. Kesmodel D (2008) Oversight 'Flaw' Led to meat recall—lax monitoring seen at plant's cattle pens before closure. Wall Str J, 11 March, B1.
17. Kilgore M (2003) Mitigating supply chain risks. White paper, Chainalytics LLC
18. Kleindorfer PR, Saad GH (2005) Managing disruption risks in supply chain. Prod Oper Manag J 14:53–68
19. Lakenan B, Boyd D, Frey E (2001) Why Cisco fell:outsourcing and its peril. strategy+business. www.strategy-business.com/article/19984. Accessed 17 Jan 2011
20. Lee H, Padamanabhan P, Whang S (1997) Information distortion in supply chain: The bullwhip effect. Manag Sci 43:546–558
21. Lunsford JL (2007) Jet Blues: Boeing scrambles to repair problems with new plane—layers of outsourcing slow 787 production; 'Hostage to Suppliers'. Wall Str J, 7 December, A1
22. Milner JM, Kouvelis P (2002) On the complementary value of accurate demand information and production and supplier flexibility. Manuf Serv Oper Manag 4:99–113
23. Narasimhan R, Jayaram J (1998) Casual linkages in supply chain management: an exploratory study of North American manufacturing firms. Decis Sci 29:579–605
24. Radjou N (2002) Adapting to supply network change. Forrester Research Inc, Cambridge MA
25. Rosenzweig ED, Roth A, Dean JW (2003) The influence of an integration strategy on competitive capabilities and business performance: an exploratory study of consumer products manufacturers. J Oper Manag 21:437–456
26. Sheffi Y (2005) The resilient enterprise: overcoming vulnerability for competitive advantage. MIT Press, Massachusetts
27. Shin H, Collier DA, Wilson DD (2000) Supply management orientation and supplier/buyer performance. J Oper Manag 18:317–333
28. Smyrlis L (2006) Risky Business: there has been a significant increase in external risk during the past five years and supply chain is the major concern, New survey shows. Can Transp and Logist, June 1, 14
29. Taylor TA (2002) Supply chain coordination under channel rebates with sales effort effects. Manag Sci 48:992–1007
30. Tang C (2006) Robust strategies for mitigating supply chain disruptions. Int J Logist Res Appl 9:33–45
31. Tomlin B (2006) On the value of mitigation and contingency strategies for managing supply chain disruptions risks. Manag Sci 52:639–657

Chapter 2
Mitigating the Impact of Disruptions in Supply Chains

Wallace J. Hopp, Seyed M. R. Iravani and Zigeng Liu

2.1 Introduction

Supply chain risk management is of growing importance, as globalization extends supply chains and makes them more vulnerable to a wide range of disruptive events. Supply interruptions can be the result of large-scale natural diasters, terrorist attacks, plant fires, electrical blackouts, financial or political crises, and many other scenarios. Some well-known examples of supply chain disruptions include:

- The 1999 earthquake in Taiwan had a dramatic impact on the global semiconductor market. At the time, Taiwan was the third largest supplier of computer peripherals in the world, so the earthquake caused a temporary global shortage of semiconductor components with production down times that ranged from 2 to 4 weeks. Production and sales of many firms were profoundly affected by this shortage [8].
- In 1999, although Hurricane Floyd struck many miles away from Daimler–Chrysler's minivan plant in Windsor, ON, Canada, it led to an indirect shutdown when another Daimler–Chrysler plant located in Greenville, NC, was flooded as a result of the hurricane, causing a shortage of suspension parts. Consequently, seven of the company's other plants across North America, including the Windsor plant, were shut down for days [46].

W. J. Hopp
Stephen M. Ross School of Business, University of Michigan,
Ann Arbor, MI, USA

S. M. R. Iravani (✉)
Department of Industrial Engineering and Management Science,
Northwestern University, 2145 Sheridan Road, C210 Tech.
Evanston, IL 60208-3119, USA
e-mail: s-iravani@northwestern.edu

Z. Liu
Wisconsin School of Business, University of Wisconsin-Madison,
Madison, WI, USA

H. Gurnani et al. (eds.), *Supply Chain Disruptions: Theory and Practice of Managing Risk*, DOI: 10.1007/978-0-85729-778-5_2,

- On March 17, 2000, a random lightning bolt struck a Royal Philips Electronics semiconductor plant in Albuquerque, New Mexico, sparking a 10-min fire. Although it was quickly extinguished, several thousand chips for mobile phones were destroyed. Worse, the fire had a large impact on the cleanroom environment in the semiconductor plant, effectively shutting it down for weeks. Nearly half of the plant's output was destined for two of Europe's biggest cell phone makers, Nokia and Ericsson. The sudden loss of a critical part had a relatively minor impact on Nokia but was disastrous for Ericsson [35, 60].
- Following the 9/11 terrorist attack, the US government temporarily suspended truck movements between the US and Canada/Mexico. The resulting delays of part shipments reduced output at Ford Motor Company during the fourth quarter of 2001 by around 13% [61].
- The Iraq War dramatically impacted global oil supply chains. As the war, as well as production disruptions in Russia, pushed crude oil prices above $45/barrel, gasoline and jet fuel prices in the United States increased rapidly, with disastrous consequences for airlines. Shortly after the start of the war, three US airlines—United, US Airways, and Hawaiian Airlines, were in bankruptcy protection—while United, Northwest, Continental, and Air Canada were announcing job cuts or temporary furloughs of employees, as well as schedule cuts [50].
- Hurricanes Katrina and Rita in 2005 threatened oil and gas production platforms on the US Gulf Coast, the home for 23% of the nation's refining capacity. Combining the effects of both Katrina and Rita, 1.3 million barrels/day of refining, representing about 8% of national capacity, was shut down [38]. On September 21, 2005, Valero Energy Corp, the nation's largest refiner, stated that Rita caused gasoline prices to rise well above $3/gal, at a time when the US average price was $2.77/gal [57].

Supply chain disruptions resulting from events like these can result in serious financial consequences. Hendricks and Singhal [28] studied a sample of 827 reported supply disruptions between 1989–2000 and found that the affected firms achieved stock returns that were on average 40% below that of comparable firms not reporting disruptions for a period extending from 1 year prior to the disruptions report and 2 years after it. While this suggests that many types of disruptive events can have significant business impacts, it is important to recognize that events like those listed above vary greatly in terms of likelihood and severity. For example, the 9/11 attacks were (we hope) a very unlikely, but very severe, event, while the fire in the Philips plant was a much more likely, but less dramatic, event. If we think of risk in terms of total impact (i.e., the product of frequency times consequence), then a high-frequency/low-consequence event, such as a routine demand fluctuation, will be viewed as similar to a low-frequency/high-consequence event, such as the fire in Philips' plant. However, these events will have vastly different qualitative effects on the company. Furthermore, they are amenable to very different interventions. For instance, steps for mitigating the effects of routine variability (e.g., safety stocks) may be poorly suited for addressing exceptional events, such as extended supply disruptions.

In this chapter we present a summary of strategies that can be used to reduce the likelihood of a supply chain disruption, or to mitigate its impact. The objective of this chapter, by no means, is to present a thorough review of the literature on supply chain disruption. We only use some of the existing literature to provide examples of different risk mitigating strategies. Specifically, in Sect. 2.2, we first present the existing classifications of supply chain risks in operations management and operations research literature. In Sect. 2.3 we describe our general framework for strategies to mitigate the impact of supply chain disruption. These strategies are further discussed in Sects. 2.4–2.7 We provide some managerial insights in Sect. 2.8, and point out some promising future research directions in Sect. 2.9.

2.2 Classification of Risk and Mitigating Strategies

There has been a fair amount of efforts devoted to describing and categorizing supply chain risks and the strategies for mitigating them. While the resulting studies do not provide a means for quantifying risks or striking an economic balance of response strategies, they are useful as a starting point for understanding supply chain risks.

Johnson [34] divided supply chain risks into two categories: (a) *demand risks*, including seasonal imbalances and new product adoption, and (b) *supply risks*, such as manufacturing and logistics capacity limitations, currency fluctuations, and supply disruptions from political issues. Chopra and Sodhi [12] further refined the supply risks into a taxonomy of risks faced in supply chains and qualitatively discussed different strategies for mitigating them. They identified the following categories of risks: (1) disruption, (2) delays, (3) system risks, (4) forecasting risks, (5) intellectual property risks, (6) procurement risks, (7) receivables, (8) inventory risks, and (9) capacity risks.

Simchi-Levi et al. [62] discussed how to build robust and competitive supply chains in the face of risk from uncertainties by using four approaches: (1) hedge strategies, (2) flexible strategies, (3) collaboration and outsourcing, and (4) "what if" analysis.

Rice [58] classified firms' responses to disruptions into four levels: Level 1 refers to "basic initiatives", which include protective activities, such as physical security, information security, and freight security; Level 2 refers to "reactive initiatives" corresponding to greater awareness of security vulnerabilities and establishment of a supply continuity plan; Level 3 refers to " proactive initiatives", which include security and resilience actions beyond normal requirements, leading to a business continuity plan; Level 4 refers to "advanced initiatives", which take the form of progressive security and resilience practices, such as customer–supplier collaboration, formal security strategy, and emergency control center.

Christopher and Peck [13] discussed how to design resilient supply chains in qualitative terms by emphasizing on recognizing the nature of supply chain risks. Christopher and Lee [14] suggested that a key element for mitigating supply chain risk is improved end-to-end visibility.

Peck [51] suggested that based on the different drivers and sources of risk, supply chain systems can be analyzed at four levels: at the first level, supply chain risks mainly refer to inefficient supply chain performance, such as lacking of the ability to react to demand uncertainty and the market changes; at the second level, whether a supply chain network is resilient or not is evaluated with respect to the implications of link failures, node failures, and the loss of other essential operating assets; at the third level, supply chain systems are viewed as inter-organizational networks, in which strategic outsourcing decisions are critical; at the fourth level, risk factors associated with the natural environment, political environment, economic situation, social conditions, and technology are considered.

Elkins et al. [20] divided risk management responsibility into internal operations, external suppliers, current business, and future business, to produce a 2×2 matrix. By examining the four regions of this matrix, they generated 18 practices (which are assigned to the four organizational areas of the matrix) to help companies to build resilient supply chains.

Tang [66] discussed the benefits under normal situations, and benefits after disruptions of nine robust supply chain strategies. He pointed out that well-designed strategies can manage supply chain uncertainties efficiently during the normal days and also make the supply chain resilient in the face of a disruptive event.

Kumar and Stecke [37] developed a catastrophe classification scheme, offered comprehensive mitigating strategies, and provided tables to help supply chain managers decide which components within the supply chain network are under the risk from catastrophes and what kinds of mitigating strategies should be implemented.

Tang and Tomlin [67] studied five stylized models (which correspond to five types of flexibility strategies) and suggested that flexibility can be used as a powerful defensive protection mechanism to mitigate supply chain risks.

From a practice standpoint, these various classifications provide checklists that can help mangers ensure that they do not overlook important categories of risk or protection strategy. From a research perspective, they enumerate the issues that need modeling support in order to guide balanced economic decisions for managing supply chain risks.

2.3 General Framework

A supply chain disruption causes cessation or restriction of production by one or more nodes in this network for an extended period of time. A node in a supply chain network may be a supplier of raw material or components, a manufacturer or an assembly plant, a distribution center, or a retailer. Without some kind of response, such a disruption could lead to disruption of shipments of finished products to consumers (see Fig. 2.1) . In the short-term, this will result in lost sales, and hence lost revenue. In the long-term, if customers whose demands are not filled during the disruption shift their future business to competitors, it could also result in lost market share. Figure 2.2 graphically illustrates these short-term (tactical) and long-term (strategic) losses from a supplier disruption.

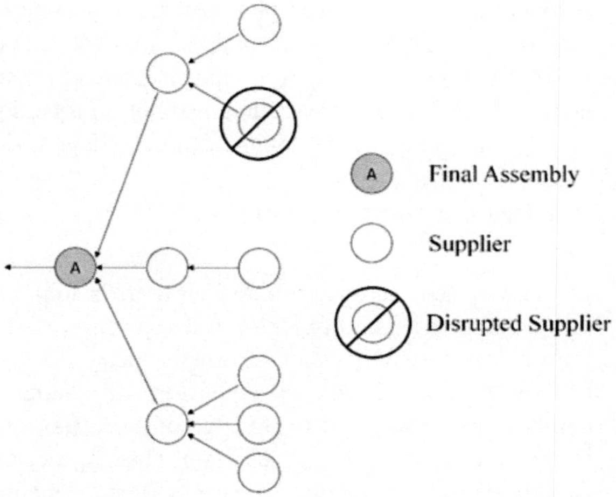

Fig. 2.1 Whether and when a disruption of a supplier will disrupt the final assembly operation depends on the duration of the outage, the amount of inventory in the supply chain between the disruption and the customer, and the ability to bring on backup capacity to replace the disrupted supply

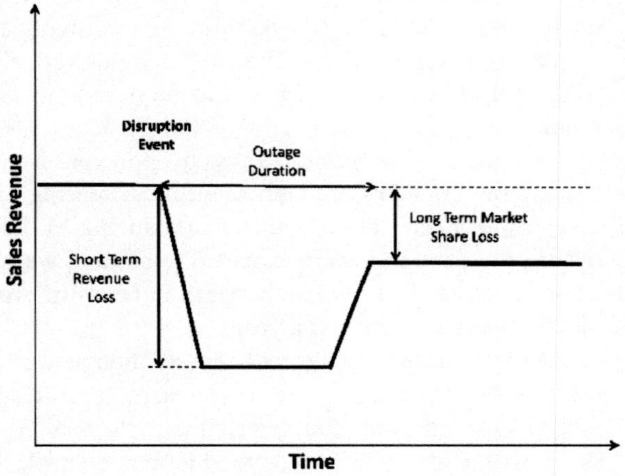

Fig. 2.2 A supply disruption that disrupts delivery of finished products to customers may reduce sales revenue in the short-term (a tactical loss) and result in a long-term reduction in market share (a strategic loss)

We can use this conceptual model to identify a range of strategies for reducing supply chain risk. For instance, the expected loss of short-term revenue and long-term market share are increasing in the likelihood of a disruption. So, *prevention* policies that make disruptions less likely can reduce risk. If a disruption does occur, then the economic consequences can be influenced by the speed and effectiveness of the firm's *response*. For example, quickly finding an alternate supplier who can provide a substitute supply for the disrupted node could minimize the ultimate impact on the customer. While the firm is waiting for some kind of resolution to the supply problem, the extent to which the disruption propagates to the final assembly node depends on the amount of *protection* that exists in the system. For example, if there is enough downstream inventory (i.e., between the disrupted node and the customer) in the supply chain, then the disruption may not affect customers. If customers are affected, then the amount of damage the firm will incur will be influenced by its *recovery* actions. For example, customers who defected to competitors during the disruption may be won back by appropriate actions during and after the event.

This suggests that supply chain risk reduction strategies can be grouped into a general framework of prevention, response, protection, and recovery policies, which will be discussed in the remainder of this chapter.

2.4 Prevention Strategies

In maintenance and repair operations there is an important distinction between reliability and maintainability. *Reliability* represents the ability of the equipment to avoid failure, and is usually characterized by mean time to failure (MTTF). The larger the MTTF, the more reliable the equipment. In contrast, *maintainability* represents the ability of the equipment to get back on line after a failure, and is usually characterized by mean time to repair (MTTR). The shorter the MTTR, the more maintainable the equipment. Availability (percent of time the equipment is up) can be increased either by increasing MTTF or by decreasing MTTR. However, as shown in [31], production efficiency is improved more by improving availability by reducing MTTR than by increasing MTTF. The reason is that reducing MTTR reduces the amount of protective inventory needed in the system to maintain flow during an equipment failure.

This same distinction exists in policies for reducing supply chain risk. Prevention strategies are equivalent to increasing MTTF of a supply node, while recovery strategies are equivalent to reducing MTTR of a supply node. Since serious supply disruptions are expected to be rare, we typically do not use MTTF as a measure. Instead, we typically speak in terms of event likelihood (e.g., probability of a disruption in any given year).

At first glance, it may not seem as though we can influence the likelihood of many of the events that cause supply disruptions. Hurricanes, earthquakes, lightning strikes, fires, etc., are often referred to as "acts of God". But even for an event, such as an earthquake, which is beyond human control, there are interventions that can influence the likelihood that the event triggers a supply disruption. For instance, sourcing supply from facilities in less earthquake prone regions of the world will reduce the likelihood of an earthquake-induced disruption.

With this perspective in mind, we can divide prevention strategies into *forecasting* and *risk reduction* strategies.

2.4.1 Forecasting

To prepare for disruptive events, firms must first identify events that are likely to result in disruptions. Although terrorist attacks get extensive coverage in the press, most supply chain disruptions are caused by more mundane scenarios, such as fires, natural disasters such as earthquakes and hurricanes, and business failures due to labor unrest, financial instability, or quality control problems. Starting with a list of these generic categories and supplementing it with special scenarios that apply to the environment in question, a firm can identify the range of events that could lead to supply disruptions.

Since firms cannot prepare individually for every event, they should also evaluate the likelihood of events, so that they can focus on events whose probability and severity warrant investment of resources. Firms can use historical data to estimate the likelihood of certain classes of events (natural disasters, fires, economic failures, etc.) [27, 70].

To evaluate the economic consequences of events or classes of events, firms can use bill of material information to trace the consequences of component disruptions on product availability. For example, one component may be used in several different products. For purposes of ranking which components are most critical for protection, firms do not need to make highly accurate forecasts of disruption occurrence frequency. Instead, they can group events into fairly coarse categories (e.g., high, medium, and low) as described in Hopp et al. [29].

2.4.2 *Risk Reduction*

Once failure modes have been identified, steps can be taken to reduce the likelihood of some of them. For events and products for which intervention is possible and makes economic sense, firms can reduce the likelihood of high-impact supply chain disruptions by increasing security. For example, firms can choose a facility location that is less likely to a terrorist object to reduce the likelihood of terrorist attacks, or that is not under the impact of unions to reduce the likelihood of labor disruptions, or that is not with earthquake-hurricane-prone areas to reduce the likelihood of nature disasters. Firms can also track the financial health of their suppliers in an effort to reduce their vulnerability to economic failures. For example, after the technology bubble burst in the early 2000s, carefully reviewing the financial viability of a software vendor became a necessary evaluation step when firms considered purchasing software products [61].

Many firms learn the value of reducing risks only after they have suffered through a disruption. But, when possible, learning from the experience of others is much more efficient. Reading about cases like those cited in this chapter is one way firms can increase their knowledge of risks and mitigation strategies. Forming consortiums and sharing information with suppliers, customers, and each other, is another way to improve their knowledge of supply chain risks. Of course, since information and preparation can provide a competitive advantage, firms may not be willing to share their knowledge [29].

Overviews of risk reduction strategies are given in [13] and [60].

2.5 Response Strategies

As prevention policies can be viewed as increasing the MTTF of a supply node, response strategies can be seen as decreasing the MTTR. The straightforward reason is that the faster a firm gets started on resolving the problem, the sooner it will be able to resume production. But a more subtle reason is that a supply disruption may put the firm in competition with other firms for limited supplies of backup capacity. Hopp et al. [29] modeled the competition between firms for such capacity as a noncooperative game. In situations where quick and decisive response to a disruption enables a firm to procure whatever supplies are available, the difference between being first and second can be huge. The "winner" of the competition to obtain the backup supplies may be able to avoid a serious disruption of its customers, while the "loser" may be unable to supply its customers for an extended period.

Hopp et al. [29] also pointed out that competitions for backup supplies can lead to even more extreme outcomes. Firms who are not affected by a supply disruption, either because they were able to procure alternate supplies or because their suppliers were not disrupted, can exploit the situation to gain market share at the disrupted firm's expense. Because of the long-term market share implications, the speed with which a firm responds to a supply disruption is an essential element of a strategy for reducing supply chain risk. Since a speedy response involves first detecting the disruption and then acting on this realization, we can divide response strategies into *detection* and *speed* strategies.

2.5.1 Detection

Detecting a disruption would seem to be a simple task—if supplies do not show up, there is a disruption. But in large organizations with complex global supply chains, it is not this easy for three reasons:

1. The firms must distinguish a disruption from normal day-to-day variation. Supplies are often late, damaged, incomplete or otherwise imperfect. But usually these variations are small and short term. Setting up a system under which staff follow-up immediately on every late arrival may be impractical. But there should be a system for following up if the delay lasts more than a short time. For example, in Philips fire case because Nokia had installed a production monitoring system, it was aware of Philips' supply problem even before it received a phone call from Philips. This enabled Nokia to start planning its response more quickly than Ericsson which did not have such a system.
2. The firm must characterize the disruption. A late delivery could be due to a manufacturing problem, a transportation problem, a border control problem, a receiving problem, etc. For major events, such as natural disasters, the root cause may be obvious (e.g., a major hurricane), but the extent and duration of the problem may not. For example, it took days, or even weeks, for the full extent of the damage from Hurricane Katrina to be known. Some firms found that they had second- and third-tier suppliers in the Gulf region that they did not even know about. Even when firms get direct information from suppliers about the problem, such information may not be accurate. For instance, when Philips initially called Nokia and Ericsson to inform them about the fire, it was believed that the event would be limited and short. By the time everyone

realized that it would take longer than expected to get operations back up and running, it was too late for Ericsson to locate backup supplies.

3. Information about the event must reach the people who need to act. While this is largely an issue of culture, related to the flow of information across the organization, it has serious operational implications. While a low-level receiving clerk may realize that supply of a component is disrupted, the VP Operations who must authorize use of a new supplier may not. Nokia was noted for cultivating a company-wide culture of awareness and communication, which encouraged individuals to swiftly notify their superiors about emerging situations and act upon them. Ericsson, in contrast, did not have such a culture and was very slow about communicating information about the Philips outage to senior management [61]. Consequently, Ericsson as a firm did not detect the problem until it was too late.

Addressing all three of these challenges requires a combination of effective information, communication, and decision-making processes. Consequently, there is no simple turnkey fix for improving ability to respond to a supply disruption. Nokia's success in identifying and understanding the Philips disruption was the result of a coordinated effort in several dimensions.

2.5.2 Speed

Once an event has been detected, the consequences can depend on how quickly the firm reacts. As we noted above, Nokia not only detected the problem at Philips more quickly than Ericsson, but it also reacted more quickly. An event manager at Nokia followed up frequently and even offered to come and help manage the situation at Philips [61]. Even before they knew the extent of the problem, Nokia purchasing people were on the phone seeking alternate suppliers, just in case they needed them. The net effect was a swift and effective response that enabled Nokia to lock up the available backup supply, to their benefit and the detriment of Ericsson.

In addition to promoting detection and response speed from the within firms, firms can improve reaction speed to an external disruptive event by strengthening relationships with suppliers. Enhancing the firm's reputation with suppliers can make it more likely that these suppliers will share information with the firm and work with it to resolve a problem. For instance, the 1997 fire at Aisin (which made 99% of Toyota's p-valves) provided an example of a successful business relationship between a manufacturer and a supplier, which led to a rapid collaborative response to a disruptive situation (see [49, 55] for discussions).

2.6 Protection Strategies

In a production system, failure of a single machine may or may not result in a loss of throughput. If there is enough extra capacity and downstream inventory to keep final assembly working, then the consequences of the failure will not be felt beyond the plant. The same is true in a supply network. Backup inventory and/or capacity could allow downstream nodes to continue working during the disruption. Even if these do not entirely isolate the disruption, they can serve to shorten the interval felt by customers.

In addition to protection of physical supplies, firms must also be concerned with protecting information systems, which are needed to coordinate flows and match supplies with demand. We summarize research addressing inventory, capacity, and information protection below. We also examine the structure of the supply network itself as a protective mechanism.

2.6.1 Inventory Protection

Inventory is ubiquitous as a response to routine variability. Retail inventory buffers against demand variability. Work in process inventory buffers against fluctuations in production rates. Raw materials inventories buffer against delivery delays. But as we noted in Sect. 2.2, we are concerned with risks of extended disruptions. While qualitatively similar to routine variability (e.g., a disruption can be thought of as a really, really late delivery), such disruptions are so quantitatively different that they require a separate set of response strategies. For instance, Hopp et al. [30] appealed to the observation of Chopra et al. [11] that protection against routine variability and catastrophic disruptions are two separate problems to justify ignoring fluctuations in production and demand rates, as well as the inventory used to buffer them (i.e., routine safety stocks are separated from protection stocks). They developed an analytical model for striking a balance between the costs of inventory (and/or capacity) protection and the costs of lost sales in an arborescent assembly network subject to disruption and showed that, under certain conditions, it is optimal to locate inventory (or capacity) protection at no more than one node along each path to the customer.

Hu et al. [32] considered a firm producing in two locations and facing risk from production capacity uncertainty (e.g., due to unexpected downtime). They assumed that, although inventory can be held from period-to-period, unsatisfied demand is lost. They examined how a central planner with full access to the inventory status at the two locations should transship inventory from one location to another to maximize the expected discounted joint profits. That transshipment strategy can significantly lower the risk of capacity uncertainty.

Liu et al. [41] considered how supply reliability influences the performance of a retail firm under joint marketing and inventory decisions. They investigated the value of higher supply reliability and showed that the optimal stocking quantity does not necessarily increase or decrease as risk is reduced due to the improvement of supply reliability. But they showed that there are conditions under which a retailer is willing to pay a higher unit price for higher supply reliability.

The overall conclusions from this stream of research are: (a) inventory can mitigate the consequences of a supply disruption, (b) the details of how to use inventory as a protective mechanism vary depending on the environment, and (c) inventory is not sufficient by itself as a protective mechanism because catastrophic failures can last too long to be covered economically by inventory.

2.6.2 Capacity Protection

A second form of protection against a supply disruption is capacity that can take the place of the disrupted supplier. This can either be *real* capacity, which has already been developed or *virtual* capacity that could be developed if needed. An example of non-contractual virtual capacity backup occurred in 1997 when a fire in an Aisin plant disrupted supply of p-valves used in almost all Toyota vehicles. By working flexibly with its supply base, Toyota was able to tap a number of firms to produce p-valves to replace the disrupted supply. Real capacity could take the form of *multi-sourcing*, so that if one supplier is disrupted another can pick up the slack, or *capacity contracts*, in which non-active suppliers agree to provide supplies if needed. For example, Boston Scientific has redundant production lines for some of their most important products to prevent competitors from poaching market share during the outages.

Researchers have investigated various actions that firms can take in advance of a disruption to enhance their ability to provide capacity protection.

Tomlin and Wang [69] considered a firm producing multiple products and facing risk from unreliable supply chains. The firm addresses this risk by striking a balance between mix-flexibility and dual-sourcing. They compared four different networks: single-source dedicated (SD), single-source flexible (SF), dual-source dedicated (DD), and dual-source flexible (DF). They showed that a DD network is preferred when a resource is less reliable, and a DF network is more preferred when the range of contribution margin is wider. In addition, they observed that, in an unreliable supply chain network, the DD design strategy can dramatically reduce a firm's downside risk exposure due to its diversification characteristics.

Wang et al. [72] also studied the supplier reliability problem and considered risk from both capacity and yield uncertainties. In their model, there are multiple suppliers and one buyer, who can dual source and/or expend effort to improve the reliability of its supplier. They found that, in the face of risk from capacity uncertainty, the buyer prefers improvement over dual sourcing as supplier cost heterogeneity increases, but prefers dual sourcing over improvement in the high-reliability heterogeneity case. In the face of risk from yield uncertainty, the buyer prefers dual sourcing over improvement as supplier cost heterogeneity increases, but prefers improvement over dual sourcing in the high-reliability heterogeneity case.

Tomlin [68] considered a firm producing one single product and having two potential suppliers, a "cheap" one that is unreliable and capacity constrained and an "expensive" one that is more reliable and has volume flexibility. He found that a supplier's uptime percentage and disruption length play a important role in determining the optimal mitigation strategy. He showed that, given an uptime percentage, if disruptions are less frequent but with longer duration, the firm prefers the strategy of sourcing exclusively from the expensive but reliable supplier over the strategy of sourcing exclusively from the cheap but unreliable supplier and carrying inventory to mitigate disruptions and the reason is that significant quantities of inventory need to be carried for extended periods without a disruption, which results in expensive holding cost.

Dada et al. [19] considered a system with multiple suppliers and each of them is either perfectly reliable (i.e., can deliver an amount equal to the desired amount) or unreliable (i.e., delivers an amount strictly less than the desired amount with some probability) and investigated the impact of the changes in supplier cost or reliability on the ordering decisions and customer service level. Their results indicated that, although one supplier's reliability affects the quantity ordered by a newsvendor, cost is a relatively more important factor to make the choice among suppliers and even a completely reliable supplier can be disqualified if that supplier's cost is considered to be not competitive.

Identifying possible outside suppliers is another important strategy for mitigating supply chain risk. Hopp et al. [29] modeled the impact of regional supply disruptions on competing supply chains. They described generic strategies that consist of two stages: (1) *preparation*, which involves investment prior to a disruption in dedicated backup capacity and/or measures that facilitate quick detection of a problem, and (2) *response*, which involves post-disruption purchase of shared (non-dedicated) back-up capacity for a component whose availability has been compromised. They characterize the conditions that pose the greatest risk and suggest ways to reduce the risk exposure.

Several research studies (e.g., [21, 22, 26, 64]) have indicated that higher levels of organization flexibility enable firms to respond to disruptions more successfully than their non-flexible counterparts. Van Mieghem [71] investigated how resource allocation flexibility can mitigate financial risk. Taking a view from both operational and financial perspectives, he studied the impact of risk attitude of newsvendors and network configuration on mitigating risk by using the strategic placement of operational resources. There are different ways to achieve parts, process, or plant flexibility within an organization, standardization is the most popular one. Standard parts, standard processes, and standard plants make firms to make the parts of their products compatible with a wider range of suppliers, move labor around, and relocate production across their plants.

Another way to increase a firm's flexibility is through contracting with suppliers. Firms can choose short- and long-term contracts. Long-term contracts have the advantage of guaranteeing a fixed cost pattern even in the event of a supply disruption. However, short-term contracts are usually less expensive. Cohen and Agrarwal [17] modeled the problem of selecting the right mix of short-and long-term policies for a firm sourcing components from external suppliers. Under their multi-period model, the buyer makes the selection decision based on the evaluation of the tradeoff between the capacity, flexibility and low-initial investment associated with short-term contracting and the price certainty, and high-initial investments associated with long-term contracting. The insight from their work is that different conditions (e.g., a firm's risk attitude to supply chain disruptions and the magnitude of required initial contract investment) lead to different optimal mixes of contracts.

Swinney and Netessine [65] studied a similar contracting problem as a two-period game. In their modeling framework, there are two suppliers and one buyer. The buyer chooses a supplier and then selects the type of contract, short or long-term, to use. They assumed that revenues from the buyer at the beginning of the game determine whether or not supplier failure occurs. They found that the buyer prefers short-term contracts over long-term contracts without the possibility of failure, but prefers long-term contracts in the presence of failure. They further considered dynamic contracts in which the contract price is tied to production costs (e.g., commodity prices of raw materials) so that the buyer will share risk from cost uncertainty with the supplier. They found that this kind of contract can coordinate the supply chain system in the face of risk from supplier failure. Actually, instead of having a fixed contract (short- or long-term), more and more firms prefer flexible contracts with their suppliers.

Barnes-Schuster et al. [4] studied the role of options in a buyer-supplier system. They illustrated how a flexible contract benefits the overall supply chain system (i.e., both the buyer and the supplier). During a disruption, flexible contracting enables firms to obtain desired amount of capacity to replace the disrupted supply within the outage from non-disrupted contracted suppliers.

2.6.3 Information Protection

As information technology becomes increasingly integrated into core business operations, firms are increasingly at risk of a disruption to an interruption of their information systems. For example, widespread power outages such as those that affected the northeastern and midwest United States in August 2003 [23], can severely impact business continuity by compromising information systems. Similarly, after the 9/11 terrorist attacks on the World Trade Center, the entire local information technology infrastructure lay in ruins. Merrill Lynch, which had a backup information technology center and redundant trading floors near New York city, quickly shifted its critical operations to the redundant facility within a few minutes [3]. In general, some type of backup information system is a key element of a supply chain protection plan.

Information redundancy is also a ubiquitous practice. Even the most casual computer user is encouraged to backup data in case of a failure. Firms most certainly back up customer data, supplier data, bill-of-material data, etc. But the functioning of supply chain requires the ability to update, process, share these data, and do so across firm boundaries. So information protection requires more than banks of backup hard drives. It requires backup of the system itself. If IT resources are physically damaged (as occurred in the 9/11 attack), compromised by a virus, or brought down for some other reason, the firm must be able to replace them somehow. Failure to do so can bring business to a halt just as effectively as a disruption of physical supply.

We noted earlier that Merrill Lynch maintained a complete redundant system for its critical IT operations that were destroyed on 9/11. This permitted them to resume operations within minutes. Firms that do not need such rapid response could provide information protection with less complete redundancy. For instance, an offsite IT group could take over operations in the event of a disruption, provided that they have been given sufficient training, hardware, software, and access to data [24]. Another approach is to outsource IT operations to a third party, which must guarantee redundancy for protection [63].

2.6.4 Supply Chain Structure

A final form of supply chain protection is the structure of the network itself. The number of levels, nodes at each level, types of nodes, location of nodes and the coordination of them can influence the performance of a supply chain in the face of a disruptive event.

2.6.4.1 Facility Location Decisions

A significant stream of research on supply chain network structure related to disruption management focuses on facility location decisions. By using facility location models in which facilities in the network are subject to disruption, these studies characterize the short- and long-term impacts and means for mitigating them.

Traditional models for the strategic design of supply chain networks focus primarily on cost-efficiency of the system. Because they often assume that every element in the supply chain will always perform as planned, many such models suggest that just-in-time supply chains are appropriate. Indeed, such supply chains have become common in many industries. But if elements do not perform as planned due to a disruption, then just-in-time supply chains may be seriously vulnerable. This has led researchers view supply chains in terms of robustness and reliability, as well as cost.

Bundschuh et al. [8] studied the problem of finding the designs that improve reliability and robustness of supply chain networks that are under the risk of supplier failures. They considered an integrated inbound supply chain network of a manufacturer and introduced two models that combine reliability with robustness: the expected-service level model and the reliability-contingency supply model. They numerically showed that both of these models can lead to significant improvement with respect to the performance of supply chain networks, but considering the cost, the second model is more promising for strategic design of inbound supply chain networks.

Santoso et al. [59] studied a global supply chain network design problem considering scenario-based uncertainty that involves two decisions: (1) decide where to build facilities (which they refer to " major configuration decisions") and (2) what machines to build at each facility (which they refer to "minor configuration decisions"). The objective was to minimize the total expected cost, including a penalty in case the constructed capacity is insufficient to meet the realized demand. They developed a practical solution methodology, which integrates the sample average approximation method with an accelerated decomposition scheme. They showed that the proposed methodology can efficiently identify the candidate design solutions.

Berman et al. [6] considered the possibility that some facilities are subject to disruptive events. Facility failure causes customers to look for alternative facilities from the remaining ones, taking into account the increase in total travel costs. They found that the patterns of the optimal locations dependent on the probability of failure: when the probability of facility failure increases, facilities are located more centralized (to provide better support in the face of a disruption).

Qi et al. [52] considered an integrated supply chain network design problem in which the supplier and retailers are subject to disruptive events that may cause failure of either the supplier or the retailers. The objective was to find the optimal locations of retailer facilities and the assignments of customers to retailers that minimize the expected total costs. They formulated the problem as a nonlinear integer programing model and investigated the impact of the disruptions on retailer location and customer allocation decisions. They numerically showed that the more a retailer is likely to be disrupted, the fewer the customers will be assigned to this retailer. Furthermore, they numerically showed that taking supply chain disruptions into considerations when initialing a supply chain network design can significantly save cost.

After a supply disruption, some facilities become unavailable during the outage, one of the most important questions is how to keep satisfying those customers who originally purchase from the disrupted facilities? One solution is to relocate them, which may result in excessive transportation costs as customers previously served by these facilities must now be served by more distant ones.

Lee and Wolfe [39] reconsidered this problem under the assumption that customers are initially assigned to a hierarchy of facilities, and "higher-level" assignments are used only when the original facilities fail because of a disruption. They used extended p-median and fixed-charge location models to select facility locations that minimize cost, including the extra transportation cost caused by facility failures. They also examined the tradeoff between traditional operation cost and the cost when disruptions are taken into account.

Berman et al. [7] studied a facility location problem where facilities are subject to disruptive events. They assumed that customers always travel to the closest facility, but since they do not know whether or not a given facility is operational in advance, they may have to visit multiple facilities before finally finding an operational one. They investigated the optimal locations of facilities to minimize the total expected cost of customer travel. They also evaluated which part of the supply chain system would benefit the most from improvements in travel cost, reliability, and/or advance information.

While none of these papers provides a simple or general solution to the problem of how to locate facilities within a supply network to achieve optimal protection against the consequences of a disruption, they do provide insights. At a high level, the main insight is that considering the possibility of disruptions has a significant impact on facility location decisions. At a more detailed level, the conclusion is that the right structure depends on detailed specifics of the environment in question.

2.6.4.2 Protection Device Location Decisions

Another stream of research on supply chain network structure related to disruption focuses on protection devices (e.g., inspection devices, detection devices). For example, Gendreau et al. [25] consider the flow interception problem (FIP) on a transportation network with an objective of optimally locating inspection stations within the network so as to maximize risk reduction. James and Salhi [33] focused on the problem of reducing the total outage time by optimally placing protection devices within electrical supply networks. Carr et al. [9] presented a series of related models for optimizing the placement of sensors in water supply networks to detect contaminants. Due to data uncertainties, they considered a restricted absolute robustness criteria. Kumar et al. [36] aimed at different objectives for sensor placement in water supply networks with a goal of minimizing the time to detection.

Although this literature is not directly related to protection of a supply chain against disruptions, it does offer some insights that could be carried over to the supply chain protection problem. Most significantly, instead of looking for the optimal placement of protection devices within a supply chain network, a firm should seek to identify the most critical system components to protect with its limited protection resources. That is, the question of which elements to protect is more important than the question of how to protect them.

2.7 Recovery Strategies

The vast majority of operations management research on supply chain risk has focused on decisions prior to a disruptive event. While an emphasis on preparedness is certainly appropriate, we should not let this overlook the opportunities for mitigating the consequences of a disruption that take place after the disruption has taken place. As we noted earlier, the prospect for losing market share as a result of failing to fill customer orders during a disruption means that the consequences of the disruption could persist long beyond the event itself. Hence, action plans that lessen the post-disruption impacts can be an important part of a supply chain risk reduction strategy.

2.7.1 Customer Loyalty

The consequences of a supply disruption can persist long after the disruption event has ended. If customers shift away from a firm during a disruption, they may not come back immediately, leading to market share and cash flow problems.

For example, Hendricks and Singhal [28] noted that stock returns were substantially depressed for a period 2 years after announcement of a supply disruption. Ericsson suffered depressed market share well beyond the resolution of the supply problem caused by the above-mentioned fire at a Philips plant [61].

Even if a firm may not be able to stop customers from defecting to competitors during a supply disruption, it can minimize the long-term loss of market share by investing in customer loyalty. Of course, cultivating customer loyalty is (or should be) part of business as usual; the more appealing the product, the more likely the customers are to return after a disruption. But there are steps firms can take during a disruption to enhance customer loyalty. For instance, the 1999 Taiwan earthquake caused shortages both in dynamic random access memory (DRAM) and microprocessors that affected many firms, including Dell and Apple. In response, Apple chose to ship pre-ordered computers with a less powerful chip without lowering prices [54], which led to many complaints and order cancelations [44]. In contrast, Dell immediately informed customers and gave them the choice of receiving computers with less memory for the same price or paying a premium for computers with more memory [61]. As a result, Dell increased its market share in spite of the disruption, while Apple lost market share [15]. Evidently, treating customers sensitively and fairly during a disruption can have a significant effect on loyalty.

A long-term customer relationship is based on high customer satisfaction. Such satisfaction yields benefits during normal operations, as well as in recovery situations. Studies indicate that investments in customer retention yield returns of 30–150% [73]. Moreover, it has been estimated that the cost of acquiring a new customer is six to ten times higher than the cost of maintaining a repeat customer [39]. Hence, firms should first seek to hang on to customers affected by a supply disruption. Regaining them later on is likely to be much more difficult and expensive.

Examples of ways a firm can cultivate stronger customer relationships include [29]:

- *Provide Discounts/Rebates*. Discounts and rebates for repeat customers during normal production can be an effective way to retain customers. Similarly, firms can offer discounts or rebates to customers who cannot be satisfied during a supply chain disruption but are willing to return after the outage. Also, if a firm has similar products available as substitutes for a disrupted product, it can discount these products during a disruption to encourage customers of the disrupted product to buy the alternatives.
- *Exceed customer expectations*. Disruptions are bound to happen, but preventing those disruptions due to mistakes from happening again and again, and immediately addressing problems that impact customers, will minimize damage to the firm in the short run and may also present an opportunity for the firm to transform an upset customer into a happy and more loyal one in the long-term. Nothing surprises customers more than an employee going the extra mile to help them. Delivering more than what customers expect is one of the most powerful ways to build a healthy customer relationship. For example, Lexus had a recall early in its history in the United States. They called customers, arranged to pick up their cars, and

left identical cars as loaners for the day in their places. When they returned the car that evening, it was fixed, cleaned, and had a full tank of gas and a coupon for a free oil change. Lexus wisely used this disruption as an opportunity to impress its customers [61]. Thinking in similar terms may provide firms with ideas for keeping customers happy and loyal in spite of a supply disruption.

- *Pay great attention to unhappy customers*. It is impossible for a firm to keep 100% of its customers happy, so strategies are needed to identify unhappy customers who may be inclined to post negative comments (or even videos) on the web. A firm should consider giving unhappy customers an outlet for their comments, so that it can monitor and respond to them as soon as possible. As an example of unsuccessful response to unhappy customers occurred in 1999, Intel was notified by a customer about a floating-point-operation defect of its Pentium 486DX. When Intel failed to respond, the customer posted this problem on a web site, where it rapidly generated public attention. However, Intel's initial response, which offered to replace the 486DX chip for users who could prove a need for using floating-point operations, served to further aggravate customer unhappiness and led to a complete recall of the 486DX. The recall cost Intel more than 400 million dollars, and even more importantly, damaged its reputation and customer relationship [48, 61]. The following is an example of a successful response occurred in 2001, after the 9/11 attacks shut down all US air freight traffic, Continental Teves (a supplier of automotive, industrial, and agricultural products) picked out shipments deemed most important to its customers, promised to reroute them via ground transportation, and tried to deliver them on time. Although this action was costly, Continental Teves successfully impressed its customers [45].
- *Know the required level of customer service*. Understanding customer requirement will help a firm define new market opportunities and drive innovation and revenue growth in every aspect of the organization. For example, Sony predicted the average user of the PlayStation would be a 24-year-old male who wanted a level of customer service that was sophisticated and engaging, so Sony decided its customer-service strategy would involve two customer lines: one for general customer care queries; another only for game enthusiasts that offers hints, tips, and even cheating methods for the specific game.

Of course, there are many other ways to increase customer loyalty, by providing good service before, during and after a supply disruption. Since these measures strengthen the firm in ways that go beyond simply reducing supply chain risk, they should be considered as important parts of an overall strategy.

2.7.2 Insurance

In contrast to "operational hedges" (e.g., inventory, capacity) against supply chain risks, business continuity insurance is a financial hedge. It does nothing to prevent or correct a disruption, but it can mitigate cash flow problems that could plague a firm in the wake of a serious disruption. Unfortunately, since 9/11, the cost of such insurance has increased, and some situations that were once insurable are no longer so. Hence, insurance is generally a supplementary, rather than a primary, risk mitigation strategy.

The finance literature is replete with hedging strategies (some of which involve insurance) but the operations literature has made far less use of these. While there has been some work at the interface of OM and finance, which examines the role of financial hedging in some settings (see e.g., [10]), none of these have explicitly focused on insurance options.

Martha and Subbakrishna [44] categorized insurance against supply chain disruptions into the following two main forms: (1) "all risk" insurance, which hedges the risk of "contingent business interruption" due to disruptive events, such as natural disasters, social disasters, or terrorist attack, and (2) "insurance for civil authority", which hedges the risk from a government's closing of critical infrastructure. While such insurance can be useful in specific settings, the cost has risen since 9/11 and some situations that were once insurable are no longer so. Furthermore, even if insurance is available, it can only mitigate the loss of revenue. It cannot replace customers who impatiently turn to the competitors during a disruption, nor can it restore a loss of reputation. Consequently, insurance and other financial hedges should only be considered as limited parts of an overall supply chain risk mitigation strategy.

The consequences of a supply disruption can persist long after the disruption event has ended. If customers shift away from a firm during a disruption, they may not come back immediately, leading to market share and cash flow problems. For example, [28] noted that stock returns were substantially depressed for a period 2 years after announcement of a supply disruption. Ericsson suffered depressed market share well beyond the resolution of the supply problem caused by the above-mentioned fire at a Philips plant [61]. Two ways a firm can mitigate these long-term consequences are cultivating customer loyalty and buying business continuity insurance.

2.7.3 Post-Disruption Recovery Plan

The term "Disruption Management" (DM) has been used in the literature to refer to recovery plans that aim at preserving, as much as possible, the original plan that was set before disruption (e.g., [16, 42, 77, 78]). This is particulary important in situations (such as airline flight schedules) where changes in the original plan significantly impact customer satisfaction. Following the recovery plan, firms may be able to avoid significant costs associated with breaking away from the original plan, even though the recovery plan might be sub-optimal solution to the original objective function.

Yu [77] investigated how to manage disruptions by using such recovery plans in the airline industry and provided customized solutions for scheduling disruption problems. Yu and Qi [78] investigated different scenarios of DM, which include:

(1) flight scheduling problems, (2) airline crew scheduling problems, (3) machine scheduling problems, (4) logistics scheduling problems, (5) economic production quantity based inventory management problems, (6) discrete production planning problems, (7) supply chain coordination problems, and (8) project scheduling problems. They emphasized that having effective and efficient recovery plans to deal with supply chain disruptions is critical for reducing the economic impact.

Xia et al. [75] focused on economic production quantity-based inventory management problems, and considered disruption recovery plans for a two-stage production and inventory system. They studied two problems: (1) *fixed setup epochs*, for which they found that, when the penalty function for quantity change is linear or quadratic, the best recovery plan can be obtained by solving a quadratic mathematical programming problem; (2) *flexible setup epochs*, for which they found that, when the penalty function for quantity change is convex, the best recovery plan is to just have the same quantities for both stages in a lot-for-lot system.

Yang et al. [76] studied the DM problem faced by a single-product manufacturing plant and investigated the problem of how to build a recovery plan, which can not only work effectively after a disruptive event (such as power failure and supply shortage), but also be close to the original plan, in order to maintain customer satisfaction and reduce the impact on downstream-stages operations. They found that firms can increase the quality of a recovery plan by enlarging recovery times. They also noted that some conditions (i.e., firms have spare production capacity and inventory costs are relatively low) makes DM more valuable as a strategy.

Qi et al. [53] considered the machine schedule updating problem in the face of a disruption. In contrast to other rescheduling analyses, they took the cost due to switching from the original plan to the recovery plan after a disruptive event into account. They focused on cases in which the shortest processing time (SPT) rule is optimal for the original problem, and found that, in many cases, the recovery plan obtained after a disruptive event still maintains the SPT order in some form.

2.8 Managerial Insights

In this section, we appeal to the above research streams to identify a list of policies that firms can pursue to mitigate the consequences of supply chain network disruptions. To categorize these, we make use of Fig. 2.3, which characterizes disruptions according to their likelihood and severity. We associate the regions of this figure with four strategies: *Do Nothing*, *Redundancy*, *Contingency Planning*, *and Crisis Management*. "Do Nothing"refers to ignoring the event, both in the preparation and response stages, which is only appropriate when the consequences are smaller than the cost of action. "Redundancy" refers to strategies that duplicate resources (e.g., inventory or capacity) as protection against disruptions. "Contingency Planning" strategies specify a course of action in advance of a disruptive event. Finally, "Crisis Management" strategies develop a plan of action only after the event has occurred.

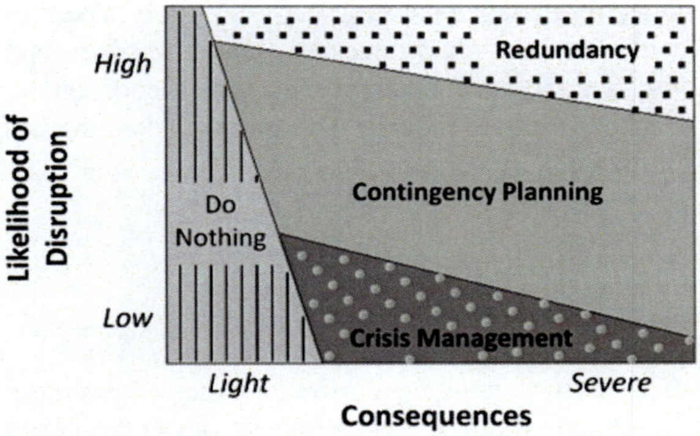

Fig. 2.3 Strategies for dealing with risks depend on likelihood and severity of disruptive event

2.8.1 Do Nothing

It might seem that doing nothing is never an appropriate way to address the risk of a supply disruption. But since the resources are finite, while the potential causes of disruptions are unlimited, firms cannot possibly prepare individually for every type of disruption that could arise. Hence firms have two options: (a) prioritize products and events so that resources are spent preparing for the scenarios with the most serious economic consequences, or (b) make use of strategies that reduce risk of broad categories of scenarios. An effective strategy must make use of a combination of both of these.

Prioritizing scenarios involve prioritizing both events and parts. The reason for prioritizing events is to focus protection attention on events for which such attention can make a difference. For instance, fires are often preventable events. So a firm may be able to reduce the risk of disruption by auditing suppliers with regard to fire prevention practices. But for a large firm with hundreds of suppliers it may not make economic sense to audit all of them. Hence, firms should prioritize their efforts to focus on parts for which disruptions will have the largest impact.

Hopp et al. [29] described the factors that should be taken into consideration when prioritizing parts. These include the likelihood and expected duration of a supply disruption, the value of a unit of market share, customer loyalty, the amount of backup capacity available and the ability of competitor to steal market share, firm's profitability, and the premium cost of a unit of backup capacity. These can be estimated by using a relatively lean set of data and provide a systematic means for identifying the parts that present the greatest risk of lost revenue due to a supply disruption.

Either type of prioritization can lead to the conclusion that doing nothing is the best option for a particular scenario. For instance, an event that will produce a disruption of sufficiently short duration that it can be covered by routine safety stocks does not need a separate risk reduction strategy. Similarly, a part for which a disruption would affect only a low margin product may not be worth exceptional effort to protect. However, for the part/event scenarios for which consequences are significant, firms should make use of redundancy, contingency planning or crisis management strategies to mitigate these consequences.

2.8.2 Redundancy Planning

The most obvious form of redundancy is inventory. Firms can deliberately hold a combination of component inventories and finished goods to protect the supply of product to its customers. For example, the Iraq War had profound consequences on global oil supply chains, but the impact was uneven. India, with inventory protection, announced, "The country will not face shortage as stocks of crude oil and petroleum products are enough to meet 2 months' requirement..." [56]. Without inventory protection, in Russia, crude oil prices were pushed above $45/barrel [18]; in the United States, gasoline and jet fuel prices increased rapidly, with disastrous consequences for airlines who were struggling with heavy debt loads and stiff competition [50].

While inventory is a very effective form of protection against a supply disruption because it is instantly available, it is expensive. Carrying enough inventory to protect against a disruption that could last weeks or even months is completely antithetical to the lean strategies pursued by firms in many industries. Moreover, studies have shown that if the premium cost of using a reliable supplier is comparable with the cost of holding inventory against supply disruptions, it is preferable to make use of reliable suppliers [41]. The reason is that lowered inventory levels reveal problems more quickly and hence are conducive to problem resolution and process improvement. Because of this, inventory protection should be used as protection against disruptions of limited duration, primarily as a means of providing continuity until a backup capacity supply can be brought online [30].

The other major form of redundancy protection against supply chain disruptions is capacity. For example, in October 2001 the United States Postal Service (USPS) shut down the brentwood and processing and distribution center in Washington D.C. after two workers there died of anthrax inhalation [43]. Fortunately, there was redundant capacity available at two other distribution centers in capitol heights and gaithersburg in Maryland and mail delivery was back to normal within a day after the closure [5].

Fully owned redundant capacity is effective because it is the quickest type of capacity to bring on line. But it is also the most expensive form of backup capacity. As we mentioned, a less expensive, but slower form of redundant capacity is virtual capacity at suppliers. For instance, contracts that obligate suppliers to increase supply by a specified percent within a specified time are one way to build virtual capacity into the system. However, because suppliers will have to reserve capacity to live up to these contracts, there will be a cost, generally in the form of higher unit purchase cost. While this cost will not be as large as that for internal backup capacity (because the excess can be pooled among multiple customers for efficiency) it is still costly.

The essential point is that idled capacity is fast but expensive. For scenarios where holding inventory is expensive and a disruption would be costly in terms of lost revenue and market share, it may be economically effective to use a capacity redundancy strategy.

When the high cost of a redundancy strategy is not justified by the potential consequences of a disruption, less expensive contingency planning or crisis management options may be appropriate.

2.8.3 Contingency Planning

A contingency plan is a course of action devised in advance of a particular event. For example, instead of holding dedicated backup capacity in reserve as protection against a disruption (i.e., a redundancy strategy) a firm could develop a plan for securing backup capacity on the open market. To do this, it might assess the suppliers capable of providing such capacity and pre-qualify them.

In addition to a plan for locating backup capacity, the firm should include in its contingency planning a means for identifying the disruption quickly, so that it will be able to secure the available backup capacity before it is bought up by other firms affected by the disruption. Installing monitoring systems on material flows can help spot disruptions quickly and to better distinguish a true disruption from day-to-day variations. For example, Nokia had installed a supply management system that enabled it to identify the Philips supply problem very quickly. As a result, Nokia was able to lock up the available backup capacity before Ericsson (which had no such monitoring system) was aware of the extent of the problem.

Contingency planning can go beyond developing a simple plan to secure backup capacity in the event it is needed. For instance, developing relationships with suppliers can enhance information sharing and make it more likely that a firm will find out about and understand a disruption early on. Similarly, enhancing the firm's reputation beyond its supply base (e.g., by giving some business to potential backup suppliers) can make it more likely that these suppliers will sell their excess capacity to the firm rather than its competitors.

Although a plan for securing backup capacity is only one type of contingency planning (others are plans to keep customers happy during a disruption, plans for maintaining financial solvency, plans for using disruptions to gain market share, etc.), it is an important one. Establishing such a plan involves several steps [29]:

- Identify potential backup suppliers (who could be approached in the face of a disruption) in advance of a supply chain disruption.
- Contractually obligate suppliers through flexible contracts to be able to deliver more than the normal amount within a specified amount of time. This can ensure that non-disrupted suppliers could cover the disrupted supply.
- Design products to be more flexible with regard to their constituent components. For example, in the previously cited case, Nokia did this prior to the Philips

fire and was able to purchase chips from suppliers who were not previously suitable [61].

- Cultivate process and organizational flexibility. For instance, the previously cited case in which a host of suppliers were tapped by Toyota to produce p-valves to replace the supply disrupted by the fire at an Aisin plant is an illustration of the power of having an organizationally flexible supply chain. Working together in task force mode, the many firms in Toyota's keiretsu quickly expanded p-valve capacity that did not exist prior to the incident (see [49, 55] for discussions).

All supply chain risk strategies should involve contingency plans for scenarios that are sufficiently likely to warrant the cost of preparing them but are not sufficiently costly to justify an expensive redundancy policy. But, because the number of disruption scenarios is unlimited, it is not possible to prepare for all of them with redundancy or contingency planning. We already noted that there are some scenarios for which doing nothing is the best strategy. But there are scenarios where advance strategies such as redundancy and contingency planning are too expensive (e.g., because the disruption event is very unlikely), but the cost of not responding at all is also too expensive. For these, a crisis management strategy is the only response left.

2.8.4 Crisis Management

A crisis is a rare event, or series of events, that has severe consequences. Research shows that a firm on average can expect to face a crisis every 4–5 years [47]. Examples include the Asian currency crisis, the 9/11 terrorist attacks, and Hurricane Katrina [2]. Managing a crisis is different from contingency planning because it involves dealing with a scenario after it has occurred rather than before. As such, it is a very flexible strategy, since a plan can be developed for any conceivable scenario. However, executing such a plan is more difficult than executing a contingency plan, because the plan must be developed and implemented at the same time.

The basic steps in crisis management as a response to a supply disruption are:

- Recognize that a crisis has occurred and initiate a response. For instance, in emergency departments, mass casualty events (e.g., major accidents that lead to a spike in patient arrivals) are formally recognized and alternate procedures (e.g., using physicians instead of nurses to conduct triage) are adopted. Toyota used a similar pronouncement to declare a state of crisis after the Aisin fire.
- Communicate with all potentially involved parties, including both suppliers and competitors, to construct a crisis management team and come up with an action plan, which addresses topics such as business-continuity-related issues and customer-related issues. Because response actions must be evolved quickly, it is important for the management team to be able to communicate very effectively. This is why Toyota adopted a war-room protocol for communicating with its major suppliers. The face-to-face communication this protocol provided helped to avoid

delays and errors that could have occurred in business-as-usual communication modes.

- Develop an initial plan quickly so that response actions can be started. For instance, if the plan is to approach outside suppliers to provide backup capacity, then purchasing managers can start communications right away with likely candidate suppliers.
- Revise the plan as additional information becomes available. For example, the suppliers initially thought to be good candidates may turn out not to have sufficient capacity, the needed capabilities, or acceptable quality levels. So an alternate plan will need to be evolved. In many situations, the response will need to be revised frequently, so organizational flexibility is a major virtue in effective crisis management. Such flexibility can be cultivated in advance of an actual crisis by means of simulation exercises in which management teams are presented with an evolving crisis and asked to develop a dynamic response.
- Learn from each crisis scenario so that the organization becomes more skilled at responding to future crises. For example, although Toyota was able to resume full production within 4 days after the Aisin fire, management reviewed the response afterward to find ways to make an even faster response possible.

The specifics of the plans created during a crisis will vary greatly. Of course, details of supply management, such as who will supply what by when, are critical. But a host of other issues, such as product pricing, allocation of limited supplies, communication to manage customer expectations, advertising to maintain brand image, etc., can be important parts of managing a supply crisis. As we noted earlier, poor decisions in such conditions can have negative ramifications. For example, Apple's decision to downgrade the CPU of their computers without reducing prices in response to the supply disruption caused by the 1999 Taiwan earthquake turned out to be a bad decision. Presumably the company learned from this error and would handle a similar scenario differently in the future.

2.9 Future Research Directions

In this chapter, we have described a wide range of research work on supply chain networks subject to disruptions. The various streams of research suggest a number of practical insights for managing supply chain risks. But our overview also suggests gaps that offer opportunities for further research.

The bulk of analytic research that explicitly models supply disruptions considers the impact on a single end product. But many disruptions of a component affect multiple final products (e.g., the Aisin p-valve problem affected virtually all of Toyota's vehicles). So it is interesting and relevant to investigate the situation where a firm may have many final products that rely on the disrupted key component. A related scenario is one in which a disruption affects a specific area (e.g., as occurred in the Taiwan earthquake that disrupted semiconductor suppliers) and a firm has many final products that rely on many components (from different suppliers) within this area.

Most existing models for supply chain network risk management use variations of a minimum total expected cost objective. But for very rare but very severe events, expected cost may not be appropriate. For example, for emergency services and post-disaster backup supplies, availability, and reliability are clearly much more important than cost. Incorporating these objectives in an appropriate way will broaden the applicability of supply chain network models.

Although the observation by Chopra et al. [11] that safety stocks to buffer day-to-day variability are not effective protection against catastrophic supply chain network disruptions, the presence of such safety stocks certainly impacts the ability to respond to a disruption. It is easy to see that if a disruption occurs when safety stock levels are high, then the supply chain system may recover with much less loss of sales or delay of orders than if the disruption occurs when safety stocks are low. The degree of oscillation of these stocks could therefore be a relevant factor in establishing an effective protection policy.

Empirical evidence suggests that disruptions are often accompanied by price shocks. As an example, world banana prices fell markedly in the first half of the 1990s, especially in Europe, and partly recovered in 1998 as supply was curtailed by adverse climatic conditions in Latin America [1]. Hence, an important future research problem is how to account for dynamic pricing in a competitive environment. By incorporating the stochastic relationship between price and time in the wake of a disruption, we could address many interesting questions, such as: What impact do price changes have on customer brand loyalty, competition, and optimal supply chain robustness policies? Which part of a supply chain network is most sensitive to varying price? Which kind of delivery channel—make-to-order or make-to-stock—is more sensitive to price changes?

Finally, the competitive environment in which a firm operates clearly has a significant impact on the firm's preparation strategies. For a firm, a key objective could be making sure that its situation after a disruption is no worse than the majority of its competitors. Embedding the above objective for individual players in a game theoretic environment is a potentially interesting extension of the resear stream on supply chain risk in competitive environments.

References

1. Arias P, Dankers C, Liu P et al (2002) The world banana economy 1985–2002. FAO. http://www.fao.org/docrep/007/y5102e/y5102e00.ht. Accessed 18 Jan 2011
2. Augustine NR, Sharma A et al (2000) Harvard business review on crisis management. Harvard Business School Press, Cambridge
3. Ballman J (2001) Merrill lynch resumes business critical functions within minutes of attack. Disaster Recover J http://www.drj.com/special/wtc/1404-04.htm. Accessed 18 Jan 2011
4. Barnes-Schuster D, Bassok Y, Anupindi R (2002) Coordination and flexibility in supply contracts with options. Manuf Serv Oper Manag 4(3):171–207
5. Beaudette M (2001) Maryland post office gets mail, after anthrax closes facility in Washington. Capital News Service
6. Berman O, Krass D, Menezes M (2007) Facility reliability issues in network p-median problems: strategic centralization and co-location effects. Oper Res 55(2):332–350

7. Berman O, Krass D, Menezes M (2009) Locating facilities in the presence of disruptions and incomplete information. Decis Sci 40(4):845–868
8. Bundschuh M, Klabjan D, Thurston DL (2003) Modeling robust and reliable supply chains. Working paper, University of Illinois at Urbana-Champaign
9. Carr RD, Greenberg HJ, Hart WE et al (2000) Robust optimization of contaminant sensor placement for community water systems. Math Prog 107:337–356
10. Chod J, Rudi N, Van Mieghem JA (2009) Operational flexibility and financial hedging: complements or substitutes? Working paper, Carroll School of Management, Boston College
11. Chopra S, Reinhardt G, Mohan U (2007) The importance of decoupling recurrent and disruption risks in a supply chain. Nav Res Logist 54(5):544–555
12. Chopra S, Sodhi MS (2004) Managing risk to avoid supply-chain breakdown. MIT Sloan Manag Rev 46:53–61
13. Christopher M, Peck H (2004) Building the resilient supply chain. Int J Logist Manag 15(2): 1–13
14. Christopher M, Lee H (2004) Mitigating supply chain risk through improved confidence. Int J Phys Distrib Logist Manag 34(5):388–396
15. Christopher BP (2003) Strategies for maximizing supply chain resilience: learning from the past to prepare for the future, Master of engineering in logistics thesis, June 2003, MIT press, Cambridge
16. Clausen J, Hansen J, Larsen J et al (2001) Disruption management. ORMS Today, October, pp 40–43
17. Cohen M, Agrarwal N (1999) An analytical comparison of long and short term contract. IIE Trans 31(8):783–796
18. ConsumerAffairs (2004) Gas prices keep climbing ... or do they? http://www.consumeraffairs.com/news04/gas_prices2.html. Accessed 18 Jan 2011
19. Dada M, Petruzzi NC, Schwarz LB (2007) A newsvendor's procurement problem when suppliers are unreliable. Manuf Serv Oper Manag 9(1):9–32
20. Elkins D, Handfield RB, Blackhurst J, Craighead CW (2005) 18 Ways to guard against disruption. Supply Chain Manag Rev 9(1):46–53
21. Fawcett SE, Calantone R, Smith SR (1996) An investigation of the impact of flexibility on global reach and firm performance. J Bus Logist 17(2):167–196
22. Fredericks E (2005) Infusing flexibility into business-to-business firms: a contingency theory and resource-based view perspective and practical implications. Ind Mark Manag 34(6): 555–565
23. Fried I, Hu J (2003) Power outage hits Northeast. cnet news. http://news.cnet.com/2100-1011_3-5063997.htm. Accessed 18 Jan 2011
24. GAO (2004) Information system controls at the federal deposit insurance corporation. http://www.gao.gov/new.items/d04630.pdf. Accesed 18 Jan 2011
25. Gendreau M, Laporte G, Parent I (2000) Heuristics for the location of inspection stations on a network. Nav Res Logist 47:287–303
26. Goldsby TJ, Stank TP (2000) World class logistics performance and environmentally responsible logistics practices. J Bus Logist 21(2):187–208
27. Handfield R (2007) Reducing the impact of disruptions to the supply chain. http://www.sas.com/resources/asset/sascom.pdf. Accessed 18 Jan 2011
28. Hendricks KB, Singhal VR (2005) An empirical analysis of the effect of supply chain disruption on long-run stock price performance and equity risk of the firm. Prod Oper Manag 14(1):35–52
29. Hopp WJ, Iravani SMR, Liu Z (2008) Strategic risk from supply chain disruptions. Working paper, Department of Industrial Engineering and Management Sciences. Northwestern University
30. Hopp WJ, Liu M, Liu Z (2010) Protecting supply chain networks against catastrophic failures. Working paper Wisconsin School of Business, University of Wisconsin–Madison
31. Hopp WJ, Spearman ML (2007) Factory Physics. McGraw-Hill, New York

32. Hu X, Duenyas I, Kapuscinski R (2008) Optimal joint inventory and transshipment control under uncertain capacity. Oper Res 56(4):881–897
33. James JC, Salhi S (2002) A tabu search heuristic for the location of multi-type protection devices on electrical supply tree networks. J Comb Optim 6:81–98
34. Johnson ME (2001) Learning from toys: lessons in managing supply chain risk from the toy industry. Calif Manag Rev 43(3):106–124
35. Kharif O (2001) Will ericsson hang up on cell-phones altogether? http://www.businessweek.com/bwdaily/dnflash/feb2001/nf20010212_208.ht. Accessed on 18 Jan 2011
36. Kumar A, Kansal ML, Arora G (1999) Detecting accidental contaminations in municipal water networks. J Water Resour Plan Manag 125(5):308–310
37. Kumar S, Stecke K (2008) Sources of supply chain disruptions, factors that breed vulnerability, and mitigating strategies. J Mark Channels 16(3):193–226
38. Kumins L, Bamberger R (2006) Oil and gas disruption from hurricanes katrina and rita. https://www.hsdl.org/?abstract\&doc=62277\&coll=limited. Accessed 18 Jan 2011
39. LeBoeuf M (2000) How to win customers and keep them for life. Berkley Trade
40. Lee HL, Wolfe M (2003) Supply chain security without tears. Supply Chain Manag Rev 7(1):12–20
41. Liu S, So KC, Zhang F (2009) The effect of supply reliability in a retail setting with joint marketing and inventory decisions. Manuf Serv Oper Manag (forthcoming) 12(1):19–32
42. Løve M, Sørensen KR, Larsen J et al (2002) Disruption management for an airline—rescheduling of aircraft, Lecture notes in computer science vol 2279, pp 289–300
43. Mairesse M (2002) Did the government okay the anthrax attacks? http://www.hermes-press.com/anthrax_atrocities2.ht. Accessed 18 Jan 2011
44. Martha J, Subbakrishna S (2002) Targeting a just-in-case supply chain for the inevitable next disaster. Supply Chain Manag Rev September/October, 18–23
45. Martha J, Vratimos E (2002) Creating a just-in-case supply chain for the inevitable disaster. Mercer Manag J 14:73
46. McGillivray G (2000) Commercial risk under JIT. Can Underwriter. January 1
47. Merdinger I (2010) Crisis Management. http://global.marsh.com/risk/crisis/. Accessed 18 Jan 2011
48. Mueller S, Soper ME (2001) Microprocessor types and specifications. http://www.informit.com/articles/article.aspx?p=130978\&seqNum=28.. Accessed 18 Jan 2011
49. Nishiguchi T, Beaudet A (1998) The Toyota group and the Aisin fire. MIT Sloan Manag Rev 40(1):49–59
50. Padley K (2003) Iraq war could cost airlines 10 billion us dollars. http://english.people.com.cn/200303/23/eng20030323_113832.shtm. Accessed 18 Jan 2011
51. Peck H (2005) Drivers of supply chain vulnerability: an integrated framework. Int J Phys Distrib Logist Manag 35(4):210–232
52. Qi L, Shen ZJM, Snyder LV (2009) The effect of supply disruptions on supply chain design decisions. Working paper, Department of Industrial and Systems Engineering. Lehigh University
53. Qi X, Bard JF, Yu G (2006) Disruption management for machine scheduling: the case of SPT schedules. Int J Prod Econ 103(1):166–184
54. Quinlan T (1999) PC Prices taking an unusual turn: up. San Jose Mercury News. 20 October
55. Reitman V (1997) To the rescue: Toyota's fast rebound after fire at supplier shows why it is tough. Wall Str J. 8 May, A1
56. Rediff (2003) Iraq war not to impact oil supplies to India. http://www.rediff.com/money/2003/mar/20war2.htm. Accessed 18 Jan 2011
57. Reuters (2005) Rita shuts down US refineries. http://www.theage.com.au/news/world/rita-shuts-down-us-refineries/2005/09/22/1126982154325.html Accessed 18 Jan 2011
58. Rice J (2003) Supply chain response to terrorism: creating resilient and secure supply chains. Interim report of progress and learning, supply chain response to terrorism project, MIT center for transportation and logistics, Cambridge

59. Santoso T, Ahmed S, Goetschalckx M et al (2005) A stochastic programming approach for supply chain network design under uncertainty. Eur J Oper Res 167:96–115
60. Sheffi Y (2001) Supply chain management under the threat of international terrorism. Int J Logist Manag 12(2):1–11
61. Sheffi Y (2005) The resilient enterprise: overcoming vulnerability for competitive advantage. MIT Press, Cambridge
62. Simchi-Levi D, Snyder L, Watson M (2002) Strategies for uncertain times. Supply Chain Manag Rev 6(1):11–12
63. Stowers D (2005) Outsourcing IT could help cut costs, while improving reliability, security. http://www.cyrusone.com/grafx/50060-017.pd. Accessed 18 Jan 2011
64. Swafforda PM, Ghoshb S, Murthyc N (2006) The antecedents of supply chain agility of a firm: scale development and model testing. J Oper Manag 24(2):170–188
65. Swinney R, Netessine S (2009) Long-term contracts under the threat of supplier default. Manuf Serv Oper Manag 11(1):109–127
66. Tang CS (2006) Robust strategies for mitigating supply chain disruptions. Int J Logist Res Appl 9(1):33–45
67. Tang C, Tomlin B (2008) The power of flexibility for mitigating supply chain risks. Int J Prod Econ 116:12–27
68. Tomlin B (2006) On the value of mitigation and contingency strategies for managing supply chain disruption risks. Manag Sci 52(5):639–657
69. Tomlin B, Wang Y (2005) On the value of mix flexibility and dual sourcing in unreliable newsvendor networks. Manuf Serv Oper Manag 7(1):37–57
70. Tomlin B, Wang Y (2009) Operational strategies for managing supply chain disruption risk, intended for Kouvelis P, Boyabatli O, Dong L et al (eds) Handbook of integrated risk management in global supply chains, Wiley, new York (to be published)
71. Van Mieghem JA (2007) Risk mitigation in newsvendor networks: resource diversification, flexibility, sharing, and hedging. Manag Sci 53(8):1269–1288
72. Wang Y, Gilland W, Tomlin B (2009) Mitigating supply risk: dual sourcing or process improvement? Manuf Serv Oper Manag. doi:10.1287/msom.1090.0279
73. Wreden N (2009) How to recover lost customers. http://www.fusionbrand.com/pdf/how_to_recover_lost_customers.pdf. Accessed 18 Jan 2011
74. Xia YS, Yang MH, Golany B et al (2004) Real time disruption management in a two-stage production and inventory system. IIE Trans 36(2):111–125
75. Yang J, Qi XT, Yu G (2005) Disruption management in production planning. Nav Res Logist 52(5):420–442
76. Yu G (1998) Operations research in the airline industry. Kluwer, Boston
77. Yu G, Qi X (2004) Disruption management: framework, models, and applications. World Scientific Publishers, Singapore
78. Zhu G, Bard JF, Yu G (2005) Disruption management for resource-constrained project scheduling. J Oper Res Soc 56:365–381

Chapter 3
Sourcing Strategies to Manage Supply Disruptions

Amanda J. Schmitt and Brian Tomlin

3.1 Introduction

Supply chains depend on the successful flow of material in order to function and satisfy customer demand. When that flow is interrupted at a particular facility, alternate sources can be considered to keep material moving through the chain. We explore such alternatives in this chapter, focusing primarily on two sourcing strategies: diversification and emergency backup sourcing. Diversification means a firm uses multiple supply sources on an ongoing basis, which lowers the impact of any one source becoming unavailable. However, this requires ongoing investment in multiple supplier relationships and logistics. Emergency backup sourcing, on the other hand, is used only if a disruption occurs and thus may cost less on an ongoing basis; however, in the event of a disruption that emergency source may be more costly to use and have a slower response time than a routine supplier.

In the chapter we introduce some simple models to analyze these considerations. We focus on developing insights as to which strategies are appropriate in various settings. Our models reflect many important aspects of disruption mitigation but do not reflect all the possible complexities that might influence strategy implementation in a given firm's setting. We would recommend a more tailored analysis for detailed implementation purposes.

While the term "supplier" typically refers to external sources, we use it throughout this chapter to mean either an external or an internal source of supply. Internal

A. J. Schmitt
Center for Transportation and Logistics, Massachusetts Institute of Technology, Cambridge, MA 02139, USA

B. Tomlin (✉)
Tuck School of Business at Dartmouth, 100 Tuck Hall, Hanover, NH 03755, USA
e-mail: brian.tomlin@tuck.dartmouth.edu

H. Gurnani et al. (eds.), *Supply Chain Disruptions: Theory and Practice of Managing Risk*, DOI: 10.1007/978-0-85729-778-5_3,

sources refer to supply chains with multiple levels within a single firm, where downstream locations depend on upstream sources for their material. Firms should evaluate sourcing strategies not only for external suppliers, but also for mitigating disruptions to their own upstream network locations.

As the focus of this chapter is on sourcing strategies to manage disruption risk, we ignore inventory in this chapter unless otherwise stated. We refer the reader to Chap. 5 for a full treatment of the inventory strategy. For the sake of completeness, we include a brief analysis of the inventory strategy in the appendix to this chapter.

The remainder of the chapter is organized as follows: in Sect. 3.2 we cover the acceptance strategy (no proactive mitigation) and introduce our modeling approach. In Sect. 3.3 we discuss diversification strategies, and in Sect. 3.4 we discuss emergency backup strategies. We compare these approaches along with acceptance and inventory in Sect. 3.5 We discuss other considerations for sourcing strategies in Sect. 3.6, and we summarize the chapter's insights in Sect. 3.7. The appendix presents the mathematical development of the profit expressions shown in this chapter.

3.2 Acceptance Strategy

We start our analysis by studying the acceptance strategy in which the firm sources from a single supplier (prone to disruptions) and does not attempt to mitigate supply risk through inventory, supplier diversification, or backup supply. The acceptance strategy can stand as a benchmark against which other strategies can be measured. Also, it may be an appropriate strategy if disruption risk is low or if the mitigation costs are very high.

In what follows we introduce our modeling approach and develop an expression for the long-run average profit the firm obtains under the acceptance strategy.[1] The other strategies will build upon the model introduced here. We assume that all unsatisfied demand is lost, but discuss backlogging of demand in Sect. 3.6. We adopt the following notation: each unit sold yields a profit margin of *MARG*, unsatisfied demand incurs a penalty cost of *LOST* per unit, and the supplier's exogenous percentage uptime is denoted by *UP*, where $0 \leq UP \leq 1$. For example, if a supplier has a probability of being available of 95%, then *UP* would be 0.95. We assume demand is constant and normalize the demand to be 1 in every period.[2]

The acceptance strategy gives a per-period profit of *MARG* in a non-disrupted period, but in a disrupted period a penalty cost of *LOST* is incurred. Thus the acceptance strategy profit, denoted ACC_PROF, is equal to $MARG \times UP - LOST(1 - UP)$.

[1] By "long-run average profit" we mean the average profit per period the firm would obtain if there were an infinite number of periods. For the sake of brevity we will drop the term "long-run average" and simply refer to profit throughout the remainder of the chapter.

[2] If demand is constant at some $d \neq 1$, then we can scale all profit expressions in this chapter by multiplying by d.

Equivalently, this can be presented as:

$$ACC\_PROF = MARG - (MARG + LOST)(1 - UP) \quad (3.1)$$

If the supplier was perfectly reliable, i.e., 100% uptime, the profit would be *MARG*. Therefore the lost profit associated with the lack of reliability is $(MARG + LOST)(1 - UP)$, and this lost profit increases (linearly) as the supplier becomes less reliable, i.e., as *UP* decreases.

3.3 Supply Diversification

Diversifying supply sources is a logical way to manage the risk of supply disruptions. Diversification means that multiple sources are used for the same product on a regular basis. Thus some material flow will still continue in the event of a disruption if at least one supplier is still operating. Diversification takes time and effort, though. Contracts and relationships with all external suppliers must be created and maintained, or investments must be made to create and maintain multiple internal sources for a product. Operational capability for receiving or picking up material from multiple sources must be put in place and maintained. There are clear savings available from making a firm's supply base more lean (reducing the number of suppliers); thus the risk-mitigation benefits of supplier diversification need to be weighed against the cost of expanding the firm's supply base.

Diversification is not a panacea for disruptions. Issues can arise if problems simultaneously disrupt multiple facilities or if spare capacity is not sufficient. Kellogg Co. recently experienced disruptions at two of its four plants that produce Eggo brand frozen waffles. The plant in Atlanta, Georgia, experienced a bacterial contamination (and unrelated flooding) that shut down the facility for most of September and October 2009 [6]. In the mean time, multiple production lines at the plant in Rossville, Tennessee, were closed for repairs. Because the other two plants did not have sufficient capacity to make up for the disrupted production, Kellogg forecasted that there would be a shortage of Eggo waffles on retail shelves until the middle of 2010 [11]. While the supplier diversification strategy helped Kellogg mitigate the disruptions it did not completely insulate the business from their effects: "The temporary supply disruption contributed to the 3% sales decline in the North American frozen and specialty channels unit last quarter, Kellogg chief executive officer David Mackay said on an October 29 earnings conference call" [3].

In what follows, we explore the impact of the number of suppliers, disruption correlation and spare capacity on the performance of the diversification strategy.

3.3.1 Number of Suppliers

Suppose the firm uses multiple suppliers, with the number of suppliers denoted by *NUM_SUP*. Assume that all suppliers have the same percentage uptime *UP*. To start, assume that supplier disruptions are not correlated and that suppliers have infinite

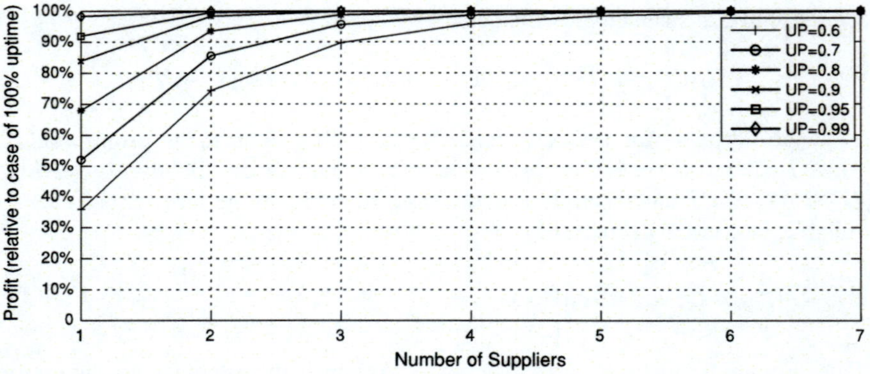

Fig. 3.1 Diminishing returns to adding suppliers

capacity (that is, any supplier can instantly satisfy 100% of the demand). Because any single supplier can satisfy all demand, the firm's supply is only disrupted if *all* suppliers are simultaneously down, which happens with probability $(1-UP)^{NUM\_SUP}$. Therefore, the diversification strategy profit is given by DIV_PROF_u (where the u subscript denotes uncorrelated suppliers):

$$DIV\_PROF_u = MARG - (MARG + LOST)(1 - UP)^{NUM\_SUP} \tag{3.2}$$

We have purposely ignored any ongoing fixed costs associated with maintaining a supplier. A full accounting would subtract such costs from this expression. If $NUM\_SUP = 1$ then, $DIV\_PROF_u = ACC\_PROF$, i.e., Eq. 3.2 reduces to Eq. 3.1. Increasing the number of suppliers increases the profit, as we would expect. An important characteristic of the profit function is that it exhibits diminishing returns to the number of suppliers, that is, the incremental value of adding another supplier decreases as the number of suppliers increases.

We illustrate this using a particular numerical example, setting the profit margin, *MARG*, at 5, and the lost sales parameter, *LOST*, at 3. For different *UP* values, Fig. 3.1 shows the profit as *NUM_SUP* increases. (The profit is reported as a percentage of the profit if there are never any disruptions, i.e, *MARG*, which is 5 for this case). Observe the rapidly diminishing returns to adding suppliers. The benefits of diversification are largely achieved with a relatively small number of suppliers. Even at extremely low uptime percentages, e.g., $UP = 0.6$, any incremental risk-mitigation benefit of going beyond four suppliers will likely be outweighed by the additional fixed costs of maintaining more suppliers. For reasonable uptime percentages, firms may find that using two or three suppliers strikes an appropriate balance between risk mitigation and supply-base rationalization.

3.3.2 Correlated-Supplier Disruptions

The underlying cause of a supplier disruption may not be specific to the affected supplier. For example, a bankruptcy at one supplier may be due to a global recession, and so other suppliers may also have a heightened risk of bankruptcy. Or if suppliers are located in the same geographic region, then a natural disaster in that region may impact multiple suppliers. Indeed, as illustrated by the Kellogg example above, multiple suppliers might be simultaneously disrupted by different events. It is important to consider the impact of disruption correlation when evaluating the diversification strategy.

Let *CORR* denote the (pair-wise) correlation coefficient for supplier disruptions. Our model allows for no correlation or positive correlation, that is $0 \leq CORR \leq 1$. $CORR = 0$ indicates independent disruptions and, at the other extreme, $CORR = 1$ indicates perfectly positively correlated disruptions, i.e., if one supplier is down then all are down. The diversified profit (allowing for correlation, denoted with the subscript c) is:

$$\begin{aligned} DIV\_PROF_c &= MARG - (MARG + LOST) \\ &\quad \times \left(1 - \frac{UP\left(1 - [(1 - CORR)(1 - UP)]^{NUM\_SUP}\right)}{1 - (1 - CORR)(1 - UP)}\right) \end{aligned} \tag{3.3}$$

The development of this expression (and all other expressions) can be found in the appendix at the end of this chapter. We note that if $CORR = 0$, this expression collapses to our earlier expression Eq. 3.2 for DIV_PROF_u. The profit decreases as *CORR* increases, that is, diversification provides less value as disruption correlation increases. At $CORR = 1$, $DIV\_PROF_c = MARG - (MARG + LOST)(1 - UP)$ which is the acceptance profit given by Eq. 3.1, and so diversification provides no value if disruptions are perfectly positively correlated. We also note that DIV_PROF_c exhibits decreasing reductions to correlation, that is, the incremental profit reduction is lower at higher values of *CORR*; however, the profit decreases in an almost straight-line fashion unless the uptime percentage is low.

We illustrate these general findings in Fig. 3.2 which shows the profit (again as a percentage of the no-disruption profit) as a function of correlation for two different percentage uptime values: 0.85 and 0.95. As before we set $MARG = 5$ and $LOST = 3$. Observe that as the correlation increases, the profit decreases and eventually equals the profit of a single-supplier strategy at $CORR = 1$. (When only one supplier is used, then obviously correlation has no affect on the profit). As before, we see that adding just a few suppliers to diversify sourcing can offer significant increases in profits, but the value of additional suppliers is lower at higher correlations.

Disruption correlation is not readily measured and firms cannot design their supply chain to "dial in" a precise correlation number. However, firms can qualitatively assess the degree of correlation in their current or proposed supply base. Suppliers located in the same region may be prone to correlated natural-hazard or socio-political-related disruptions. Suppliers using the same raw material input pose the

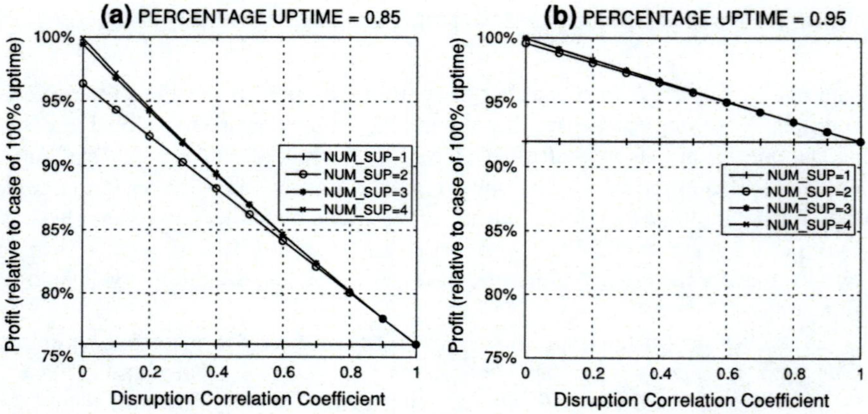

Fig. 3.2 Profit decreases as disruption correlation increases

risk of simultaneous contamination-induced disruptions. As correlation reduces the value of diversification, firms can and should design their supply chains to reduce correlation when possible, e.g., by not concentrating suppliers in one region.

3.3.3 Limited Spare Capacity

To this point we have assumed that each supplier had sufficient capacity to produce the total demand if needed. While one would expect that suppliers can provide *some* additional capacity beyond their normal volume, it may not be reasonable to expect that they can produce the full demand amount. We now expand our model to consider capacity constraints at the suppliers. In doing so, we assume that order quantities are evenly split across the available suppliers and that all suppliers have the same spare capacity. We introduce the parameter *SP_CAP* to denote how much capacity any supplier can provide in excess of their normal order quantity; if 10 units are normally ordered from a supplier and its spare capacity is $SP\_CAP = 0.20$, then the maximum volume that supplier can provide is 12 units. When suppliers can provide some spare capacity, but not infinite capacity, the mathematical formulation becomes complex. We include it in the appendix for reference, but discuss and demonstrate the impact of spare capacity here.

First, we consider the case in which there is *no* spare capacity at the suppliers. The expected profit is $DIV\_PROF_{ns} = MARG - (MARG + LOST)(1 - UP)$, with the subscript *ns* denoting no spare capacity. Perhaps surprisingly, this is the same as the acceptance strategy profit given in Eq. 3.1. In other words, diversification provides no value if there is no spare capacity. While diversification does provide protection if a supplier is down (because only a fraction of supply is lost), the likelihood of at

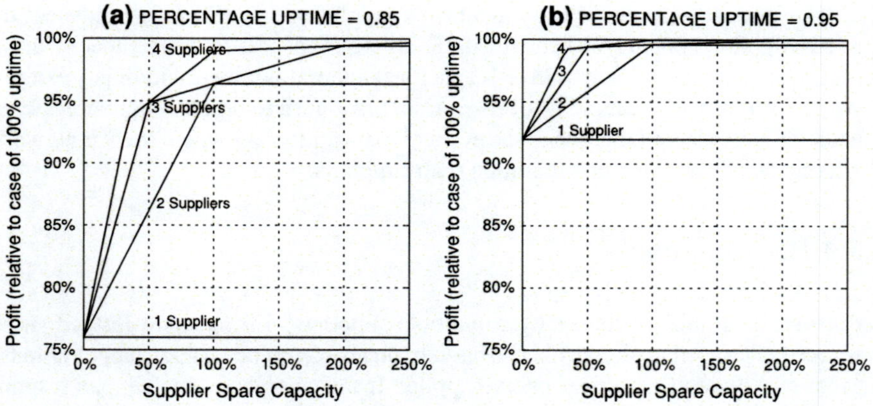

Fig. 3.3 Diminishing return to increasing spare capacity

least one supplier being down is higher when there is more than one supplier. These counteracting factors balance each other to eliminate any value.[3]

Next we consider the case in which suppliers have some spare capacity. The profit increases in the spare capacity but it exhibits diminishing returns, that is, the incremental value of adding more spare capacity is lower at higher spare capacities. These general findings are illustrated for two different percentage uptime values (0.85 and 0.95) in Fig. 3.3. As before we set $MARG = 5$ and $LOST = 3$. Observe that spare capacity has a large impact on profit and that there are diminishing returns to spare capacity.

Recall that SP_CAP is the percentage a supplier can provide above and beyond its normal order quantity. For example, if demand is 100 units and SP_CAP is 50%, then with two suppliers each can provide at most 75 units (1.5 times their normal order quantity of 50), and with three suppliers each can provide at most 50 units (1.5 times 33.33). For a given SP_CAP value, the total capacity in the supply base is the same regardless of the number of suppliers, e.g., 150 in the example just given. Looking at the two-supplier case in Fig. 3.3, we see that the profit increases until the spare capacity reaches 100% (where either supplier can fully back each other up) and then remains level. When one supplier has enough spare capacity to meet all demand, then there is no additional value to adding more capacity. With three suppliers, the profit increases until each supplier has 50% spare capacity (at which point if one supplier goes down the other two can provide full demand coverage) and then increases at a lower rate until each supplier has 200% spare capacity (at which point each supplier can fully back up both other suppliers). The 4-supplier curve has a similar form, with distinct changes in slope when a capacity is reached where more suppliers may be disrupted without reducing material availability.

[3] We are assuming a risk neutral objective function, i.e., long-run average profit. Diversification would provide value in a risk-averse setting even if there is no spare capacity.

The amount of spare capacity in a firm's supply base significantly influences the risk mitigation benefits of diversification. To ensure access to spare capacity, firms may need to invest in some "safety" capacity at internal suppliers and/or pay external suppliers to provide volume flexibility. Also, firms need to rapidly communicate the need for increased production in the event of a disruption elsewhere in their network. All this entails ongoing collaboration with suppliers.

3.4 Backup Supply

Rather than routinely source from multiple suppliers, a firm might instead single source under normal circumstances but rely on an emergency backup supplier in the event of a disruption to its primary supplier. If the emergency backup can respond rapidly when called upon, then an adequate flow of material can be maintained. Single sourcing eliminates the complexities and costs associated with routinely using multiple suppliers. However, emergency backup sourcing has its own set of complexities. Assuming that a backup source exists without validating availability and response time can leave a firm vulnerable to disruptions. Firms that rely on a spot market for emergency supply may discover that their competitors do also, causing issues with price and material availability in the event of an industry-wide disruption. Even similar plants within the firm's own network may not be able to back each other up if they do not have available capacity, or if labeling, customs, or overlooked plant differences cause issues.

For example, we learned from a large consumer packaged goods firm that they had issues implementing their backup plan when customs went on strike in a South American country in which one of their plants was located. Managers had assumed that they would be able to source material from another plant in their network which made the same product and had some capacity flexibility. However, when trying to implement the plan, it was quickly realized that packaging and labeling was not identical for the two plants and therefore the emergency backup could not be put into effect. The firm lost over a million dollars while it waited for the strike to end so that the South American plant could receive raw materials again and try to catch up on backlogged demand. Effective backup sourcing requires proactive planning, and the firm is working to outline better plans to provide backups for certain critical facilities if they are disrupted. It put one such advance plan into action in 2009 when a tornado disrupted one of its distribution centers (DCs) and another DC was able to cover the disrupted DCs' demand within a day or two.

In what follows, we explore the impact of the backup supplier cost, response time and capacity on the performance of the backup strategy.

3.4.1 Modeling Backup Supply

As sourcing from an emergency supplier typically costs more than a routine supplier we assume the unit profit margin when using the backup supplier, *EM_MARG*, is less than unit profit margin, $MARG$, when using the regular supplier. The backup supplier will only be used if it makes economic sense, that is, using the backup supplier must be better than incurring a penalty for not filing demand, i.e., $EM\_MARG > -LOST$; otherwise the backup strategy is not viable. We assume that it takes time for the backup supplier to come on stream. This response time, EM_RES, reflects the time taken to alert the emergency backup and to get material flowing from the backup. If a disruption occurs, the backup can only provide supply after EM_RES periods have passed. Also, we allow for the fact that the backup supplier might not have sufficient capacity to cover all demand; we denote the backup's capacity limit as EM_CAP, where $0 \leq EM\_CAP \leq 1$ is the percentage of demand that the backup supplier can fill.[4] Because we allow for a delayed response, the percentage uptime UP does not fully capture the disruption attributes relevant to the backup strategy (as it did for the acceptance and diversification strategies modeled earlier). In addition to the percentage uptime we also need to characterize the length of disruptions. For simplicity, we assume that (1) when the primary supplier is up there is a constant probability of a disruption occurring, and (2) when the primary supplier is down there is a constant probability of the disruption ending. That is, disruptions are geometrically distributed. We denote the average disruption length as DIS_LEN.

The profit in a period when the regular supplier is up is $MARG$. If the regular supplier has been down EM_RES periods or less, then the profit in that period is $-LOST$ as the backup supply has yet to come on stream. If the regular supplier has been down for more than EM_RES periods, then the profit in that period is $EM\_MARG \times EM\_CAP - LOST \times (1 - EM\_CAP)$, as the backup can only supply EM_CAP. This leads to the following profit for the emergency backup strategy, EM_PROF :

$$EM\_PROF = MARG - (MARG + LOST)(1 - UP) + (EM\_MARG + LOST)(1 - UP)(EM\_CAP)\left(1 - \frac{1}{DIS\_LEN}\right)^{EM\_RES} \tag{3.4}$$

As one would expect, the profit decreases (in a straight line) as the emergency supplier becomes more expensive, i.e., as EM_MARG decreases. Interestingly, the profit increases as the average disruption length increases, i.e., as DIS_LEN increases. This is because for a given percentage uptime, the frequency of disruptions decreases as the average disruption length increases. Also, the fraction of the disruption time for which the backup is not producing (due to the delayed response) is shorter for longer disruptions, e.g., 50% for a disruption of length 6 and 25% for a disruption of length 12 if the response time is 3. Therefore, the backup strategy is more effective at mitigating rare-long disruptions than short-frequent ones, and this is why the profit increases as the average disruption length increases (for a given percentage uptime).

---

[4] Because we assume unfilled demand in lost rather than backlogged, there is no additional value to having emergency capacity that exceeds demand, i.e., $EM\_CAP > 1$. Therefore we restrict attention to $0 \leq EM\_CAP \leq 1$. If demand is backlogged then there can be additional value for $EM\_CAP > 1$, see Sect. 3.6

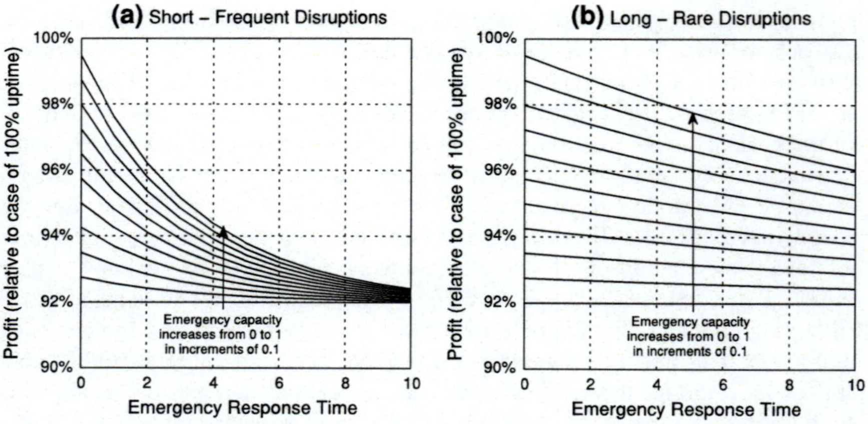

Fig. 3.4 Increasing returns to reducing response time

We now turn our attention to the role that response time and emergency capacity play in determining the effectiveness of the backup strategy.

3.4.2 Emergency Response Time

The backup profit decreases in the response time, and firms should make every effort to eliminate unnecessary delays in detecting the disruption and activating the emergency supply. Managers should ensure there is an agreed plan in place that sets out the actions and responsibilities required for rapid response. Emergency suppliers should be selected and validated in advance of any disruption. Effective planning can minimize unnecessary delays but it may take time to ramp up capacity at the backup and there may be physical constraints that preclude immediate production.

The backup strategy profit exhibits increasing returns to response time reductions; that is, the incremental benefit to reducing response times is higher the faster the response time. This is illustrated in Fig. 3.4 for two different disruption profiles: short-frequent and long-rare. The percentage uptime was set at $UP = 0.95$ and, as before, we used $MARG = 5$ and $LOST = 3$. The average disruption length was 4 periods for the short-frequent cast and was 20 periods for the long-rare case. Because the uptime was the same for both cases, disruptions are much more common in the short-frequent case - hence the name.

Looking at Fig. 3.4 we see that the nature of the disruption (short-frequent or long-rare) plays a crucial role. The profit falls off rapidly in the case of short-frequent disruptions; if the emergency supplier cannot respond very quickly then the backup strategy is not effective at mitigating short-frequent disruptions. The profit falls off much more slowly in the case of long-rare disruptions. In negotiating with a potential

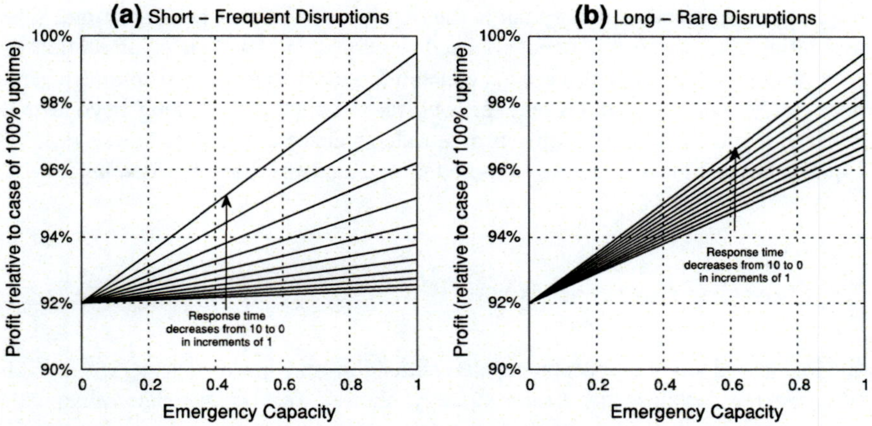

Fig. 3.5 Profit increases (linearly) in emergency capacity (until $EM\_CAP = 1$)

backup supplier, there may be a tradeoff between cost, response time, and capacity. In the case of long-rare disruptions it may make sense to sacrifice some response time to gain on the other dimensions. For short-frequent disruptions, the firm has to gain very significant concessions on cost and capacity to make up for response time degradation.

3.4.3 Emergency Capacity

The backup profit increases in the emergency capacity. Ensuring adequate surge capacity at an internal source (if that is the backup) may necessitate additional capital investment or a flexible workforce. Ensuring additional capacity at an external backup source might require the firm to pay an ongoing fee to reserve a desired level emergency capacity. If no one supplier can guarantee sufficient capacity, then the firm might contract with multiple suppliers to provide backup capacity.

We illustrate the impact of emergency capacity on the backup profit in Fig. 3.5, using the same parameters as used for Fig. 3.4. The profit increases in a straight line until the emergency capacity, EM_CAP, equals 1, after which the profit would stay constant because unfilled demand is not backlogged (see earlier Footnote #4). Importantly, we again see that the nature of the disruption (short-frequent or long-rare) plays a significant role. For short-frequent disruptions, the profit is somewhat insensitive to capacity when the response time is long (e.g., 8, 9, or 10 in this figure) because additional capacity does not matter greatly if disruptions are almost over by the time the backup supplier comes on stream. In the case of rare-long disruptions, however, the profit increases significantly in capacity even at these longer response times.

Firms may face a tradeoff when selecting a backup supplier: one supplier might offer rapid response but only provide a limited capacity whereas another supplier might provide greater capacity but at the expense of a slower response time. When evaluating such tradeoffs, managers need to understand the type of disruption risk they face. Response time is a crucial concern for short-frequent disruptions whereas emergency capacity is important for long-rare disruptions. Moving beyond these generalities to explore a specific firm's tradeoff curve between response time and capacity is possible but requires a more tailored analysis than provided here.

3.5 Choosing the Strategy

Having examined the acceptance, diversification and backup strategies in isolation, we now explore the firm's strategy choice. That is, we determine which strategy best fits the firm's situation. We now explicitly account for ongoing supplier maintenance costs. The firm incurs a per-period cost of *FIXED* for each supplier in the diversification strategy, where *FIXED* represents the operational, logistical and other volume-independent costs of maintaining a supplier. In addition to primary supplier maintenance cost of *FIXED*, we assume the firm incurs an ongoing cost of *EM_FIXED* to ensure access to the backup supplier in the backup strategy.

Because there are situations in which stockpiling inventory might be the most appropriate strategy, we allow this as a fourth option in this section. We use *HOLD* to denote the inventory holding cost per unit per period. We refer the reader to the appendix (heading *Inventory model*) for a brief analysis of the inventory strategy.

Let us examine how the nature of the disruption risk (percentage uptime and disruption length/fequency) influences the strategy choice. Holding the supplier percentage uptime constant for the moment, we first consider the impact of the average disruption length or, equivalently, disruption frequency on the performance of the four strategies. Now, based on our earlier analysis, the acceptance and diversification strategy profits do not depend on the average disruption length. That is, two disruption profiles with the same percentage uptime will result in the same profits even if the average disruption length differs. See the profit expressions Eq. 3.1 and Eq. 3.3. As discussed in Sect. 3.4.1, the backup strategy profit increases as the average disruption length decreases. In contrast, the inventory strategy profit decreases (for a given percentage uptime) as the average disruption length increases. The net effect is that the average disruption length has a profound impact on the preferred strategy, with inventory favored for short (more frequent) disruptions, backup favored for long (less frequent) disruptions, and diversification favored in between.

This is illustrated in Fig. 3.6, which presents the preferred strategy as a function of percentage uptime and average disruption length.[5] For short average disruption

[5] As before, we used $MARG = 5$ and $LOST = 3$. We used three suppliers in the diversification strategy, each with a spare capacity of $SP\_CAP = 0.4$, and we set $CORR = 0$, i.e., no disruption correlation. We set $EM\_MARG = 4.5$, $EM\_RES = 4$, and $EM\_CAP = 0.75$ in the backup strategy. The cost of holding a unit of inventory for one period was set at $HOLD = 0.05$. Fixed supplier costs were set at $FIXED = 0.04$ and $EM\_FIXED = 0.5 \times FIXED$.

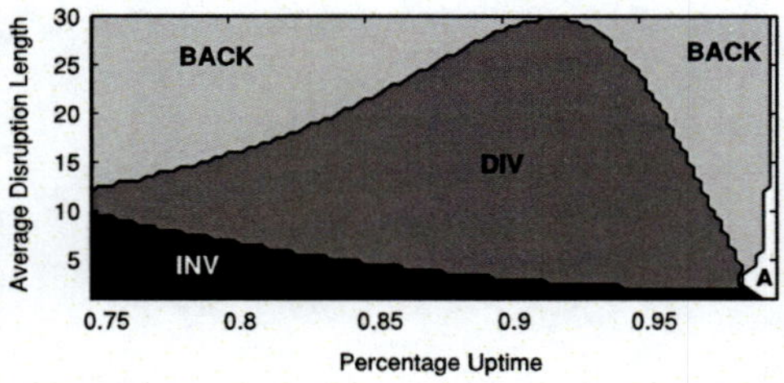

Fig. 3.6 Strategy choice depends on disruption profile

lengths, inventory (INV) is the preferred strategy unless the percentage uptime is very high. As the average disruption length increases (i.e., moving due north), inventory is initially displaced by diversification (DIV) [or acceptance (A)], but eventually the backup strategy (BACK) is preferred.

We now turn our attention to the influence of the percentage uptime. Different from the average disruption length, the percentage uptime has the same directional effect on all four strategies, with the profits all decreasing as the percentage uptime decreases. Looking at Fig. 3.6, we see that diversification is preferred over a large region but as the percentage uptime increases (i.e., moving due east), it is displaced by the backup strategy. At higher percentage uptimes the supply risk is by definition lower, and the supplier-related costs incurred by diversification to mitigate this lower risk become unattractive. Acceptance eventually becomes the preferred strategy because at very high uptimes the risk-cost tradeoff is such that it does not make economic sense to incur even the backup strategy's supplier-related costs. Moving due west, i.e., as the percentage uptime decreases, diversification becomes less attractive because the likelihood of multiple suppliers being down increases. This could be alleviated with additional suppliers, but at the expense of higher supplier-related costs. If the average disruption length is short, then diversification is displaced by inventory (as short-frequent disruption risks do not require the firm to stockpile prohibitive amounts of inventory), but for higher average disruption lengths diversification is displaced by the backup strategy.

Summarizing Fig. 3.6, acceptance is preferred only if the percentage uptime is very high, and mitigating disruption risk through sourcing (diversification or backup) is preferable to mitigation through inventory unless disruptions are short/frequent and/or supplier-related fixed costs are prohibitive. (In certain capital intensive industries, e.g., pharmaceuticals, inventory might be used to protect against long/rare disruptions due to the high cost of building multiple plants.) In comparing the sourcing strategies, backup is preferred to diversification as disruptions become lengthier and more rare.

The disruption-risk profile is not the only determinant of the preferred strategy. For example, Fig. 3.7 illustrates the impact of inventory holding costs and supplier fixed costs. Figure 3.7a (top left quadrant) is identical to Fig. 3.6, but the other figures have a higher inventory holding cost ($HOLD = 0.2$ instead of $HOLD = 0.05$) and/or higher supplier fixed costs ($FIXED = 0.08$ instead of $FIXED = 0.04$.) Comparing Fig. 3.7a–c, we see that inventory is preferred over a smaller region as the holding cost

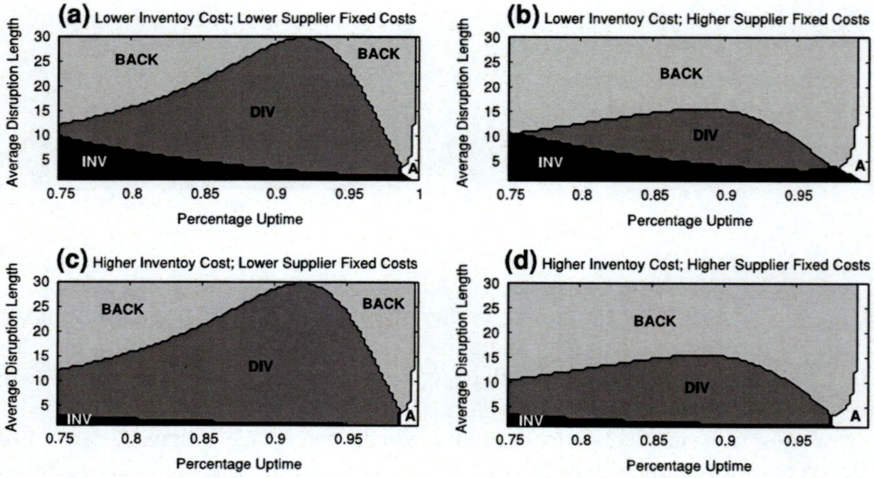

Fig. 3.7 Influence of inventory and supplier fixed costs

increases. Comparing Fig. 3.7a–b, we see that diversification is less often preferred as the fixed costs of maintaining suppliers increases. Comparing Fig. 3.7a–d, we see that acceptance is preferred over a larger region as inventory holding and fixed supplier costs increase. When choosing the appropriate mitigation strategy, managers need to account for all the significant factors that influence performance, including the disruption profile, inventory costs, the fixed and variable supplier costs, capacities, response times, and disruption correlation.

For the sake of clarity, we have discussed the strategies as if they were mutually exclusive, that is, the firm can choose only of the four strategies. In fact, there may be situations in which it makes sense to deploy a combination of strategies. For example, a firm whose backup supplier has a longer-than-desired response time might hold some inventory to use during the early stages of a disruption. Or, a firm pursuing a diversification strategy might find that instead of maintaining three routine suppliers, it is better off maintaining two routine suppliers but having a backup in place.

3.6 Additional Considerations

The intent of this chapter was to explore the role of sourcing strategies in mitigating the risk of supply disruptions. To that end, we introduced some simple models that captured many salient features of sourcing strategies. However, by design, we did not address all possible complexities, as doing so might obscure some fundamental insights. We now briefly discuss some additional considerations that may be relevant when crafting a strategy to mitigate disruption risk. In doing so, we refer the interested reader to the relevant academic literature for a deeper treatment of these issues.

Backlogging of demand: we assumed that unfilled demand was lost rather than backlogged. While this might seem like a minor distinction it does have some important implications. For example, with lost sales, there was no benefit to the backup strategy having more capacity than demand. This is not the case with backlogged demand: the backup can reduce the backlog (that accumulated during the time it took to come onstream) only if its capacity exceeds normal demand. In general, capacity concerns are amplified in both the diversification and backup strategies if unfilled demand is backlogged rather than lost. The impact of a disruption is felt for longer in the case of backlogged demand as the firm may have to work through its backlog after the disruption ends. The lower the capacity, the longer it takes to eliminate the backlog. We refer the interested reader to [8] for a treatment of the inventory, diversification, and backup strategies when demand is backlogged.

Product life cycle: the models presented in this chapter implicitly assumed that the firm sells and replenishes the product on an ongoing basis. While no product lasts forever, an "infinite-horizon" model, as adopted in this chapter, is a reasonable approximation to reality if the product life cycle is significantly longer than procurement lead times or disruptions. However, a finite horizon model may be more appropriate if the product lifecycle is not significantly longer. If product life cycles are very short relative to procurement lead times, such that the firm only has one ordering opportunity, then inventory is not a viable strategy and single-period models are needed to analyze the sourcing strategies. We refer the interested reader to [4, 9] for this type of analysis.

Type of supply risk: we modeled supply risk by assuming suppliers where either fully operational, i.e., up, or temporarily completely unavailable, i.e., down. This is a reasonable model of disruptions but not the only one. Rather than complete failures, disruptions might entail the loss of a portion of the capacity at a facility. If the possible capacity loss is deterministic, then the models used here are easily modified. If the capacity loss is uncertain, then the analysis becomes more complicated. We refer the reader to [4, 12] for a treatment of diversification in a random capacity setting. In some industries, e.g., the semiconductor industry, significant yield loss is of more (or equal) concern to supply disruptions. We refer the reader to [1, 4, 5] for a treatment of diversification in a random yield setting. Chopra et al. [2] explores the backup strategy in which the primary supplier is subject to complete failures and yield variability, showing that it is important to correctly account for the underlying drivers of supply risk. Schmitt and Snyder [7] extend Chopra et al.'s model to consider multiple periods. Yano and Lee [13] provides an excellent review of the random-yield literature.

Nonidentical suppliers: for simplicity, we assumed that suppliers were identical in the diversification strategy. While this might be a reasonable approximation for some settings, oftentimes a firm's supply base will contain suppliers that differ across costs, reliabilities, and capacities. The diversification question then moves beyond how many suppliers to use to include the questions of which suppliers to select and how much volume to allocate to suppliers. Tomlin [8] explores diversification with non-identical suppliers in an infinite-horizon setting with random disruptions. We refer the reader to [1, 4, 5, 12] for a treatment of non-identical suppliers in a single-period setting with random yield and/or random capacity.

Risk attitudes: we assumed that management's goal in crafting the disruption strategy was to maximize the long-run average profit. In other words, we assumed managers were risk neutral. If managers are risk averse, then disruption mitigation is more easily justified. More than that, risk attitudes will influence the preferred strategy as different strategies result in different profit/loss distributions. If disruptions are short and frequent, then a mean-variance framework might be reasonable for evaluating strategies. However, such a framework might not be desirable if the firm is facing low-probability, high-impact events such as rare-but-long disruptions. In this case, maximizing the average profit subject to some constraint on downside risk might be preferred. A Variance-at-Risk (Var) or Conditional Variance-at-Risk (CVaR) approach could be used. In addition to a risk-neutral approach, Tomlin [8] considers both a mean-variance and a CVaR approach in selecting a disruption strategy. In single-periods setting, Tomlin [9] explores a risk neutral objective and a loss-averse objective in which managers value profits less than they fear losses. These papers show that the preferred strategy can be heavily influenced by risk attitudes.

Multiple products: the unit of analysis in this chapter was a single product rather than a portfolio of products. With regard to the strategies presented here, this is appropriate if the products do not share any supply chain resources (e.g., inventories, facilities, or suppliers) and also provides a reasonable "first-cut" analysis if they do share resources. However, a portfolio perspective is recommended if any resources are shared. Disruption to a shared resource will impact multiple products and this interaction needs to be evaluated as it has direct implications for the spare or backup capacity available to any individual product. The issue of multiple products introduces a tension between demand and supply risks. Component commonality and flexible facilities are well-established approaches to managing demand risk by taking advantage of demand pooling. However, these approaches can also concentrate supply risk such that multiple products are now at risk if a particular component supplier or facility fails. Tomlin and Wang [10] examines this tension by exploring dual sourcing and flexibility. Firms selling multiple products may have the option of influencing customer demand in the direction of a particular product. This demand management (or shaping) can be used during a disruption to direct customers from constrained-supply products to less constrained ones. We refer the interested reader to [9] which explores demand management and its interaction with the diversification and backup strategies.

Supplier leadtimes: we were silent on procurement (production and/or transportation) lead times in this chapter. If all suppliers have the same lead times, then the analyses presented here are perfectly adequate. However, if suppliers differ in their lead times, then a more sophisticated analysis would be required to fully capture the lead time differentials.

Other situation-specific considerations might come to light when evaluating a particular firm's supply chain. While the models introduced in this chapter provide a good starting point for evaluating different mitigation strategies, we highly recommend that managers carefully determine the relevant considerations for their supply chain and conduct a rigorous analysis that adequately captures those considerations.

3.7 Conclusions

Supplier diversification and backup sourcing offer alternatives to stockpiling inventory as a means of mitigating disruption risks. The effectiveness of diversification depends largely on the number of suppliers, the possibility of disruption correlation, and the available spare capacity at suppliers. Most of the benefits are achieved with a small number of suppliers; oftentimes firms may find that using two or three suppliers strikes an appropriate balance between risk mitigation and supply chain rationalization. However, disruption correlation and spare capacity considerations can limit the mitigation benefit of diversification. If these concerns cannot be alleviated, firms may need to increase the number of suppliers to adequately mitigate disruption risk.

Backup supply may be an attractive alternative to supplier diversification as a means of mitigating disruption risk. The effectiveness of backup sourcing depends largely on the cost and availability of the backup source, with availability being measured as response time and capacity provided. Effective backup supply requires the firm to have agreed emergency plans and protocols in place before a disruption occurs. In addition to cost, potential backup suppliers should be evaluated along the dimensions of response time and capacity. While a firm will of course prefer faster response and higher capacity, it may be forced to make a tradeoff between these two dimensions and/or pay higher price to improve one or both. The nature of the disruption risk (short-frequent versus long-rare) influences the value of response time and magnitude, and so managers need to account for this when developing their backup plan.

The disruption profile (uptime and frequency/severity) plays a major role in determining the most appropriate mitigation strategy, with, for example, backup sourcing being appropriate for long/rare disruptions but inventory being appropriate for short/frequent ones. As there is no one-size-fits-all solution to mitigating disruption risk, firms should choose the strategy that best aligns with their internal and external operating environment, recognizing that this may mean different strategies for different parts of the business. Choosing and crafting the best strategy relies on sound judgment aligned with suitable analysis; guesswork is neither required nor recommended. Regardless of the chosen strategy, successful implementation depends on proactive planning. Selecting and validating additional or alternative suppliers can be time consuming and cannot be left until a disruption occurs. Likewise, disruption detection and notification protocols should be designed, agreed upon, and documented in advance.

We offer the following thoughts in closing. One, be prepared. As the adage goes, failing to plan is planning to fail. Senior executives should ensure that there is a systematic approach to identifying, evaluating, and managing supply risk throughout their organization. Two, be vigilent. Delays in detecting and responding to a disruption can dramatically amplify its impact, especially if competitors preempt any backup supply options. Ongoing supplier communication and threat monitoring can aid in rapid detection. Three, be flexible. Risk identification is an inexact science; a disruption may occur at unanticipated location due to some unexpected cause.

Supply chains that can rapidly detect anomalies and that are flexible enough to divert flows to other parts of the network are best able to react to unforeseen events.

3.8 Appendix

In this appendix we present some of the underlying analysis for the profit expressions used in this chapter.

3.8.1 Acceptance

Profit in a period when supplier is up: $MARG$
Profit in a period when supplier are down: $-LOST$
Steady-state probability supplier is up: UP
Steady-state probability supplier is down: $1 - UP$
Average profit: $ACC\_PROF = MARG \times UP - LOST(1 - UP)$
Rearranges to: $ACC\_PROF = MARG - (MARG + LOST)(1 - UP)$, i.e., Eq. 3.1.

3.8.2 Supply Diversification

3.8.2.1 Basic Model

There are NUM_SUP identical suppliers.
Each supplier has enough capacity to meet all demand.
Profit in a period when at least one supplier is up: $MARG$
Profit in a period when all suppliers is down: $-LOST$
Steady-state probability at least one supplier is up: $1 - (1 - UP)^{NUM\_SUP}$
Steady-state probability all suppliers are down: $(1 - UP)^{NUM\_SUP}$

Average profit:
$DIV\_PROF_u = MARG(1 - (1 - UP)^{NUM\_SUP}) - LOST(1 - UP)^{NUM\_SUP}$
Rearranges to:
$DIV\_PROF_u = MARG - (MARG + LOST)(1 - UP)^{NUM\_SUP}$, i.e., Eq. 3.2.

3.8.2.2 Correlated Disruptions

We model correlated disruptions as follows. Let there be two types of disruptions: systemic disruptions and supplier-specific disruptions. A systemic disruption causes all suppliers to be down. For example, a natural diaster might shut down all suppliers if they are located in the same region. A supplier-specific disruption only causes that specific supplier to be down. For example, an internal strike might disrupt one supplier but not disrupt other suppliers. Denote the steady-state probability that the overall system is up as SYS_UP. Denote the steady-state probability that an individual supplier is down due to a supplier-specific disruption as $1 - SUP\_UP$. Therefore, all suppliers are down if either the system is down or the system is up but all suppliers are down. The probability of this event is $1 - SYS\_UP + SYS\_UP(1 - SUP\_UP)^{NUM\_SUP}$. Following a similar logic to that presented for the basic model above, we have $DIV\_PROF_c = MARG - (MARG + LOST)\left(1 - SYS\_UP + SYS\_UP(1 - SUP\_UP)^{NUM\_SUP}\right)$. Now, any particular supplier is up only if the system and the supplier is up. Therefore, the steady-state probability of a supplier being up (i.e., the percentage uptime) is $UP = SYS\_UP \times SUP\_UP$. The correlation coefficient, denoted as $CORR$, for any two suppliers being up can be shown to be $CORR = \frac{SUP\_UP(1-SYS\_UP)}{1-SYS\_UP \times SUP\_UP}$. Using these expressions to write the profit expression in terms of UP and $CORR$ instead of SYS_UP and SUP_UP gives the DIV_PROF_c expression Eq. 3.3.

3.8.2.3 Limited Spare Capacity

We assume that order quantities are evenly split across the available suppliers. Therefore the normal volume (i.e., when all suppliers are up) sourced from a supplier is $1/NUM\_SUP$. (Recall that we scale demand to equal 1.) Each supplier has spare capacity to produce more than its normal volume. Spare capacity, denoted by $SP\_CAP \geq 0$, is measured as the percentage a supplier can provide above and beyond its normal order quantity. Therefore, the maximum quantity that a supplier can produce is given by $(1/NUM\_SUP)(1 + SP\_CAP)$. Let n denote the number of suppliers that are up. If $n \geq \frac{NUM\_SUP}{1+SP\_CAP}$, then these n suppliers together can meet all demand, otherwise only $\frac{n(1+SP\_CAP)}{NUM\_SUP}$ of demand can be filled, and the remainder is lost. Define $\hat{n} = \left\lceil \frac{NUM\_SUP}{1+SP\_CAP} \right\rceil$ where $\lceil x \rceil$ is the smallest integer greater than or equal to x (i.e., round up x to the nearest integer). Therefore, the profit in a period for which $0 \leq n < \hat{n}$ suppliers are up is $MARG \times \frac{n(1+SP\_CAP)}{NUM\_SUP} - LOST \times \left(1 - \frac{n(1+SP\_CAP)}{NUM\_SUP}\right)$, and the profit in a period for which $n \geq \hat{n}$ suppliers are up is $MARG$. As before (see correlated disruptions above), the steady state probability that $n = 0$ suppliers are up is $1 - SYS\_UP + SYS\_UP(1 - SUP\_UP)^{NUM\_SUP}$. The probability that exactly $0 < n \leq NUM\_SUP$ of the NUM_SUP suppliers are up is $SYS\_UPNUM\_SUPn(SUP\_UP)^n(1 - SUP\_UP)^{NUM\_SUP-n}$. (These probabilities can be expressed in terms of UP and $CORR$ instead of SYS_UP and SUP_UP by using the transformations shown above.) Knowing the probability of there being exactly n suppliers up and knowing the associated profit, the average profit can then be evaluated. A closed-form expression for the profit will not typically exist, however.

Consider the special case where $SP\_CAP = 0$. Then, $\hat{n} = NUM\_SUP$, and when $0 \leq n \leq NUM\_SUP$ suppliers are up, the profit is $MARG \times \frac{n}{NUM\_SUP} - LOST \times \left(1 - \frac{n}{NUM\_SUP}\right)$, or equivalently, $\frac{n}{NUM\_SUP}(MARG + LOST) - LOST$. The long-run average profit is then

$$DIV\_PROF_{ns} = \sum_{n=0}^{NUM\_SUP} \left(\frac{n}{NUM\_SUP}(MARG + LOST) - LOST \right) P[n],$$

where $P[n]$ is the probability of n suppliers being up. Rearranging terms gives $DIV\_PROF_{ns} = \left(\frac{MARG+LOST}{NUM\_SUP}\right) \times \sum_{n=0}^{NUM\_SUP} (nP[n]) - LOST$. Now, $\sum_{n=0}^{NUM\_SUP} nP[n]$ is the average number of suppliers up, and this equals $NUM\_SUP \times UP$. Substituting this into the profit expression, we then have $DIV\_PROF_{ns} = (MARG + LOST)UP - LOST = MARG - (MARG + LOST)(1 - UP)$.

3.8.3 Backup Supply

The backup supplier is not available until *EM_RES* periods of a disruption have passed, after which the backup can provide a capacity of $0 \leq EM\_CAP \leq 1$. Units sourced from the backup supplier give a profit of $EM\_MARG < MARG$, where *MARG* is the profit obtained using the regular supplier. Therefore, the profit in a period when the regular supplier is up is *MARG*. If the regular supplier has been down *EM_RES* periods or less, then the profit in that period is $-LOST$ as the backup supply has yet to come on stream. If the regular supplier has been down for more than *EM_RES* periods, then the profit in that period is $EM\_MARG \times EM\_CAP - LOST \times (1 - EM\_CAP)$, as the backup can only supply EM_CAP. Let $P[t]$ represent the steady-state probability that the regular supplier has been down for $t = 1, 2, 3, \ldots$ periods. Then, the average profit is

$$EM\_PROF = MARG \times UP - \sum_{t=1}^{EM\_RES} LOST \times P[t]$$
$$+ \sum_{t=EM\_RES+1}^{\infty} (EM\_MARG \times EM\_CAP - LOST \times (1 - EM\_CAP))\, P[t].$$

To evaluate this profit expression, we need to know $P[t]$, the steady-state probabilities of being up and of being down for $t = 1, 2, 3, \ldots$ periods. This requires us to make an assumption about the disruptions. For simplicity, we assume that (1) when the primary supplier is up there is a constant probability, denoted by *FAIL*, of a disruption occurring, and (2) when the primary supplier is down there is a constant probability, denoted by *RECOVER*, of the disruption ending. With these assumptions, it can be shown that $UP = \frac{RECOVER}{RECOVER+FAIL}$ and $P[t] = (1 - UP)(RECOVER)(1 - RECOVER)^{t-1}$. Substituting these into the profit expression above, followed by some algebra, gives

$$EM\_PROF = MARG - (MARG + LOST)(1 - UP) +$$
$$(EM\_MARG + LOST)(1 - UP)(EM\_CAP)\,(1 - RECOVER)^{EM\_RES},$$

which is the same as Eq. 3.4 because the average disruption length is given by $DIS\_LEN = 1/RECOVER$.

3.8.4 Inventory Model

In the inventory strategy, the firm single sources but stockpiles a quantity of inventory, denoted by *INV*, to use during a disruption. There is a per-unit cost of *HOLD* to hold inventory for a period and the inventory is accounted for at the end of a period. For simplicity we assume the supplier has infinite capacity and, therefore, the inventory returns to *INV* as soon as a disruption ends. Therefore, the firm has *INV* units of inventory at the start of any disruption. The firm can fill demand from production if the supplier is up or from inventory if the supplier has been down for less than or equal to *INV* consecutive periods. (Recall that we scale demand equal to 1.) Demand cannot be filled if the supplier has been down for more than *INV* consecutive periods. Therefore, the profit in a period in which the supplier is up is $MARG - HOLD \times INV$. The profit in a period in which the supplier has been down for $t \leq INV$ periods is $MARG - HOLD \times (INV - t)$ as t units of inventory will have been used to fill demand. The profit in a period in which the supplier has been down for $t > INV$ periods is $-LOST$ as the inventory is gone and demand cannot be filled. Then, the average profit is

$$INV\_PROF = (MARG - HOLD \times INV)UP + \sum_{t=1}^{INV} (MARG - HOLD \times (INV - t))\, P[t] - \sum_{t=INV+1}^{\infty} LOST \times P[t].$$

To evaluate this profit expression, we need to know $P[t]$, the steady-state probabilities of being up and of being down for $t = 1, 2, 3, \ldots$ periods. For simplicity, we will assume geometric disruptions as we did for the backup strategy. Therefore, $UP = \frac{RECOVER}{RECOVER+FAIL}$ and $P[t] = (1 - UP)(RECOVER)(1 - RECOVER)^{t-1}$. Substituting these into the profit expression above, followed by some algebra, gives

$$INV\_PROF = MARG - HOLD \times INV + (1 - UP)\Bigg(HOLD \times DIS\_LEN - (MARG + LOST + HOLD \times DIS\_LEN)\left(1 - \frac{1}{DIS\_LEN}\right)^{INV}\Bigg), \quad (3.5)$$

where we have also used the fact that $DIS\_LEN = 1/RECOVER$. In deciding how much inventory to stockpile, the firm faces a tradeoff between increasing sales during a disruption by carrying more inventory and incurring the additional cost for holding more inventory. Using the *INV_PROF* expression above, the value of *INV* that maximize the firm's profit is given by $OPT\_INV = \max\left\{0, \lceil \frac{lnA}{lnB} \rceil\right\}$ where

$A = \frac{HOLD}{(1-UP)\left(HOLD+\frac{MARG+LOST}{DIS\_LEN}\right)}$, $B = 1 - \frac{1}{DIS\_LEN}$, and $\lceil x \rceil$ is the smallest integer greater than or equal to x (i.e., round up x to the nearest integer).

References

1. Agrawal V, Nahmias S (1997) Rationalization of the supplier base in the presence of yield uncertainty. Prod & Oper Manag 6(3):291–308
2. Chopra S, Reinhardt G, Mohan U (2007) The importance of decoupling recurrent and disruption risks in a supply chain. Nav Res Logist 54(5):544–555
3. Clothier M (2009) Kellogg predicts Eggo waffle shortage next year. http://www.boston.com/lifestyle/food/articles/2009/11/15/kellogg_predicts_eggo_waffle_shortage_next_year/
4. Dada M, Petruzzi N, Schwarz L (2007) A Newsvendor's procurement problem when suppliers are unreliable. Manuf Serv Oper Manag 9(1):9–32
5. Federgruen A, Yang N (2008) Selecting a portfolio of suppliers under demand and supply risks. Oper Res 56(4):916–936
6. Hartman B (2009) Eggo Waffle Shortage: bacteria first forced plant closure. http://www.abcnews.go.com/Business/eggo-waffle-shortage-bacteria-forced-plant-closure/story?id=9117059. Accessed 18 Jan 2011
7. Schmitt AJ, Snyder LV (2009) Infinite-horizon models for inventory control under yield uncertainty and disruptions. Working paper, P.C. Rossin College of Engineering and Applied Sciences, Lehigh University, Bethlehem, PA
8. Tomlin B (2006) On the value of mitigation and contingency strategies for managing supply chain disruption risks. Manag Science 52(5):639–657
9. Tomlin B (2009) Disruption-management strategies for short life-cycle products. Nav Res Logist 56:318–347
10. Tomlin B, Wang Y (2005) On the value of mix flexibility and dual sourcing in unreliable newsvendor networks. Manuf Serv Oper Manag 7(1):37–57
11. Turner D (2009) Leggo my Eggo! Kellogg fights waffle shortage. http://www.boston.com/business/articles/2009/11/18/leggo_my_eggo_kellogg_fights_waffle_shortage
12. Wang Y, Gilland W, Tomlin B (2009) Mitigating supply risk: dual sourcing or process improvement? Working Paper, Forthcoming in Manuf Serv Oper Manag
13. Yano CA, Lee HL (1995) Lot sizing with random yields: a review. Oper Res 43(2):311–334

Chapter 4
Supply Chain Management Under Simultaneous Supply and Demand Risks

Awi Federgruen and Nan Yang

4.1 Introduction

Standard supply chain management texts discuss the benefits of consolidating the set of suppliers in the chain. These benefits include economies of scale in the production costs as well as statistical economies of scale due to the pooling of demand risks. Recently, many corporations and governments, alike, have recognized a variety of risks associated with external disruptions of the supply process. These provide a powerful argument *against* (maximal) consolidation. Such disruptions may arise because of "natural" disasters, e.g. fires in production plants or the need to shut down a facility because of violations of quality regulations or standards. Disruptions may also occur because of labor strikes, or planned acts of sabotage, resulting from terror attacks among others. While these disruptions may be rare, their consequences can be catastrophic for an individual firm as well as for a region or a country as a whole.

In the private sector, "Planning for Disaster" has become one of the foci of supply chain planning, see e.g. [25]. This conference report describes, e.g., a case study of Ericsson, which, in contrast to Nokia, suffered major and long-term losses in profits and market shares for its cellular phone business, due to its unhedged dependence on a single chip supplier in New Mexico and its lack of preparedness to switch to alternative suppliers in response to a major fire disabling this chip supplier. Terrorist generated disasters targeted at such universally critical component suppliers as chip manufacturers may have a crippling effect, not

---

A. Federgruen
Graduate School of Business, Columbia University,
New York, NY 10027, USA
e-mail: af7@columbia.edu

N. Yang (✉)
Olin Business School, Washington University in St. Louis,
St. Louis, MO 63130, USA
e-mail: yangn@wustl.edu

H. Gurnani et al. (eds.), *Supply Chain Disruptions: Theory and Practice of Managing Risk*, DOI: 10.1007/978-0-85729-778-5_4,

just on this industry sector, but on many other major sectors of the economy as well.

Similarly, in the Fall of 2004 the United States saw half of its supply of flu vaccines cut out, when the Chiron plant in Liverpool had to be closed down because of violations of FDA standards. Even in a year without vaccine shortages, no less than 36,000 deaths—twelve times the number of 9/11 victims—and 200,000 hospitalizations are attributed to influenza and its complications. In terms of productivity, between $11 and $20 billion is lost annually due to influenza. The sudden elimination of one of only two manufacturers and half the national vaccine supply was hardly an unforeseeable or rare event, as numerous Senate testimonies and General Accounting Office reports have documented recurring supply problems with this and other vaccines, (see e.g. [20, 21, 22]). In 2004, the Centers for Disease Control and Prevention (CDC) identified a target population of 100 million individuals who should have been vaccinated with the flu vaccine. Remarkably, the United States was dependent on just *two* suppliers, while England, with a target population of only 14 million, had its supply spread over *six* suppliers. Moreover, the vulnerability experienced with respect to the flu vaccine is hardly unique. Similar problems have arisen repeatedly over the last decade with respect to other, perhaps even more crucial vaccines such as those required to immunize the children's population against highly contagious diseases.[1]

As a final example, oil is arguably the most critical commodity for the functioning of our economy. Its supply is primarily limited by existing refinery capacity. In the past 20 years, as the real valued US Gross Domestic Product grew by 86.5%, the number of refineries decreased by more than 50%. This consolidation occurred because various types of economies of scale drove smaller refineries out of the market; other refineries identified the above mentioned benefits of pooling capacity and of running refineries at near 100% utilization. (In July and August of 2004, US refineries were operating at 97% of available capacity.) Moreover, in case of a domestic supply disruption, little recourse can be expected from overseas refineries: The push of oil prices to record high levels, this year, is generally attributed to a lack of global refinery capacity. The Department of Energy predicts that current "financial, environmental and legal considerations make it unlikely that new refineries will be built in the United States," see [9]. Most ominously, close to half of our capacity is located in a relatively small region on the Gulf Coast; disruption of its refinery and distribution process, the result of Katrina like hurricanes, for example, could have a crippling effect on our economy. Since the fifties, all US administrations have intervened in the market by maintaining a stockpile of Strategic Reserves so as to mitigate the impact of sudden supply

[1] "In recent years there have been many significant disruptions of vaccine supplies. Between November 2000 and May 2003, there were shortages of 8 of the 11 vaccines for childhood diseases in the United States including those for tetanus, diphtheria, whooping cough, measles, mumps and chicken pox. There have been flu vaccine shortages or miscues for four consecutive years." See [30].

problems. However, the stockpile is largely in terms of crude oil, while refinery capacity has become the true bottleneck in the system. It is also the most vulnerable part of the oil supply chain, since repairs of refinery equipment can take months to years to be completed. This puts into question whether the Strategic Reserves should not be replaced or complemented by incentives to expand the refinery base in different parts of the country.

In this chapter, we characterize how supply risks arising from the above yield uncertainties impact on supply chain management. We first consider the implications for the procurement strategies of a firm which has access to a set of potential suppliers which differentiate themselves in terms of their prices and yield distributions. The fundamental questions that arise here are (i) how many suppliers to maintain and diversify one's purchase orders amongst; (ii) how to select the desired number of suppliers from the set of potential suppliers; (iii) how to adjust one's inventory strategy to account for the supply risk, in particular how *total* purchase quantities should be set in the simultaneous presence of supply and demand risks; (iv) the final question is how aggregate orders are to be split among the selected suppliers and whether the tradeoffs between reliability and cost differentials among the suppliers can be captured in terms of simple allocation rules.

We first characterize the answers to the above questions assuming given cost and reliability profiles of the potential suppliers. In the last part of this chapter, we proceed to analyze how these competing suppliers may wish to invest in process and technology improvements so as to "optimally" affect their reliability characteristics and resulting market shares. In addressing the latter questions, we model the suppliers as engaged in a non-cooperative competition game.

The remainder of the chapter is organized as follows: In Sect. 4.2, we characterize procurement strategies in settings where shortfalls are controlled by the specification of service constraints. Section 4.3 is devoted to so-called Total Cost Models where direct and indirect cost consequences of these shortfalls are added to the cost objective. This section also contains a systematic comparison of the above two modeling approaches. Section 4.4 provides an analysis of supplier competition models in which the suppliers select or adjust their yield characteristics so as to maximize their expected market shares and expected profits based on the procurement strategies identified in Sect. 4.2. Section 4.5 completes the chapter with a brief summary and a discussion of important generalizations.

4.2 Procurement Strategies in a Single Period Setting Under a Service Constraint

We start our analyses with a planning model for a firm or public organization which needs to cover uncertain demand over a single sales season, assuming a single replenishment epoch at the start of the season. In this section, we take the perspective that demand needs to be covered with at least some prespecified probability. We refer to this model as the Service Constraint Model (SCM). In the next section, we adopt the more traditional approach in inventory and supply chain management models where the consequences of stockouts are assumed to be associated with specific stockout costs. We refer to this alternative as the Total Cost Model (TCM) approach. In traditional inventory theory with fully reliable suppliers, assigning a direct stock-out penalty for each unsatisfied unit of demand and employing a constraint on the probability of a stock-out represent the two common approaches to control the stock-out phenomenon. Much has been written about the relative merits of both modeling approaches, see e.g. [37]. Both approaches continue to be pursued in parallel, even though in classical inventory models, the two approaches are known to be equivalent: an instance of (TCM) with a given stockout cost induces the same optimal inventory strategy as an instance of (SCM) with a corresponding permitted shortfall probability, and vice versa. See [4] for a recent discussion of this equivalency in classical inventory models. The equivalency breaks down under multiple suppliers with unreliable yields, adding to the need to pursue both planning approaches in parallel.

The firm has access to N potential suppliers. Yield and reliability risks are reflected by the fact that only a random *fraction* of any given order becomes available as useable or sellable items. In the literature, this representation of yield uncertainty is referred to as "Stochastically Proportional Yield Models". See [36] for a survey of alternative yield models as well as an excellent review of planning models, prior to 1995. (Almost all of these papers deal with a single supplier or at most two suppliers, i.e., dual sourcing.) Thus, each supplier faces a random yield with a general probability distribution on the unit interval. An important special case is where this distribution has a positive mass at zero, representing the possibility of a complete shutdown due to an unplanned disruption. Demand during the sales season is uncertain as well, but characterized by a known demand distribution. The planning problem amounts to selecting which of the given set of suppliers to retain, and how much to order from each, so as to minimize total procurement costs, while ensuring that the uncertain demand is met with a given probability. The total procurement costs consist of *variable* costs which are proportional to the total quantity delivered by the suppliers and a *fixed* cost, for each participating supplier. We therefore assume that the potential suppliers may differ from each other, in terms of their units prices, their fixed procurement costs and their distributions of the random yield factor. The yield factors at different suppliers are assumed to be independent of the season's demand which is described by a general probability distribution. Thus, let:

N = the number of all available suppliers,
c_i = unit price charged by supplier i for every useable unit, $i = 1,\ldots,N$;
K_i = fixed cost when purchasing from supplier i, $i = 1,\ldots,N$;
X_i = the random yield factor at supplier i's facility, with cdf $G_i(\cdot)$, mean p_i, variance ς_i^2, and coefficient of variation $\gamma_i = \varsigma_i/p_i$, $i = 1,\ldots,N$;
D = the uncertain demand during the season, with a strictly increasing continuous cdf $F(\cdot)$, complementary cdf $\overline{F}(\cdot)$, inverse cdf $F^{-1}(\cdot)$, mean μ, variance σ^2, coefficient of variation $\gamma_D = \sigma/\mu$ and finite moments;
I^0 = initial inventory, at the beginning of the season;
α = maximum permitted probability of a shortfall ($\leqslant 0.5$);
$p_{\min}\,[p_{\max}] = \min_i p_i\,[\max_i p_i]$

4.2.1 Identical Suppliers

We first consider the special case where all suppliers share identical cost parameters and yield distributions. We therefore omit the subscript i from the parameters and distributions. Since the suppliers are indistinguishable, assume we place identical orders of size y with each of a selected set of $n \leqslant N$ suppliers, for a total order $Y = ny$. (As shown in [13], with identical suppliers, it is often optimal to place identical orders, for example when the demand distribution has a decreasing pdf.) The service constraint can thus be expressed as:

$$\Pr\left(I^0 + y\sum_{i=1}^{n} X_i \geqslant D\right) = \int_0^n F(I^0 + yu)\,dG^{(n)}(u) \geqslant 1 - \alpha, \tag{4.1}$$

where $G^{(n)}(\cdot)$ denotes the n-fold convolution of the $G(\cdot)$ distribution. The probability to the left of $(I^0 + yu)$ clearly increases with y, so that the order size which minimizes variable procurement costs is given by:

$$y^*(n) = \min\left\{y : \int_0^n F(I^0 + yu)\,dG^{(n)}(u) \geqslant 1 - \alpha\right\}. \tag{4.2}$$

Let $Y^*(n) \stackrel{\text{def}}{=} ny^*(n)$ denote the minimum total order size.

The following theorem shows that it is optimal to place orders if and only if the starting inventory I^0 is below a given threshold level. In this base case model, the threshold level corresponds with the known fractile of the demand distribution in the standard inventory model without supply risks. Assuming I^0 is below the threshold level, and a complete supply disruption may occur with positive probability at any of the suppliers, a feasible solution exists if and only if the number of retained suppliers is in excess of a given minimum number. The expected aggregate effective supply,

$pY^*(n)$, may both be larger or smaller than the optimal order quantity in the standard model without supply risks but approaches the latter when the number of suppliers goes to infinity.

Theorem 1 *Assume the suppliers have identical characteristics and receive identical orders.*

1. *If* $I^0 \geqslant F^{-1}(1-\alpha)$, $y^* = 0$.
2. *Assume* $I^0 < F^{-1}(1-\alpha)$. *Let* $\underline{n} \overset{\text{def}}{=} [\ln \overline{F}(I^0) - \ln \alpha]/[-\ln G(0)]$, *if* $G(0) > 0$ *and* $\underline{n} \overset{\text{def}}{=} 0$, *otherwise. If* $n < \underline{n}$, *no optimal order quantity exists. Conversely, if* $n > \underline{n}$, *an optimal order quantity exists.*
3. *Assume* $I^0 < F^{-1}(1-\alpha)$. *For* $n > \underline{n}$, y^* *is decreasing in* n.
4. *Assume* $I^0 < F^{-1}(1-\alpha)$. $\lim_{n \to +\infty} Y^*(n) = [F^{-1}(1-\alpha) - I^0]/p$

4.2.2 Non-Identical Suppliers

As pointed out above, under general demand distributions, it may be optimal, or even necessary, to place different size orders with the various retained suppliers, even if they share the same yield and cost characteristics. In this case, the necessary and sufficient condition for a feasible procurement strategy becomes considerably harder than the simple criterion $n > \underline{n}$, in Theorem 1. The same applies, a fortiori, to the characterization of the *optimal* solution, assuming a feasible solution does exist. An additional level of complexity arises when the suppliers have different yield- and cost- characteristics. In this subsection, we assume that the suppliers are differentiated on the basis of their yield distributions $\{G_i(\cdot)\}$ and fixed costs $\{K_i\}$, while charging the same unit price $c_i = c$ for all useable units. The most general case where the unit purchase prices $\{c_i\}$ may vary by supplier as well, is addressed in the next subsection. Without loss of generality, assume that the suppliers are ranked in increasing order of their yield volatility, i.e., $\gamma_1 \leqslant \gamma_2, \leqslant \cdots \leqslant \gamma_N$.

To characterize feasible and optimal procurement strategies with non-identical suppliers, one needs to employ an approximation for the shortfall distribution $\left(D - I^0 - \sum_{i=1}^{n} y_i X_i\right)$. Federgruen and Yang [13] develop two such approximations, one based on a Large Deviation Technique and one based on a Central Limit Theorem. Here, we confine ourselves to the latter. Let

$y_i =$ the absolute order size placed with supplier i, $i = 1, \ldots, N$;
$w_i = \frac{y_i}{\sum_{j=1}^{N} y_j} =$ the relative order size placed with supplier i, $i = 1, \ldots, N$;
$Y_E = \sum_{i=1}^{N} p_i y_i =$ the expected effective supply resulting from the order vector $\mathbf{y}$.

The basic idea is to replace the shortfall distribution by a Normal with matching mean and variance, i.e.,

$$\left(D - I^0 - \sum_{i=1}^{N} y_i X_i\right) \sim \text{Normal}\left(\mu - I^0 - \sum_{i=1}^{N} p_i y_i, \sigma^2 + \sum_{i=1}^{N} y_i^2 \varsigma_i^2\right) \quad (4.3)$$

Thus, under the Normal approximation, a set of orders $\mathbf{y}$ satisfies the service constraint if and only if

$$\frac{Y_E + I^0 - \mu}{\sqrt{\sigma^2 + \sum_{i=1}^N y_i^2 \varsigma_i^2}} \geqslant z_\alpha = \Phi^{-1}(1-\alpha),$$

which is equivalent to the following pair of inequalities:

$$(Y_E - \mu + I^0)^2 - z_\alpha^2 \left(\sum_{i=1}^N \varsigma_i^2 y_i^2 \right) - z_\alpha^2 \sigma^2 \geqslant 0 \tag{4.4}$$

$$Y_E \geqslant \mu - I^0 + z_\alpha \sigma \tag{4.5}$$

(When $I^0 \geqslant \mu + z_\alpha \sigma$, $Y_E = 0$ is feasible and optimal.) (4.4) can be written as: $(Y_E - \mu + I^0)^2 - z_\alpha^2 Y_E^2 \left(\sum_{i=1}^N \varsigma_i^2 w_i^2\right) / \left(\sum_{i=1}^N p_i w_i\right)^2 - z_\alpha^2 \sigma^2 \geqslant 0$, or

$$Y_E^2 \left(1 - \frac{z_\alpha^2}{R(\mathbf{w})}\right) - 2Y_E(\mu - I^0) + (\mu - I^0)^2 - z_\alpha^2 \sigma^2 \geqslant 0, \tag{4.6}$$

$$\text{where } R(\mathbf{w}) \stackrel{\text{def}}{=} \frac{\left(\sum_{i=1}^N w_i p_i\right)^2}{\left(\sum_{i=1}^N w_i^2 \varsigma_i^2\right)}$$

Thus, if the pair $(Y_E, \mathbf{w})$ results in a feasible solution, i.e., satisfies (4.5) and (4.6), so does $(Y_E, \mathbf{w}^*)$, where

$$w_i^* = \frac{p_i / \varsigma_i^2}{\left(\sum_{j=1}^N p_j / \varsigma_j^2\right)}, \quad i = 1, \ldots, N. \tag{4.7}$$

The latter follows from the fact that the allocation vector $\mathbf{w}^*$ is easily verified to maximize R($\mathbf{w}$), i.e., $\overline{R} \stackrel{\text{def}}{=} R(\mathbf{w}^*) \geqslant R(\mathbf{w})$ for all allocation vectors $\mathbf{w}$. This allows us to state the feasibility conditions as a pair of inequalities in the single variable Y_E only, i.e., (4.5) in conjunction with

$$Y_E^2 \left(1 - \frac{z_\alpha^2}{\overline{R}}\right) - 2Y_E(\mu - I^0) + (\mu - I^0)^2 - z_\alpha^2 \sigma^2 \geqslant 0 \tag{4.8}$$

This pair of inequalities already indicates that a feasible procurement strategy can be achieved by selecting a high enough expected effective supply Y_E, but only if $\overline{R}$ is sufficiently large.

Theorem 2 (General Suppliers with Common Unit Prices) *Under the CLT-approximation:*

1. *A feasible solution exists if and only if it exists under* theallocation scheme $\mathbf{w}^*$.
2. *A feasible solution exists if and only if condition (F) is satisfied:*

$$(F)\quad (i)\quad If\ \ I^0 \leqslant \mu,\quad \overline{R} = \sum_{i=1}^{N} \gamma_i^{-2} > z_\alpha^2;$$

$$(ii)\quad If\ \ I^0 > \mu,\quad \overline{R} = \sum_{i=1}^{N} \gamma_i^{-2} \geqslant z_\alpha^2 - (I^0 - \mu)^2/\sigma^2$$

3. *Assume all unit prices* $c_i = c, i = 1, \ldots, N$. *Assume* $I^0 < \mu + z_\alpha \sigma \Leftrightarrow Y_E^* > 0$. *Under (F), the set of orders which minimizes the variable procurement costs is given by the allocation scheme* $\mathbf{w}^*$ *and the expected effective supply level:*

$$Y_E^* = \begin{cases} (1 - \frac{z_\alpha^2}{\overline{R}})^{-1} \left[(\mu - I^0) + z_\alpha \sqrt{\frac{(\mu - I^0)^2}{\overline{R}} + \sigma^2 \left(1 - \frac{z_\alpha^2}{\overline{R}}\right)} \right], & if\ \overline{R} = \sum_{i=1}^{N} \gamma_i^{-2} \neq z_\alpha^2 \\ \left[z_\alpha^2 \sigma^2 - (I^0 - \mu)^2 \right] / \left[2(I^0 - \mu) \right], & if\ \overline{R} = \sum_{i=1}^{N} \gamma_i^{-2} = z_\alpha^2\ and\ I^0 > \mu \end{cases} \tag{4.9}$$

Proof See the proof of Theorem 5 in [13]. □

Theorem 2 shows that a feasible procurement strategy exists if and only if procurements can be diversified over sufficiently many or sufficiently reliable suppliers. More specifically, whether a set of suppliers allows for a *feasible* solution, depends not just on its cardinality but also on each supplier's predictability, as measured by the coefficient of variation (c.v.) of his yield distribution. Define a (hypothetical) supplier with a c.v. value of one as a "Base Supplier". A supplier with c.v. $= \gamma$ represents γ^{-2} Base Supplier Equivalents (*BSE*). A set of suppliers is feasible if its total number of BSE's is in excess of a critical number, given by a simple function of the permitted shortfall probability α, only. When the initial inventory level exceeds the mean demand by s standard deviations of demand, this minimum threshold is reduced by s^2. In particular, whether a set of suppliers is feasible or not does not depend on any of the demand distribution's moments, the mean and standard deviation included, as long as the starting inventory is below the mean. (If the starting inventory is above the mean, feasibility of a set of suppliers depends on the demand distribution, but only via the single measure s.) It follows that the minimal number of required suppliers is given by the smallest number n for which the total number of *BSE*'s among the *first n* suppliers exceeds the critical threshold. The number of required suppliers can be reduced by improving their reliability; moreover, as a supplier with c.v. $= \gamma$ contributes γ^{-2} BSE's, the benefits of reductions of the c.v. value of a yield distribution become progressively larger. This gives support to management philosophies like "Six Sigma," which advocate that companies should strive for near perfection, rather than terminating their quality improvement program when a "reasonable" level of quality or reliability is reached. The allocation scheme which splits the aggregate order in proportion to the suppliers' mean-to-variance ratio of their yield distribution has the best chance of enabling feasibility: if a feasible solution fails to exist under *this* scheme, it fails to exist under *any*. Note, under this scheme, each supplier's share in the expected effective supply of useable units (Y_E) is determined by the square of the reciprocal of the coefficient of variation of his yield distribution. The same allocation scheme also optimizes the variable procurement costs.

Additional BSE's beyond the minimum number help to reduce the variable procurement costs. We refer to this surplus as the *"suppliers' safety margin"*. Under the CLT approximation, the minimal procurement cost for a given set of suppliers can be given as a closed form expression which depends on the set of suppliers and their yield distributions via a *single* measure, i.e. the number of BSE's. Moreover, as follows from the proof of Lemma 1 that the cost minimizing value of the expected effective supply Y_E is, under a minor assumption, a convex decreasing function of the number of BSE's which converges to $\mu + z_\alpha \sigma - I^0 = \mu(1 + z_\alpha \gamma_D) - I^0$, the classical optimal order quantity under a single reliable supplier. To appreciate the price paid for dealing with unreliable suppliers, consider, for example, the case where the starting inventory is zero. Here, the asymptotic lower bound needs to be adjusted in *two* ways: first the mean μ needs to be increased by a factor given by (#BSE's)/(the suppliers' safety margin); second the coefficient of variation γ_D needs to be increased to $\gamma_D' = \sqrt{\gamma_D^2 + (1 - z_\alpha^2 \gamma_D^2)/(\#BSE's)}$.[2] The correction factor for the mean and the correction term for γ_D' decrease to one and zero, respectively, as the number of BSE's grows.

The above results are obtained by approximating the shortfall distribution by the Normal distribution in (3) with matching mean and variance. This approximation is of course exact if all yield and demand distributions are Normal, themselves. Moreover, the Normal approximation turns out to be remarkably accurate even when the underlying yield and demand distributions are rather differently shaped. Indeed, the approximation is supported by the following asymptotic properties:

For any total order size Y and allocation scheme $\mathbf{w}$, let $P_n(Y, \mathbf{w})$ denote the *exact* shortfall probability and $\widetilde{P}_n(Y, \mathbf{w})$ denote the shortfall probability obtained on the basis of the Normal approximation. When dealing with an ever larger set of potential suppliers, the share of the aggregate order that is assigned to any given supplier will, under a given allocation scheme, depend on the total number of suppliers. We therefore write $w_{i,n}$ to represent supplier i's share when considering the set of suppliers $\{1, \ldots, n\}$. For example, the allocation scheme which splits the aggregate order equally among all potential suppliers, has $w_{i,n} = 1/n$ for all $i = 1, \ldots, n$. The following asymptotic accuracy result applies to any allocation scheme $\{w_{i,n}\}$ with

$$\frac{\max_i w_{i,n}}{\min_i w_{i,n}} \leqslant A \quad \text{for some constant } A,$$

[2] This assumes $\mu - z_\alpha \sigma > 0$, i.e. the probability of the Normal demand distribution adopting a negative value is itself less than α.

i.e. the ratio of the largest and the smallest order size remains bounded, as $n \to \infty$. Assume uniform positive lower bounds $\underline{p}$ and $\underline{\varsigma}$ such that $p_i \geqslant \underline{p}$ and $\varsigma_i \geqslant \underline{\varsigma}$ for all $i = 1, 2, \ldots$.

$$\lim_{n\to\infty} \left[P_n(Y, w_{1,n}, \ldots, w_{n,n}) - \widetilde{P}_n(Y, w_{1,n}, \ldots, w_{n,n})\right] = 0 \quad \text{for all } Y. \tag{4.10}$$

$$\lim_{n\to\infty} Y_E^*(n|\mathbf{w}) = F^{-1}(1-\alpha) - I^0 = \lim_{n\to\infty} \widetilde{Y}_E(n|\mathbf{w}). \tag{4.11}$$

See Theorem 4 in [13] for a proof. In addition to these asymptotic properties, a numerical study in [13] has confirmed the accuracy of the CLT-based approximation, even when the number of suppliers N is moderate or small, and even when the yield distributions are anything but Normal, for example when they represent Bernoulli random variables. See ibid for details.

We now address the problem of selecting the best possible set of suppliers, considering the *total* of fixed and variable operating costs. Proposition 1 in [13] shows that the problem is NP-complete even if the total variable cost is given by the CLT-based approximation. Nevertheless, [13] designs a simple heuristic and proves its remarkable *worst-case* optimality gap, while demonstrating its even more remarkable *average* performance on the basis of an extensive numerical study.

For the sake of brevity, assume $I^0 = 0$. In view of (4.9), the problem of selecting the optimal set of suppliers $S^* \subseteq \{1, \ldots, N\}$ may be formulated as:

$$\min_S \{z(S) : S \subseteq \{1, \ldots, N\}\}, \tag{4.12}$$

where $z(S) \stackrel{\text{def}}{=} \sum_{i\in S} K_i + C\left(\sum_{i\in S} \left(\gamma_i^{-2}\right)\right)$ and

$$C(R) \stackrel{\text{def}}{=} \begin{cases} \dfrac{c\mu}{\left(1-\frac{z_\alpha^2}{R}\right)} \left(1 + z_\alpha\sqrt{\gamma_D^2\left(1-\frac{z_\alpha^2}{R}\right) + \frac{1}{R}}\right), & \text{if } R > z_\alpha^2 \\ \infty, & \text{if } 0 < R \leqslant z_\alpha^2 \end{cases}$$

More generally, we may wish to limit the number of suppliers to some maximum $\widehat{N} \leqslant N$, in which case the problem can be formulated as $\min_S \{z(S) : S \subseteq \{1, \ldots, N\}$ and $|S| \leqslant \widehat{N}\}$. Federgruen and Yang [13] show that the selection problem (4.12) is a problem of minimizing a supermodular set function. (A set function $h : 2^{\{1,\ldots,N\}} \to \mathbb{R}$ is supermodular if $h(T \cup \{j\}) - h(T) \geqslant h(S \cup \{j\}) - h(S)$, for all $S \subset T$ and $j \notin T$).

Lemma 1 *Assume* $I^0 = 0$. *(a) The function* $C(\cdot)$ *is decreasing.*

(b) Assume $\gamma_D \leqslant 2\sqrt{3}/z_\alpha$. *The function* $C(\cdot)$ *is convexly decreasing, and the set function z(S) is supermodular.*

It is easily verified that the condition $\gamma_D \leqslant 2\sqrt{3}/z_\alpha$ is without any loss of practical generality.

The class of combinatorial optimization problems which can be formulated as the minimization of a supermodular set function is broad and has been studied intensively, see e.g. [6, 27, 28, 29]. The class includes, for example, the uncapacitated plant location problem and more generally the problem of finding a maximum weight independent set in a matroid. As such, the class is NP-complete. In our case, the set function *z(S)* is of the special type:

$$z(S) = f\left(\sum_{i \in S} a_i\right) + g\left(\sum_{i \in S} b_i\right) \tag{4.13}$$

for a given sequence of positive pairs $\{(a_1, b_1), \ldots, (a_N, b_N)\}$ and with $f\colon \mathbb{R} \to \mathbb{R} \cup \{\infty\}$ and $g\colon \mathbb{R} \to \mathbb{R} \cup \{\infty\}$ convex. Federgruen and Groenevelt [11] refer to this structure as "generalized symmetric" set functions. Proposition 1 in [13] shows that even this subclass is NP-complete.

Since the problem is NP-complete, no exact polynomial time procedure for the selection problem can be expected. However, [6] and [29] show that a simple greedy procedure has a low worst-case optimality gap; in practice, it comes within a few percentage points of optimality.

Normally, the greedy procedure for a set selection problem operates as follows: starting with the empty set, in each iteration an element is added to the set whose addition results in the biggest cost reduction. The greedy procedure terminates after $\widehat{N}$ iterations or in an earlier iteration if no cost saving can be achieved by adding a new element to the set. In our case, a slight modification is required since $z(S) = \infty$ if the set S contains insufficiently many BSE's. Thus, the greedy procedure must be initiated from a minimum set of suppliers enabling a feasible solution. We define a set S to be *minimally feasible* if $\sum_{i \in S} \gamma_i^{-2} > z_\alpha^2$ while $\sum_{i \in T} \gamma_i^{-2} \leqslant z_\alpha^2$ for all $T \subsetneq S$. Let

$$\bar{n} = \min\left\{ n : \sum_{i = N-n+1}^{N} \gamma_i^{-2} > z_\alpha^2 \right\}.$$

Clearly, all minimally feasible sets have a cardinality below $\bar{n}$.

When, as usual, $p_i \geqslant 0.5$, $\gamma_i \leqslant 1$ and $\bar{n} \leqslant \lceil z_\alpha^2 \rceil$. In practice, $\bar{n}$ is much smaller. For example, if all $p_i \geqslant 0.9$, it can be shown that $\bar{n} \leqslant \lceil \frac{z_\alpha^2}{9} \rceil \leqslant 2$ with $\alpha \geqslant 1.1045 \times 10^{-5}$; if all $p_i \geqslant 0.8$, $\bar{n} \leqslant \lceil \frac{z_\alpha^2}{4} \rceil \leqslant 2$ with $\alpha \geqslant 0.0023$. There are at most $\binom{N}{\bar{n}}$ minimally feasible sets, since every set of cardinality $\bar{n}$ is feasible, but several such sets may contain the same minimally feasible set. The bound is tight in case all $\gamma_i = \gamma$.

Our proposed solution method consists of enumerating the list $\mathcal{L}$ of all minimally feasible supplier sets and applying the above described greedy procedure to each of them. The Supplier Selection Algorithm in [13] employs an efficient creation of the list $\mathcal{L}$. Starting from a given minimally feasible set $S \in \mathcal{L}$, the greedy procedure requires $O(N^2)$ elementary operations and square root calculations, since in each of up to N iterations, up to N potential supplier sets have to be evaluated. Since there are at most $\binom{N}{\bar{n}}$ minimally feasible sets, the Algorithm's complexity is $O(N^{\bar{n}+2})$.

(As mentioned, usually $\bar{n} \leqslant 2$ and the Algorithm is of complexity $O(N^3)$ or $O(N^4)$.) Let Z^* denote the total cost of the optimal set of suppliers and Z^G the cost associated with the set of suppliers generated by the greedy-type Algorithm. Let ρ denote the *maximum* cost deterioration due to the addition of a single supplier to an existing set. By the convexity of the function $C(\cdot)$,

$$\rho = \left[\max_i \left\{ K_i + \left(C\left(\sum_{j=1}^{N} \gamma_j^{-2} \right) - C\left(\sum_{j \neq i} \gamma_j^{-2} \right) \right) \right\} \right]^+ .$$

The following optimality gap is immediate from Theorem 4.2 in [29].

$$\frac{[Z^G - Z^*]}{\min\limits_{S \in \mathcal{L}} [z(S) - Z^* + \widehat{N}\rho]} \leqslant \left(1 - \frac{1}{\widehat{N}}\right)^{\widehat{N}} \leqslant e^{-1}$$

Disregarding the correction term $\widehat{N}\rho$, the optimality gap $Z^G - Z^*/z(S) - Z^*$ is somewhat unconventional, but, as argued in [6], it may actually be more descriptive of the quality of the heuristic: it relates the absolute gap between its cost value Z^G and the optimal cost value Z^* to the span between Z^* and the cost of an arbitrary starting solution.

4.2.3 General Suppliers

We now discuss the most general case where the suppliers charge non-identical unit prices. Without loss of generality, assume that the suppliers are ranked in ascending order of their cost rate, i.e., $c_1 \leqslant c_2 \leqslant \ldots \leqslant c_N$. As in the previous subsections, we focus first on procurement strategies which minimize the expected *variable* procurement cost among all feasible strategies, and base our analyses on the CLT-based approximation.

When the cost rates $\{c_i\}$ fail to be identical, it is not necessarily optimal to choose the *smallest* possible value of Y_E: as we will show, this smallest feasible value requires the participation of all N suppliers and a cheaper solution may be obtained by enlarging the effective supply Y_E while allocating the aggregate order only to some of the less expensive suppliers. The necessary and sufficient conditions for a *feasible* set of orders continue to be given by the pair of inequalities (4.4) and (4.5). This implies that an *optimal* set of orders can be found as the optimal solution to the following mathematical program:

$$\text{(SCM)} \quad \min_{Y_E} \Psi^S(Y_E) \quad \text{s.t.} Y_E \geqslant \mu - \mathrm{I}^0 + z_\alpha \sigma, \text{ where} \tag{4.14}$$

$$\Psi^S(Y_E) \stackrel{\text{def}}{=} \min \sum_{i=1}^{N} c_i p_i y_i \tag{4.15}$$

$$\text{s.t.} \quad \sum_{i=1}^{N} p_i y_i = Y_E \tag{4.16}$$

$$(Y_E - \mu + I^0)^2 - z_\alpha^2 \left(\sum_{i=1}^{N} \varsigma_i^2 y_i^2 \right) - z_\alpha^2 \sigma^2 \geqslant 0 \tag{4.17}$$

$$y_i \geqslant 0, \quad i = 1, \ldots, N \tag{4.18}$$

Federgruen and Yang [15] show that the function $\Psi^S(\cdot)$ is strictly convex and differentiable with a unique minimum Y_E^*. Moreover, the mathematical program (4.15)–(4.18) which defines the function $\Psi^S(\cdot)$, is itself a convex program. Let λ_1 denote the unique Lagrange multiplier associated with the constraint (4.16). The following theorem, proven in [15], characterizes the optimal procurement cost associated with a given choice of Y_E, that is undominated. (We call a solution $\{Y, Y_E\}$ *undominated* if it is feasible and satisfies the service constraint as an equality. It is easily verified that any optimal solution $\{\mathbf{y}^*, Y_E^*\}$ of (SCM) is undominated.)

Theorem 3 (General Suppliers) *Let Y_E be part of an undominated solution and let $\lambda_1(Y_E)$ denote the optimal Lagrange multiplier associated with the mathematical program (4.15)–(4.18) which defines $\Psi^S(Y_E)$.*

1. *The optimal set of retained suppliers consists of the k^* cheapest suppliers, where $k^* = \max\{k : c_k < \lambda_1(Y_E)\}$.*
2. *There exists a unique optimal set of orders, corresponding with the selected level of expected effective supply Y_E, as follows:*

$$p_i y_i^* = \begin{cases} \frac{(\lambda_1 - c_i)\gamma_i^{-2}}{\sum_{l=1}^{k^*}(\lambda_1 - c_l)\gamma_l^{-2}} Y_E, & i = 1, \ldots, k^* \\ 0, & i = k^* + 1, \ldots, N \end{cases} \tag{4.19}$$

This solution implies the following optimal allocation of the aggregate order among the suppliers:

$$w_i^* \stackrel{\text{def}}{=} y_i^* \Big/ \sum_{l=1}^{N} y_l^* = \frac{[(\lambda_1 - c_i)^+] p_i / \varsigma_i^2}{\sum_{l=1}^{N} [(\lambda_1 - c_l)^+] p_l / \varsigma_l^2} \tag{4.20}$$

Thus, the optimal supplier base is always consecutive in the cost rates, and consists of all suppliers whose cost rate is below the above defined *benchmark cost rate*, $\lambda_1(Y_E)$. Also, each supplier in the selected supplier base is assigned an overall score, given by the product of *two* factors: the *first* is the mean-to-variance ratio of the supplier's yield distribution; the *second* factor is the net cost saving, relative to the benchmark cost rate. A supplier's market share is given by his overall score relative to the sum of the selected suppliers' scores.

We have assumed that the firms are required to pay only for the ordered units that turn out to be useable. In some settings, the firm may be charged for all ordered units, irrespective of whether they turn out to be useable or not. It is easily verified that all of the results in Theorem 3 continue to apply, merely replacing the suppliers' unit prices $\{c_i\}$ by their effective cost rates $\{c_i/p_i\}$. In particular, the optimal set of retained suppliers is now consecutive in the effective cost rates $\{c_i/p_i\}$. Finally, there are hybrid settings where the cost risks are shared between the firm and his suppliers, i.e., part of the procurement cost components is charged for every ordered unit and the remainder only for those units that are delivered as useable. Once again, all of the results in Theorem 3 continue to apply, merely replacing the unit prices $\{c_i\}$ by the suppliers' effective cost rates, defined as the expected variable costs incurred by the firm to obtain a single useable unit.

To complete the characterization of the optimal set of orders in Theorem 3, [15] shows that the benchmark cost rate $\lambda_1(Y_E)$ can be obtained as the unique root of a closed-form non-linear equation. As the Lagrange multiplier of constraint (4.16), the benchmark cost rate $\lambda_1(Y_E)$ may be interpreted as the marginal cost saving which can be obtained if a marginal unit of expected effective supply could be procured risk and cost free, i.e., without placing orders to any of the (unreliable) suppliers. As such, it is intuitive and indeed proven that $\lambda_1(Y_E)$ is a decreasing function of the expected effective supply level Y_E. In view of Theorem 3(1), this implies that as one increases the expected effective supply level Y_E, one faces a smaller risk of failing to meet the service level constraint, and hence, one can afford to diversify the orders among possibly fewer and less expensive suppliers. This represents the fundamental tradeoff: increase the expected effective supply so as to retain fewer and cheaper suppliers or limit the aggregate expected effective supply but diversify, possibly, over a larger number and more expensive suppliers. Recall from Theorem 2 that for a supplier set to be feasible, its number of BSE's must exceed a specific critical value. Since the optimal supplier set is consecutive in the unit cost rates, the minimum number of suppliers $\underline{k}$ can be determined as follows: starting with the cheapest supplier (supplier 1), keep adding suppliers to the supplier set until the corresponding number of BSE's exceeds the critical value specified in Theorem 2(1). It thus follows from Theorem 3(1) that critical expected effective supply levels $Y_E^N \leqslant Y_E^{N-1} \leqslant \cdots \leqslant Y_E^{\underline{k}}$ exists such that

$$k^*(Y_E) = k, \quad \text{iff} \quad Y_E^k \leqslant Y_E < Y_E^{k-1}, k = \underline{k}, \ldots, N. \tag{4.21}$$

Along with the above mentioned strict convexity and differentiability properties of the $\Psi^S(\cdot)$ function, the ability to compute the benchmark cost rate $\lambda_1(Y_E)$ as the unique root of a closed-form increasing function suggests that the globally optimal set of orders, among all feasible levels of the expected effective supply Y_E can be obtained by the following simple algorithm:

The above algorithm can be further accelerated by determining Y_E^* as the unique *root* of the increasing function $\Psi^{S\prime}(Y_E)$, since [15] shows that this derivative function can be evaluated as a simple function of the benchmark cost rate $\lambda_1(Y_E)$, the suppliers'

Algorithm 1 Algorithm SCM

<u>**STEP**1:</u> Determine a lower bound $\underline{Y_E}$ and an upper bound $\overline{Y_E}$ for all feasible solutions. ($\underline{Y_E}$ is given by the expression in (4.9), which corresponds with a solution to retain all suppliers.)

<u>**STEP**2:</u> Determine the optimal solution Y_E^* of the function $\Psi^S(\cdot)$ via a bisection search exploiting its strict convexity. To evaluate the function $\Psi^S(Y_E)$ for any candidate value Y_E, employ formula (4.19) for the optimal orders, after computation of the benchmark cost rate $\lambda_1(Y_E)$ as the unique root of a closed form non-linear function.

unit prices $\{c_i\}$ and the coefficients of variation of their yield distributions $\{\gamma_i\}$. See [15] for an even more efficient implementation of this bisection based algorithm.

As explained, retaining a smaller set of suppliers, when feasible, has the advantage of reducing the average procurement cost per unit, (even though it may come at the expense of requiring a larger aggregate order to hedge against the increased supply risks). The presence of fixed costs provides an additional incentive to pursue solutions with a smaller set of suppliers. If the *same* fixed cost K is incurred for every retained supplier, it is quite easy to incorporate the fixed costs into the analysis. Let Y_E^* denote the optimal effective supply and $\widehat{k}$ the number of associated suppliers, in the absence of fixed costs. Since by (4.21), $k^*(Y_E) \geqslant k^*(Y_E^*)$ for all $Y_E < Y_E^*$, Y_E^* continues to be preferred over all *lower* values of Y_E in the presence of fixed costs. Since $\Psi^S(Y_E)$ is increasing for $Y_E > Y_E^*$, it follows that only one of the $(\widehat{k} - \underline{k} + 1)$ values in $\{Y_E^*, Y_E^{\widehat{k}-1}, \ldots, Y_E^{\underline{k}}\}$ may arise as the total optimal supply level $Y_E^*(K)$ and $Y_E^*(K) = argmin\left\{\Psi^S(Y_E^*) + \widehat{k}K, \Psi^S(Y_E^{\widehat{k}-1}) + (\widehat{k} - 1)K, \ldots, \Psi^S(Y_E^{\underline{k}}) + \underline{k}K\right\}$. This characterization also implies:

Corollary 1 *The optimal effective supply $Y_E^*(K)$ is increasing in K, while the optimal number of suppliers is decreasing in K. In particular, $Y_E^*(K) \geqslant Y_E^*(0) = Y_E^*$, the optimal level in the case without fixed costs.*

When the fixed costs are supplier dependent, we already showed in Sect. 4.2.2 that the problem is NP-complete even in the special case where the suppliers have identical unit cost rates $\{c_i\}$, in which case the optimal set of orders for any given selection of suppliers can be determined in closed form. Nevertheless, we showed that a simple greedy type supplier selection procedure comes very close to being optimal, both empirically and in terms of a worst-case optimality gap. As discussed, these results follow from the fact that the marginal benefit associated with a new supplier is smaller when the supplier is added to a larger list of potential suppliers. (This property implies that the optimal cost value, viewed as a function of the set of potential suppliers, is submodular.) We continue to advocate the use of the supplier selection procedure described in Sect. 4.2.2, even though the set selection problem is no longer guaranteed to be submodular. In practice, this property is satisfied, except in certain extreme cases, thus, guaranteeing the worst-case optimality gap discussed in Sect. 4.2.2. Evaluating any candidate set of suppliers can, of course, be done with the Algorithm SCM.

4.2.4 The Impact of Initial Inventory, Demand and Supply Risks

In this subsection, we discuss how the optimal cost value, the (effective) order size and supplier base vary with the supply risks, the demand magnitude and risks, as well as the initial inventory. We start with the impact of supply risks.

Supply Risks

Proposition 1. (Impact of Supply Risks)

1. *If a supplier j fails to be competitive, i.e.* $y_j^* = 0$ *or* $j > k^*$, *this supplier cannot become part of the supplier base by improving* ς_j, *the reliability of his yield distribution, alone.*
2. *There exists a maximum price* $c_{k^*+1} > c^* \geqslant c_{k^*}$ *such that any of the currently uncompetitive suppliers* $k^* + 1, \ldots, N$ *receives a positive market share if and only if his unit cost rate is reduced to below* c^*. *In addition, the break-even value* c^* *is independent of the yield distributions of suppliers* $k^* + 1, \ldots, N$.
3. *The optimal cost value increases with any of the parameters* $\{\varsigma_j\}$.
4. *Assume* $I^0 \leqslant \mu \cdot k^*$ *is an increasing step function of any of the yield distributions' standard deviations* $\{\varsigma_j\}$.

While the *number* of suppliers increases when any of the yield distributions becomes more volatile, and while for *constant* input parameters, additional suppliers allow for a reduction in the expected effective supply Y_E (see (4.21)), Y_E^* fails, in general, to decrease with any of the standard deviations $\{\varsigma_j\}$, as illustrated with an example in [15]. The same example shows that both the expected safety stock and the total order may fail to increase when all of the standard deviations of the yield distributions are increased, in parallel, by the same percentage. This counterexample also implies that, the expected safety stock, as well as the total order, may fail to be monotone in any *individual* supplier's yield standard deviation. Similarly, as with the dependence of the expected effective supply Y_E^* with respect to any of the yield *standard deviations* $\{\varsigma_j\}$, Y_E^* fails, in general, to be monotone in any of the average yields $\{p_j\}$,

Proposition 1 shows that if a supplier improves the *volatility* of his yield distribution, this can only result in a contraction of the supplier base. Does the same monotonicity pattern apply when any of the suppliers improves his *average* yield? A different in [15] example disproves this conjecture.

Demand Magnitude and Risks

In the classical inventory model with a single fully reliable supplier, it is well known that the order-up-to level increases when demands have a larger mean and become more variable. (Indeed, it is $\mu + z_\alpha \sigma$, a simple linear functions of both μ and σ.)

The following Proposition shows that the expected order-up-to level and hence the expected safety stock continue to be increasing in μ, albeit that the dependence is no longer linear. When $I^0 \leqslant \mu$, the same monotonicity property applies to σ; moreover, extensive numerical studies show that this monotonicity result also applies when $I^0 > \mu$. Similarly, the optimal number of suppliers increases with μ. Perhaps more strikingly, the Proposition also shows that when $I^0 \leqslant \mu$, an increase in the demand variability can only result in the elimination of suppliers from the supplier base. In contrast, when $I^0 \geqslant \mu$, increased demand variability can only result in an *expansion* of the supplier base.

Proposition 2 (Impact of Demand Magnitude and Risk)

1. *The optimal cost value increases with μ and σ.*
2. *Both k^* and Y_E^* increase with μ.*
3. *k^*, the optimal number of suppliers (i) decreases with σ, when $I^0 < \mu$; (ii) is independent of σ when $I^0 = \mu$ and (iii) increases with σ, when $I^0 > \mu$.*
4. *If $I^0 \leqslant \mu$, Y_E^* increases with σ.*

The intuition behind the, at first counterintuitive, monotonicity results of k^* with respect to σ is as follows: (4.21) shows that the optimal order-up-to level is optimal to use the cheapest k suppliers if and only if targeting an expected supply level $Y_E^k \leqslant Y_E < Y_E^{k-1}$, $k = \underline{k}, \ldots, N$, i.e., the larger an expected supply level is sought, the fewer suppliers one should use so as to control the cost. An increase in the demand volatility σ has two opposite effects: first, as in the classical model, more safety stock is needed to cover the demand, hence a larger value of Y_E is required. (This is indeed proven in part (d) for $I^0 \leqslant \mu$; a numerical study shows that the result holds throughout.) If the critical effective supply values $\{Y_E^k\}$ were invariant with respect to σ, this would imply that k^* *decreases* with σ. However, these critical values are *increasing* in σ, so that for a given value of Y_E, the *same or a larger* number of suppliers is to be used. When $I^0 < \mu (I^0 > \mu)$, the first (second) effect dominates.

Initial Inventory

Note that, the optimal order quantities depend on μ and I^0 only via $(\mu - I^0)$. As a consequence, all monotonicity properties identified in Proposition 2 with respect to μ, imply the reverse monotonicity pattern with respect to I^0.

It appears intuitive that the safety stock requirement, in our setting with *combined* demand and supply risks, should be larger than the optimal safety stock in a setting where only demand risks prevail and all suppliers are completely reliable. Indeed, we showed that $I^0 + Y_E$, the expected inventory after ordering, satisfies (4.14), i.e., it is larger than or equal to the optimal inventory level when the suppliers are fully reliable.

4.3 Total Cost Model

In this section, we return to the more traditional approach in inventory and supply chain management models where the consequences of stockouts are assumed to be associated with specific stockout costs. We immediately analyze a general multi-period model, in which the firm faces an arbitrary sequence of uncertain demands and where it has the opportunity to place replenishment orders with any subsets of the collection of N potential suppliers, at the beginning of each period. In Subsect. 4.3.3, we'll apply our results to the special case where the planning horizon consists of a single sales season, with only a single opportunity to place replenishment orders, as in the model of Sect. 4.2, thus enabling a comparison of the qualitative behavior of the Service Constraint Model (SCM) and the Total Cost Model (TCM), analyzed here.

More specifically, we consider a planning horizon of T periods, in which the firm faces a given sequence of independent but possibly non-stationary, continuously distributed random demands. Suppliers are differentiated by their time-dependent unit prices and (general) yield distributions. As in Sect. 4.2, we initially assume that there are no fixed costs associated with the procurement orders, see, however, Sect. 4.3.2 for a treatment of this more general case. As in Sect. 4.2, we assume that the firm only pays for useable units, as opposed to all ordered units. However, as explained there, this is without loss of generality. The firm's holding and backlogging costs are proportional with the end of the period's inventory levels and backlog sizes respectively. (More generally, all of our results continue to apply when these costs are given by convex functions of the inventory levels.) We initially assume that (the useable parts of) orders placed in any period become available in time to meet that period's demand. However, as discussed in Sect. 4.3.2, all of our results can easily be extended to allow for a fixed lead time, as long as the actual yield realizations of a given period's orders are revealed to the firm by the beginning of the next period. We assume that unsatisfied demand in any given period is backlogged, at a cost which is proportional with the backlog size. In the absence of lead times, all of our results continue to apply when stockouts result in lost sales. We number the periods of the planning horizon backwards, i.e., period t is the t-th period before the end of the planning horizon. Thus, let:

$c_{it} =$ the unit price charged by supplier i in period t.
$X_{it} =$ the random yield factor of supplier i, with cdf $G_{it}(\cdot)$, mean p_{it}, variance $\varsigma_{it}^2 > 0$ and coefficient of variation $\gamma_{it} = \varsigma_{it}/p_{it}$; $i = 1, \ldots, N$; $t = 1, \ldots, T$.
$D_t =$ the uncertain demand in period t, with a general continuous cdf $F_t(\cdot)$, pdf $f_t(\cdot)$, ccdf $\overline{F}_t(\cdot)$, mean μ_t, standard deviation σ_t, where the pdf is assumed to be continuously differentiable on the interior of the distribution's support, $t = 1, \ldots, T$.[3]

[3] Demand may represent net demand, net of pre-committed and guaranteed deliveries, as under the flexible quantity contracts. In our base model, we assume that the demand distributions have the positive half line as their support. All of our results are easily extended when the support is given by a different interval, for example the full real line, as in the case of Normal distributions.

$h_t =$ the inventory cost per unit carried in inventory at the end of period t, $t = 1, \ldots, T$.
$b_t =$ the backlogging cost rate per unit backlogged at the end of period t, $t = 1, \ldots, T$.
$\alpha =$ the discount factor $(0 < \alpha \leqslant 1)$.

The yield factors $\{X_{it}\}$ and the sequence of demand volumes $\{D_t\}$ are assumed to be independent. See Sect. 4.3.2 for a relaxation of this assumption. To formulate the planning problem as a dynamic program, we need the following variables and value functions:

$I_t =$ the inventory level at the beginning of period t, $t = 1, \ldots, T$.

$y_{it} =$ the order placed with supplier i in period t.

$Y_t = \sum_{i=1}^{N} y_{it} =$ the aggregate order in period t.

$$Y_t^e = \sum_{i=1}^{N} p_{it} y_{it} = \text{the expected } \textit{effective} \text{ supply in period } t, \text{ i. e., the expected amount of useable supply resulting from the various orders.} \tag{4.22}$$

$v_t(I_t) =$ the minimum expected discounted costs in periods $t, t-1, \ldots, 1$, when starting period t with an inventory level I_t.

The inventory dynamics are described by the following recursive scheme:

$$I_{t-1} = I_t + \sum_{i=1}^{N} X_{it} y_{it} - D_t, \quad t = 1, \ldots, T, \tag{4.23}$$

and the value functions therefore satisfy the following recursive equations: (Let $x^+ \stackrel{\text{def}}{=} \max\{x, 0\}$).

$$v_t(I_t) = \min_{\mathbf{y}_t \geqslant \mathbf{0}} H_t(\mathbf{y}_t, I_t), \quad \text{where} \tag{4.24}$$

$$\begin{aligned} H_t(\mathbf{y}_t, I_t) = {} & \sum_{i=1}^{N} c_{it} p_{it} y_{it} + h_t E\left[I_t + \sum_{i=1}^{N} X_{it} y_{it} - D_t\right]^+ + b_t E\left[D_t - I_t - \sum_{i=1}^{N} X_{it} y_{it}\right]^+ \\ & + \alpha E v_{t-1}(I_t + \sum_{i=1}^{N} X_{it} y_{it} - D_t), \quad t = 1, \ldots, T, \text{ and} \end{aligned} \tag{4.25}$$

$$v_0 \stackrel{\text{def}}{=} 0. \tag{4.26}$$

Thus, [17] shows that the value functions are strictly convex and that a unique optimal order vector exists for every starting inventory level.

Theorem 4 *Fix* $t = 1, \ldots, T$.

1. *For all* $-\infty < I_t < +\infty$, $v_t(I_t) < \infty$ *and* $\lim_{I_t \to -\infty} v_t(I_t) = \lim_{I_t \to +\infty} v_t(I_t) = \infty$.
2. *The value function* $v_t(\cdot)$ *is strictly convex, and the function* $H_t(\mathbf{y}_t, I_t)$ *is jointly strictly convex and supermodular in* $(\mathbf{y}_t, I_t)$.
3. *For each starting inventory level* I_t, *there exists a unique optimal order vector* $\mathbf{y}_t^*(I_t)$.

As it is immediate from its proof, all of the structural results in Theorem 4 continue to apply when [some of] the demand distributions are discrete or mixed. Additional structural results depend on the value functions $v_t(\cdot)$ being continuously differentiable and twice differentiable almost everywhere, as shown in Lemma 1 of [17]. These analytical properties of the value functions rely on our assumption that the demand functions have continuous distributions.

In view of the joint convexity of the function $H_t(\mathbf{y}_t, I_t)$, one can show that the optimal set of orders to be used in period t is the unique solution of the following system of equations and inequalities:

$$\partial H_t(\mathbf{y}_t^*, I_t)/\partial y_{it} = c_{it} p_{it} + h_t p_{it} - (b_t + h_t) E_{\{X_{lt}\}} \left[X_{it} \overline{F}_t \left(I_t + \sum_{l=1}^{N} X_{lt} y_{lt}^* \right) \right]$$
$$+ \alpha E_{\{X_{lt}\}} \left[X_{it} \int_{-\infty}^{+\infty} v_{t-1}' \left(I_t + \sum_{l=1}^{N} X_{lt} y_{lt}^* - u \right) dF_t(u) \right]$$
$$\begin{cases} = 0, & \text{if } y_{it}^* > 0 \\ \geqslant 0, & \text{if } y_{it}^* = 0 \end{cases} \tag{4.27}$$

This characterization permits us to show that the optimal set of suppliers to be retained in any period t, is consecutive in the unit cost rates $\{c_{it}\}$. In fact, in Theorem 5, below, we show that, as in the SCM model, the optimal set of suppliers in period t consists of those whose unit cost rates fall below the following benchmark cost rate:

$$\lambda_t^E(I_t) \stackrel{\text{def}}{=} -v_t'(I_t)$$
$$= -h_t + (b_t + h_t) E_{\{X_{lt}\}} \left[\overline{F}_t \left(I_t + \sum_{l=1}^{N} X_{lt} y_{lt}^* \right) \right]$$
$$- \alpha E_{\{X_{lt}\}} \left[\int_{-\infty}^{+\infty} v_{t-1}' \left(I_t + \sum_{l=1}^{N} X_{lt} y_{lt}^* - u \right) dF_t(u) \right] \tag{4.28}$$

$$\leqslant \overline{\lambda}_t(I_t) \stackrel{\text{def}}{=} -h_t + (b_t + h_t) \overline{F}_t(I_t) - \alpha \int_{-\infty}^{+\infty} v_{t-1}'(I_t - u)\, dF_t(u) \tag{4.29}$$

As in the (SCM) model of Sect. 4.2, the benchmark cost rate may be interpreted as the expected value of the total cost saving, associated with a marginal effective unit, delivered—for free and outside of the normal procurement process—beyond those arising from the optimal set of orders; here, the total cost saving relates to current holding and backlogging costs as well as all future costs.

Theorem 5 *Fix period* $t = 1, \ldots, T$. *Renumber the suppliers such that* $c_{1t} \leqslant c_{2t} \leqslant \cdots \leqslant c_{Nt}$.

1. *The optimal set of suppliers is consecutive in the sequence of unit cost rates* $\{c_{it}\}$. *In other words, this optimal set is of the form* $\{1, \cdots, k^*(I_t)\}$ *with* $k^*(I_t) = \max\{k : c_{kt} < \lambda_t^E(I_t)\}$.
2. *The benchmark cost rate* $\lambda_t^E(I_t)$ *is a strictly decreasing, continuous function of* I_t.
3. $k^*(I_t)$, *the optimal number of suppliers, is decreasing in* I_t, *i. e., the optimal set of suppliers shrinks as a function of* I_t. *In other words, additional units of starting inventory may result in the elimination of one or more of the most expensive suppliers in the retained supplier set.*
4. *There exists an inventory level* S_t, *such that it is optimal to place orders if and only if the starting inventory* $I_t < S_t$.
5. S_t *is the unique root of the strictly increasing function* $c_{1t} + h_t - (b_t + h_t)\overline{F}_t(I_t) + \alpha \int_{-\infty}^{+\infty} v'_{t-1}(I_t - u)dF_t(u)$.

Theorem 5(1) shows that the optimal set of suppliers in any given period is consecutive in *that* period's unit cost rates. The *degree* of supplier diversification, i. e., how *many* suppliers are to be retained depends, of course, on all current and future yield and demand distributions, as well as all current and future cost rates. Strikingly, the dependence on these various distributions and cost parameters occurs via a single aggregate measure, i.e., the benchmark cost rate $\lambda_t^E(I_t)$: the retained suppliers are precisely those whose unit cost rate is strictly below this benchmark. This characterization also implies that, as in the SCM model, see Proposition 1, if a supplier fails to be part of the set of retained suppliers, he cannot become competitive by improving his yield distribution alone. In the words of [23], cost can be thought of as an "order qualifier", while reliability acts as an "order winner." The consecutiveness property of the optimal supplier set was first obtained by [1] in a two supplier, but multi-period setting, with Bernoulli yield factors, and by [7] and [15] for a single period setting with an arbitrary number of suppliers. Similarly, the monotonicity of the optimal supplier set as a function of the starting inventory generalizes the same result obtained by [1] in their two-supplier model.

Our assumption that the demand distributions are continuous, is essential for the consecutiveness property. Under discrete or mixed demand distributions, any of the functions H_t, while convex, may fail to be differentiable at the optimal solution $\mathbf{y}_t^*$. Similarly, the value function v_{t-1} may not be differentiable in countably many points, so that the (last term in the) benchmark cost rate $\lambda_t^E(I_t)$, itself, is ill defined, when the demand distribution has a positive mass in any of the points where v_{t-1} fails to be differentiable. Indeed, [32] shows that under discrete distributions, the optimal supplier set may fail to be consecutive, even in a single period setting with $N=2$ suppliers; i. e., with a discrete demand distribution, it may be optimal to order from the more expensive supplier exclusively.

Our model with multiple unreliable suppliers inherits the well known property in the classical model, that, in each period t, orders are placed if and only if the starting inventory is below a given threshold S_t. However, the threshold S_t is no longer the order-up-to level for all the inventory levels below it. The single period example in Fig. 2 of [15] exhibits the following phenomena: (i) the expected order-up-to level usually decreases, but it sometimes fails to be monotone, and may even be increasing in the starting inventory; and (ii) the expected order-up-to level is sometimes larger but sometimes smaller than the level in the corresponding classical model. Observation (ii) is somewhat surprising, as one might conjecture that the need for safety stocks would be larger when supply risks compound on demand risks. Actually, it is the *relative* magnitude of the supply risks compared to the demand risks which determines whether the order-up-to level is larger or smaller than what is optimal in the absence of supply risks. In particular, when the cost consequences of a shortage are relatively low, the additional supply risks may render it optimal to target a lower rather than a higher expected inventory level after ordering.

The optimality conditions (4.27), along with the consecutiveness property of the optimal set of suppliers, suggest the following algorithm (GTA) to find the optimal order vector $\mathbf{y}_t^*$, in any given period t: (As in Theorem 5, the suppliers are numbered such that $c_{1t} \leqslant c_{2t} \leqslant \cdots \leqslant c_{Nt} \leqslant c_{(N+1)t} \stackrel{\text{def}}{=} +\infty$).

Algorithm 2 General TCM Algorithm (*GTA*)

FOR $k := 1$ TO N DO

BEGIN

<u>**STEP 1:**</u> Find the unique non-negative solution $\{y_{1t}^*, \cdots, y_{kt}^*\}$ of the following system of k equations in k unknowns $\{y_{1t}, \cdots, y_{kt}\}$:

$$c_{it} p_{it} + h_t p_{it} - (b_t + h_t) E_{\{X_{lt}\}} \left[X_{it} \overline{F}_t \left(I_t + \sum_{l=1}^{k} X_{lt} y_{lt} \right) \right] \tag{4.30}$$
$$+ \alpha E_{\{X_{lt}\}} \left[X_{it} \int_{-\infty}^{+\infty} v'_{t-1} \left(I_t + \sum_{l=1}^{k} X_{lt} y_{lt} - u \right) dF_t(u) \right] = 0, \quad i = 1, \ldots, k$$

<u>**STEP 2:**</u> If $k < N$, set $y_{it}^* = 0$ for $i > k$. Calculate $\lambda_t^E(I_t)$ from (4.28).

IF $\lambda_t^E(I_t) \leqslant c_{(k+1)t}$ THEN

exit with $\mathbf{y}_t^*$ as the optimal order vector

ENDIF

END

(If the test in Step 2 is met, the right hand side of (4.27) is non-negative for all $i = k+1, \ldots, N$, while in view of Step 1, it equals zero for $i = 1, \ldots, k$, thus verifying that $\mathbf{y}_t^*$ is the optimal order vector. Also, the test in Step 2 is met for exactly *one* value of k, permitting an exit as soon as it is met: if it were met for two different values of k, there would be two distinct optimal solutions, while by Theorem 5(1), (4.27) would fail to have a solution if the test in Step 2 was never met.) (GTA) amounts to solving at most N systems of well behaved equations, each of which has a unique solution. The main difficulty arises from the evaluation of the multivariate expectations to the right of (4.30). For general yield and demand distributions, this is best done by a simulation technique. In Sect. 4.3.1, we describe a more efficient algorithm based on a CLT-approximation of each period's shortfall distribution. This approximate algorithm also generates insights into the impact which price- and yield-differentials among the suppliers have on their respective market shares.

Finally all of the above results have been extended to the case of an infinite-horizon model with stationary parameters and yield and demand distributions, both under the expected total discounted cost criterion and the long-run average cost criterion, see [18] for detailed analyses.

4.3.1 *Efficient Approximations and Heuristics*

The efficient determination of the optimal procurement strategy is hampered by the fact that the distribution of the end-of-the period inventory level is in general quite complex, even for a given set of orders $\mathbf{y}_t$. A fortiori, it is prohibitively difficult to use the exact distribution when determining the optimal set of suppliers, along with the aggregate order and its allocation among the selected suppliers, in any given period t. Applying the CLT-based approximation for the conditional inventory level distributions to our multi-period model allows us to derive a very efficient procedure for the determination of the optimal procurement strategy. It also enables additional insights into the factors which determine the optimal market shares each of the suppliers acquires. We thus approximate the conditional inventory distributions $(I_{t-1}|I_t)$:

$$(I_{t-1}|I_t) \sim N\left(I_t + \sum_{i=1}^{N} p_{it}y_{it} - \mu_t, \sigma_t^2 + \sum_{i=1}^{N} y_{it}^2 \varsigma_{it}^2\right) \tag{4.31}$$

where $N(\nu, \tau^2)$ denotes a Normally distributed random variable with mean ν and standard deviation τ.

Under the CLT-based method, we first show that for any given value of

$$Y_t^e = \sum_{i=1}^{N} p_{it}y_{it}, \quad t = 1, \ldots, T, \tag{4.32}$$

the corresponding set of suppliers and their market shares are easily obtained. This permits one to project the N-dimensional vector of the decision variables at each stage of the dynamic program onto the single aggregate supply measure Y_t^e. Thus, let:

$\Psi_t(Y_t^e|I_t) =$ minimum expected total cost in periods $t, t-1, \cdots, 1$, when starting period t with an inventory level I_t and when selecting an expected effective supply Y_t^e for this period.

Theorem 6 *Fix $t = 1, \ldots, T$ and $Y_t^e > 0$.*

1. *There exists a unique order vector $\mathbf{y}_t^*(Y_t^e)$.*
2. *There exists a benchmark cost rate $\lambda_t(Y_t^e|I_t) > \min_l c_{lt}$ such that*

$$p_{it} y_{it}^* = \frac{[(\lambda_t(Y_t^e|I_t) - c_{it})^+]\gamma_{it}^{-2}}{\sum_{l=1}^N [(\lambda_t(Y_t^e|I_t) - c_{lt})^+]\gamma_{lt}^{-2}} Y_t^e \tag{4.33}$$

where $\lambda_t(Y_t^e|I_t)$ can be computed as the unique root of a strictly decreasing function.

3. *Assume that the suppliers are renumbered such that $c_{1t} \leqslant c_{2t} \leqslant \cdots \leqslant c_{Nt}$. The optimal set of suppliers is given by $\{1, \ldots, k_t^*(I_t|Y_t^e)\}$ where $k_t^*(I_t|Y_t^e) \stackrel{\text{def}}{=} \max \{1 \leqslant k \leqslant N : c_{kt} < \lambda_t(Y_t^e|I_t)\}$ and $y_{it}^* / \sum_{l=1}^N y_{lt}^* =$*

$$\left\{[(\lambda_t(Y_t^e|I_t) - c_{it})^+] p_{it}/\varsigma_{it}^2\right\} \Bigg/ \left\{\sum_{l=1}^N [(\lambda_t(Y_t^e|I_t) - c_{lt})^+] p_{lt}/\varsigma_{lt}^2\right\}$$

Thus, for a given choice of the expected effective supply Y_t^e, both the set of suppliers to be retained and their respective market shares are easily characterized and determined. It suffices to compute the benchmark cost rate $\lambda_t(Y_t^e|I_t)$ as the unique root of a single-variable decreasing function. Similar to the characterization of the set of suppliers associated with the *globally optimal* solution, see Theorem 5(1), the suppliers optimally retained to achieve any expected effective supply level Y_t^e are precisely those whose unit cost rate (c_{lt}) is strictly below the $\lambda_t(Y_t^e|I_t)$-value. $\lambda_t(Y_t^e|I_t)$ can be interpreted as the cost saving incurred when a marginal effective unit is delivered—free of charge and outside of the normal procurement process—beyond the optimal orders associated with the expected effective supply Y_t^e.

As in the (SCM) model, the respective market shares of the retained suppliers are obtained by computing a supplier score, itself a product of a reliability score γ_{lt}^{-2}, and a cost saving score which measures the saving, relative to the benchmark cost rate $\lambda_t(Y_t^e|I_t)$, per effective unit, of using this supplier. By (4.33), the optimal market shares of the retained suppliers are simply proportional to the values of this supplier score. It is particularly noteworthy that the optimal market shares of the suppliers can be obtained as a simple closed-form expression in terms of the model parameters, once the (single) benchmark cost rate has been computed. Note also that the dependence of the optimal supplier set and their market shares with respect to the current starting inventory and all future cost, demand and yield considerations are aggregated in a single quantity, i.e., the benchmark cost rate $\lambda_t(Y_t^e|I_t)$.

It remains to be shown how Y_t^{e*}, the *optimal* value of the expected effective supply, can be determined efficiently. [18] shows that, in any given period $t = 1, \ldots, T$, the function $\Psi_t(\cdot|I_t)$ is strictly convex so that Y_t^{e*} is its unique (local) minimum. The determination is further simplified by the fact that the functions $\Psi_t(\cdot|I_t)$ are shown to be differentiable, with an analytical expression for the derivative in terms of the model parameters, the bench mark cost rate $\lambda_t(Y_t^e|I_t)$ and the value function $v_{t-1}(\cdot)$.

This suggests a very efficient algorithm to compute the optimal order vector in any period t and given any initial inventory I_t. As mentioned above, when $I_t \geqslant S_t$, $\mathbf{y}_t^* = \mathbf{0}$. When $I_t < S_t$, $Y_t^{e*} > 0$, and, since $\Psi_t(\cdot|I_t)$ is a strictly convex function, we can find the optimal effective supply Y_t^{e*} as this function's unique minimizer: Y_t^{e*} is the unique root of the strictly increasing function $\Psi_t'(\cdot)$ which can be found by simple bisection. The derivative function value $\Psi_t'(Y_t^e|I_t)$ is available in closed form, after computing the bench mark cost rate $\lambda_t(Y_t^e|I_t)$ which, in turn, can be evaluated as the unique root of a strictly decreasing function. Once the optimal expected effective supply Y_t^{e*} and the associated benchmark cost rate $\lambda_t(Y_t^{e*}|I_t)$ have been determined, the complete order vector follows from (4.33).

The Infinite Horizon Model with the Long-run Average Cost Criterion

We complete this subsection with a discussion of the long-run average cost criterion, in a model with stationary parameters and stationary demand and yield distributions. An optimal strategy can, in principal, be determined by a value iteration method, i.e., by solving the dynamic programming recursion (4.24), (4.25) iteratively for ever larger planning horizons $t = 1, 2, \cdots$.

This method identifies a policy whose long-run average cost approaches the optimum long-run cost value g^* with any desired level of precision. However, to achieve an optimality gap less than 1%, this method typically requires several hours of CPU time, for example when executed on a Dell Optiplex GX620 computer with Pentium D CPU of 3.00 GHz and 3.5 GB of RAM, and when employing an inventory level grid of several hundreds of points. In addition, it is somewhat challenging to implement the generated policy, in that the manager is provided with an algorithmic oracle rather than a simple, intuitive policy rule.

To address both complications, [18] develops a simple and intuitive heuristic, which can be computed in no more than 0.2 CPU seconds on the above platform. The heuristic is a generalization of the linear inflation rule proposed by [10] and [3] for the case of a single supplier: *aggregate* orders are prescribed according to a simple base stock policy, as in the case of fully reliable suppliers, however—inflated by a selected inflation factor $\beta^{-1} > 1$ to account for the imperfect random yields. In other words, for any period t, let

$$Y_t = \text{the aggregate order size} = \frac{(S - I_t)^+}{\beta}, \quad t = 1, 2, \cdots, \tag{4.34}$$

for an appropriately chosen base stock level S and inflation factor β^{-1}. (Recall that the *optimal* aggregate order policy fails to be of this base stock type.)

In our context, with multiple potential suppliers, this aggregate ordering policy needs to be complemented with an allocation heuristic to identify which suppliers are to be retained, and to distribute the aggregate orders among the selected suppliers. Our proposed allocation heuristic is motivated by the market share formula (4.33). While stated in terms of the suppliers' shares in the *expected effective* supply Y_t^e,

(4.33) is easily adapted to derive their shares in the aggregate gross order. Dividing both sides of (4.33) by p_i and summing the equations over all $i = 1, \cdots, N$, we obtain

$$Y_t = \frac{\sum_{l=1}^{N}[(\lambda^E(I_t) - c_l)^+]p_l/\varsigma_l^2}{\sum_{l=1}^{N}[(\lambda^E(I_t) - c_l)^+]\gamma_l^{-2}} Y_t^e \quad \text{and} \quad y_{it}^* = \frac{[(\lambda^E(I_t) - c_i)^+]p_i/\varsigma_i^2}{\sum_{l=1}^{N}[(\lambda^E(I_t) - c_l)^+]\gamma_l^{-2}} Y_t^e.$$

Therefore,

$$y_{it}^* = \frac{[(\lambda^E(I_t) - c_i)^+]p_i/\varsigma_i^2}{\sum_{l=1}^{N}[(\lambda^E(I_t) - c_l)^+]p_l/\varsigma_l^2} Y_t, \quad i = 1, \cdots, N; t = 1, 2, \cdots \qquad (4.35)$$

The benchmark cost rate $\lambda^E(I_t)$ depends on the period's starting inventory level; in fact, it is a decreasing function of the latter, see Theorem 5(2). In our proposed allocation heuristic, we replace the benchmark cost rate function $\lambda^E(\cdot)$ by a constant λ. (One implication of the usage of a constant benchmark cost rate is that both the set of suppliers and their market shares remain constant, while those in the optimal policy vary with the prevailing inventory level.) The proposed linear inflation heuristic thus depends on three parameters: the base stock level S, the inflation factor β^{-1} and the benchmark cost rate λ. [18] selects (S, β, λ) as the parameter triple which minimizes an approximate closed form expression of the long-run average cost under this type of policy, generalizing the derivation in [37] for the special case of a single supplier. (That derivation, in turn, was motivated by [33] and [8] which proposed linear control policies for different types of inventory systems.) The optimizing parameters S and β can be obtained as closed form functions of the constant benchmark cost rate λ, thus reducing the search for an optimal triple of parameters to that of the single variable λ.

4.3.2 Extensions

In this subsection, we discuss several important generalizations of the TCM model.

Fixed Supplier Cost; Price Benefits Associated Multi-Sourcing

In this section, we have, thus far, ignored any fixed cost K_i incurred for each supplier i that is added to the pool of potential suppliers $\mathcal{P}$ (for example, costs associated with buyers and information systems). Such fixed costs provide an incentive to limit the degree of supplier diversification. They can, of course, easily be incorporated when comparing aggregate expected costs under two or more sets of suppliers. Such comparisons may also allow us to model a second benefit of supplier diversification, i.e., the ability to negotiate better prices, when dealing with a larger pool of qualified suppliers. Returning to finite horizon models and representing the cost rates $\{c_{it}\}$ as decreasing functions of the number of suppliers in the pool $\mathcal{P}$, i.e., $c_{it} \stackrel{\text{def}}{=} c_{it}(|\mathcal{P}|)$, the overall cost of a pool $\mathcal{P}$ is given by $C(\mathcal{P}) \stackrel{\text{def}}{=} \sum_{i\in\mathcal{P}} K_i + v_T(I_T|c_{it} = c_{it}(|\mathcal{P}|))$. Identifying the optimal pool of potential suppliers $\mathcal{P}^* = \arg\min_{\mathcal{P}} C(\mathcal{P})$ is in general a complex combinatorial problem, which is NP-hard even in single period settings, Sect. 4.2.

Lead Times

Positive lead times of $L \geqslant 1$ periods can be handled in a similar manner as in the classical model with a single, fully reliable supplier, provided the actual yields for the orders placed at the beginning of a given period become known before the start of the next period: only orders in periods $t = L+1, L+2, \ldots, T$ are relevant, thereafter, the orders fail to be received during the considered planning horizon. We now use as the state variable: $IP_t = I_t + \sum_{\tau=t+1}^{t+L} y_{i\tau} X_{i\tau} =$ the *inventory position* at the beginning of period $t =$ the inventory on hand, plus the effective supply in process. Since all unsatisfied demand is backlogged, the inventory position satisfies the same recursive scheme as (4.23): $IP_{t-1} = IP_t + \sum_{i=1}^{N} X_{it} y_{it} - D_t, t = 1, \ldots, T$, while$I_{t-L-1} = IP_t + \sum_{i=1}^{N} X_{it} y_{it} - (D_t + D_{t-1} + \cdots + D_{t-L})$. Recognizing that the expected end-of-the-period holding and backlogging costs in the first L periods cannot be affected by any of the procurement decisions, these cost terms may therefore be eliminated from the dynamic program. Charging to period t the expected inventory-and-backlogging costs that incur at the end of period $t - L$, we obtain the modified dynamic program:

$$v_t(IP_t) = \min_{\mathbf{y}_t \geqslant \mathbf{0}} H_t(\mathbf{y}_t, IP_t), \quad t = L+1, \ldots, T, \quad \text{where}$$

$$\begin{aligned} H_t(\mathbf{y}_t, IP_t) = &\sum_{i=1}^{N} c_{it} p_{it} y_{it} + \alpha^L h_t E\left[IP_t + \sum_{i=1}^{N} X_{it} y_{it} - (D_t + D_{t-1} + \cdots + D_{t-L})\right]^+ \\ &+ \alpha^L b_t E\left[D_t + D_{t-1} + \cdots + D_{t-L} - IP_t - \sum_{i=1}^{N} X_{it} y_{it}\right]^+ \\ &+ \alpha E v_{t-1}(IP_t + \sum_{i=1}^{N} X_{it} y_{it} - D_t), \end{aligned}$$

$$v_L(\cdot) = 0.$$

It is easily verified that *all* of the results in this paper continue to apply. The problem is considerably more complex when the yields of a given period's orders do not become known to the purchasing firm before the start of the next period. As an extreme case, assume that the actual yields are not revealed until the orders are delivered. In this case, the inventory position at the beginning of period t is itself unknown and only partially observable. To model this situation, we need to keep track of the L order vectors in process, requiring a dynamic program with a state space of dimension $LN + 1$.[4]

Capacities

Capacity bounds often represent an additional complication. Thus, assume that a capacity limit M_{it} prevails for any order placed with supplier i in period t, i.e., $y_{it} \leqslant M_{it}(i = 1, \ldots, N; t = 1, \ldots, T)$. One easily verifies that all our results continue to apply for arbitrary capacities. In particular, in any period t, the optimal set of retained suppliers continues to be the consecutive set $\{i : c_{it} < \lambda_t^E(I_t)\}$. However, the simple market share formula (4.33) no longer applies, as the market shares are now affected by the capacity levels in addition to the yield reliabilities and cost differentials.

Correlated Yields and Demands

In some settings, supply risks may be correlated, for example when natural disasters (storms, floods) or sabotage by terrorists are likely to hit multiple facilities in a given geographic region, or when the suppliers depend on common second-tier suppliers. Similarly, the yields and demand distributions in a given period may be correlated, for example when both are dependent on weather-related factors or common economic variables, see, for example, [2] for a procurement model with multiple suppliers subject to correlated yield risks.

It can be verified that all of the results in Theorem 4 continue to apply. However, it is no longer true that the optimal set of suppliers, in any given period, is consecutive in the unit cost rates, i.e., consists of those whose unit cost rate is below a given benchmark rate.

4.3.3 Comparisons between the SCM and TCM Models

As mentioned in Sect. 4.2, the Total Cost Model (TCM) and the Service Constraint Model (SCM) are equivalent in classical inventory models. An instance of (TCM) with a given backlogging rate induces the same optimal inventory strategy as an instance of (SCM) with a corresponding permitted shortfall probability, and vice versa. See [4] for a recent discussion of this equivalency in classical inventory models. However, the equivalency breaks down under multiple suppliers with unreliable yields, even when considering single period models. First, while a feasible solution always exists in the (TCM), in the (SCM), feasibility requires a minimum number of

[4] See [14] for a reduction to a $L + 1$ dimensional state space when all distributions are Normal or when end-of-the-period inventory levels are approximated as Normals.

sufficiently reliable suppliers, as explained in Sect. 4.2. A key concept in both models is the so-called expected effective supply, i.e., the expected total number of usable units obtained from the various suppliers. We have shown that, for a *given* expected effective supply level, the optimal set of suppliers and the orders can be obtained in closed form, after determining the root of a single variable function. Also, the total cost is a strictly convex function of the expected effective supply with a unique minimum. However, in the (SCM), a larger expected effective supply is optimally assigned to the same number or fewer suppliers, see (4.21). In other words, if one is willing to place a larger aggregate order, there is less need to diversify among suppliers and one may therefore be able to retain fewer and less expensive suppliers; in the (TCM), this monotonicity property may fail to hold. In the (SCM), the safety stock (= expected inventory after ordering - expected demand) is always larger than in the classical model without supply risks. Once again, this is not always the case in the (TCM). Also, in all of our numerical experiments with (SCM) instances, the expected inventory after ordering decreases as a function of the initial stock until it hits the classical level. However, in the (TCM), the above consideration may be counterbalanced when, in the presence of relatively low stockout cost rates, the supply risks justify an expected inventory level after ordering *below* the classical level. Here, additional units of initial stock allow one to target a *higher* expected inventory level after ordering, *closer* to the optimal level in the classical model.

Most of the qualitative insights and comparative statics obtained in Sect. 4.2.4 for the (SCM) model continue to apply in the single-period (TCM) model; however, in the (TCM) model, many of these properties are based on numerical studies, as opposed to the (SCM) model where almost all these results are substantiated by analytical proofs.

4.4 Supplier Competition

In this section, we discuss how the different suppliers may improve their market shares and expected profits by targeting specific characteristics of their yield distributions. We observe that in most industries, component suppliers or original equipment manufacturers (OEMs) increasingly compete in terms of product attributes other than direct cost. Many goods have become commoditized, and gross profit margins have shrunk, making it increasingly difficult to compete on the basis of price differentials (alone). The supplier's *quality* and his *yield effectiveness and reliability*, as measured by the percentage of effectively produced units, rank among the most critical of the various dimensions along which competing firms differentiate themselves. The same applies to suppliers of consumer goods to large department stores, retail organizations or government agencies (in the latter case, e.g., vaccines or medical devices). As mentioned, the yield characteristics include the possibility of complete disruptions due to natural causes, such as fires or hurricanes, man made breakdowns (e.g. sabotage or terrorist attacks), as well as bankruptcies.[5]

To provide insights into how suppliers may impact their competitive position by specific investments in the quality or yield characteristics of their production processes, we turn to our initial (SCM) model of Sect. 4.2.2. In this model, we have assumed that suppliers charge identical unit prices and are therefore differentiated only in terms of their yield distributions. As in Sect. 4.2.2, consider a single purchasing firm or agency. However, almost all our results carry over to the general oligopsony case with an arbitrary number of buyers. Recall that, under the CLT-approximation, this model allows for closed-form expressions of the aggregate purchase order as well as the individual suppliers' shares in the aggregate sales, see Theorem 2(3), and Equations (4.7) and (4.9). It is easily verified that the optimal allocation vector $\mathbf{w}^*$ in (4.7) implies the following suppliers' shares in the expected effective supply Y_E^* :

$$\frac{p_i y_i^*}{Y_E^*} = \frac{\gamma_i^{-2}}{\sum_{j=1}^N \gamma_j^{-2}} \tag{4.36}$$

(The market shares in (4.36) also follow from (4.19), when applying this formula to the case where the suppliers charge identical unit prices.) Combining (4.36) with (4.9), we observe that the suppliers' expected revenues depend on their yield distributions only via their coefficients of variation $\{\gamma_i, i=1,\cdots,N\}$, or more specifically, via the reciprocals of the squared coefficients of variation. Thus, for $i=1,\cdots,N$, let $x_i = \gamma_i^{-2}$ denote supplier i's reliability measure. Assuming the purchasing firm's initial inventory $I_0 = 0$, we obtain that:

$$p_i y_i^* = \left(\frac{x_i}{\sum_{j=1}^N x_j}\right) Y_E^*, \quad \text{and} \quad Y_E^* = T\left(\sum_{i=1}^N x_i\right) \quad \text{where}$$

$$T(R) \stackrel{\text{def}}{=} c^{-1}C(R) = \mu\left(1-\frac{z_\alpha^2}{R}\right)^{-1}\left[1 + z_\alpha\sqrt{\frac{1}{R} + \gamma_D^2\left(1-\frac{z_\alpha^2}{R}\right)}\right]. \tag{4.37}$$

It follows from Lemma 2 that the function $T(\cdot)$ is decreasing.

[5] Even before the 2008 financial crisis, [2] describes the severity of this type of risk: "Credit rating firms report that in 2002 over 240 firms defaulted on 160 billion dollars of debt, the largest amount ever over any one year period. ... The combined volume of defaults in 2001 and 2002 exceeded the total volume of defaults in the US over the previous twenty years. What is especially striking about the current trends is the surge in the defaults of large, well-established companies. Even in the relative stable years 2000–2005, almost 50 firms with assets or liabilities exceeding one billion dollars have filed for bankruptcy." In the automobile industry, for example, many suppliers routinely incur losses, with Delphi, the largest supplier of automotive parts in the United States, residing in Chap. 11, until recently. Choi and Hartley [5] document that in this industry, purchasing managers consider the financial solvability of the suppliers a major selection criterion, along with criteria like consistency and reliability.

Depending on the source(s) behind his random yields, a supplier may improve the coefficient of variation of his yield distribution, by investing more time and effort into the design phase, or by adopting appropriate technologies, materials, manufacturing and logistical processes or a more secure financial structure, as well as improving his facilities' security. A supplier can improve his reliability by (i) increasing the *yield predictability*, via a reduction of the standard deviation of the yield factor, or (ii) increasing the *yield target*, i.e. the mean of the yield distribution, or (iii) improving *both* the yield target and its standard deviation. Consequently, we distinguish between three types of competition, which we refer to, respectively, as (I) Yield Predictability Competition, (II) Yield Target Competition, and (III) Simultaneous Yield Target and Predictability Competition.

Each supplier i's choice of his yield characteristics has important implications for his variable per unit cost rates. Let:

$\xi_i(p_i, \varsigma_i) =$ the expected cost for supplier i to procure an *effective* unit, a twice differentiable function, with $\lim_{\varsigma_i \downarrow 0} \xi_i(p_i, \varsigma_i) = \lim_{p_i \uparrow 1} \xi_i(p_i, \varsigma_i) = +\infty$.

The cost rate functions $\xi_i(p_i, \varsigma_i)$ may be derived from an underlying more primitive description of the cost structure: for example, assume first that all of supplier i's labor and material costs are incurred for every *attempted* unit, whether ultimately resulting in an *effective* unit or not, and the cost per unit is given by $\overline{\xi}_i(p_i, \varsigma_i)$. The supplier's cost associated with an order of size y_i, is then given by $\overline{\xi}_i(p_i, \varsigma_i)y_i = \xi_i(p_i, \varsigma_i)(p_i y_i)$, where $\xi_i(p_i, \varsigma_i) = \overline{\xi}_i(p_i, \varsigma_i)/p_i$ may be interpreted as the expected cost incurred for each *effective* unit that is procured. However, some cost components (e.g., packaging, warehousing and shipping cost) may be incurred only for effective units, that satisfy the required quality specifications. Assume these cost components amount to $\overline{\xi}_i^{(2)}(p_i, \varsigma_i)$ per (effective) unit. In this case, the total variable costs incurred by supplier i is given by $\overline{\xi}_i(p_i, \varsigma_i)y_i + \overline{\xi}_i^{(2)}(p_i, \varsigma_i)(p_i y_i) = \xi_i(p_i, \varsigma_i)(p_i y_i)$, where $\xi_i(p_i, \varsigma_i) = \overline{\xi}_i(p_i, \varsigma_i)/p_i + \overline{\xi}_i^{(2)}(p_i, \varsigma_i)$. Note that, if

$$\lim_{\varsigma_i \downarrow 0} \overline{\xi}_i(p_i, \varsigma_i) = \lim_{p_i \uparrow 1} \overline{\xi}_i(p_i, \varsigma_i) = +\infty,$$

the same limiting behavior applies to the cost rates $\xi_i(p_i, \varsigma_i)$ as well.

4.4.1 The Yield Predictability Competition Model (YPC)

We model the competition between the suppliers assuming they select a *predictability* level for their yield distribution. We refer the reader to [16] for a parallel treatment of the Yield Target Competition and Simultaneous Yield Target and Predicability Competition Models. Almost all of the results discussed below carry over to these alternative competition models.

A predictability level can be targeted by adopting appropriate design and technology choices or quality control processes. Since competition is restricted to the choices of the standard deviations of the yield distributions, we assume, here, that the yield targets $\{p_i\}$ are exogenously given at levels $p_i = p_i^0 \geqslant 0$.

As far as the per unit cost rate functions $\xi_i(\cdot,\cdot)$ are concerned, in this model, we merely assume

$$\frac{\partial \xi_i(p_i, \varsigma_i)}{\partial \varsigma_i} < 0 \tag{4.38}$$

to reflect the fact that a less volatile yield distribution can only be achieved by adopting better materials, technologies and quality processes, as well as higher investments in the design phase.

For any supplier i, selecting the yield standard deviation ς_i, is equivalent to selecting the c.v. value $\gamma_i = \varsigma_i / p_i^0$ or the supplier's reliability level, i.e. $x_i = \gamma_i^{-2} = (p_i^0)^2/\varsigma_i^2$. (We include the possibility of $p_i^0 = 0$ and hence $x_i = 0$ to enable the modeling of firms entering the industry.) To highlight the dependence of any supplier i's cost of manufacturing an effective unit on x_i, define:

$$\xi_i^P(x_i|p_i^0) \stackrel{\text{def}}{=} \begin{cases} \xi_i(p_i^0, p_i^0/\sqrt{x_i}), & \text{if} p_i^0 > 0 \text{ and hence } x_i > 0; \\ \xi_i(0,0), & \text{if} p_i^0 = x_i = 0. \end{cases}$$

which is clearly strictly increasing in x_i, since $\partial \xi_i(p_i, \varsigma_i)/\partial \varsigma_i < 0$. We assume, in addition, that $\xi_i^P(x_i|p_i^0)$ is decreasing in p_i^0, i.e., it is less costly to procure an *effective* unit with a given reliability measure x_i when the supplier's expected yield is larger:

$$\frac{\partial \xi_i^P(x_i|p_i^0)}{\partial p_i^0} \leqslant 0 \tag{4.39}$$

In choosing a reliability level x_i, firm i faces a natural upper limit:

$$x_i \leqslant \overline{x}_i(p_i^0) \stackrel{\text{def}}{=} \begin{cases} \max\{x_i : \xi_i^P(x_i|p_i^0) \leqslant c\} < \infty, & \text{if } p_i^0 > 0; \\ 0, & \text{if } p_i^0 = 0. \end{cases} \tag{4.40}$$

(The gross profit margin per effectively delivered unit for supplier i is given by $c - \xi_i^P(x_i|p_i^0)$; since $\xi_i^P(x_i|p_i^0)$ is continuously increasing and $\lim_{\varsigma_i \downarrow 0} \xi_i(p_i^0, \varsigma_i) = \lim_{x_i \uparrow \infty} \xi_i^P(x_i|p_i^0) = \infty$, $\overline{x}_i < \infty$ is well defined.) Note from (4.39) and (4.40) that $\overline{x}_i(p_i^0)$ is increasing in p_i^0 and c. Similarly, it is easily verified that $\varsigma_i \leqslant \sqrt{p_i^0(1-p_i^0)}$, i.e., the standard deviation of the yield distribution is maximally large when the support of this distribution is confined to the extreme values $X_i = 1$ and $X_i = 0$; see [26,page 57, Example 1.10.5]. This upper bound implies:

$$x_i \geqslant p_i^0/(1-p_i^0) \tag{4.41}$$

In addition, a lower bound $\underline{x}_i^e$, independent of the yield target p_i^0, may be imposed, either by the buyer, or by external stipulations, such as government regulations.[6] Thus, let

$$\begin{aligned}\underline{x}_i(p_i^0) &= \text{the minimum reliability level for supplier} i \\ &\stackrel{\text{def}}{=} \max(\underline{x}_i^e, p_i^0/(1-p_i^0)), i = 1, \ldots, N.\end{aligned} \tag{4.42}$$

Like the upper bound $\overline{x}_i(p_i^0)$, the lower bound $\underline{x}_i(p_i^0)$ is increasing in p_i^0 as well. Finally, to exclude situations where no feasible solution exists under some of the suppliers' choices, see Theorem 2, we assume:

$$\sum_{i=1}^{N} \underline{x}_i > z_\alpha^2 \tag{4.43}$$

(We revisit this assumption at the end of this section.) To simplify the notation, we generally, suppress the dependence of the parameters with respect to p_i^0. Supplier i's expected profit function is given by

$$\begin{aligned}\pi_i(\mathbf{x}) &= \left(c - \xi_i^P(x_i)\right) p_i^0 y_i^* = \left(c - \xi_i^P(x_i)\right) \left(\frac{p_i^0 y_i^*}{Y_E^*}\right) Y_E^* \\ &= \left(c - \xi_i^P(x_i)\right) \left(\frac{x_i}{\sum_{j=1}^N x_j}\right) T\left(\sum_{j=1}^N x_j\right),\end{aligned}$$

With $x_{-i} = \sum_{j \neq i} x_j$, it is easier to employ

$$\widetilde{\pi}_i(\mathbf{x}) \stackrel{\text{def}}{=} \log \pi_i(\mathbf{x}) = \log\left(c - \xi_i^P(x_i)\right) + \log x_i + \log\left[\frac{T(x_i + x_{-i})}{x_i + x_{-i}}\right], \text{ with} \tag{4.44}$$

$$\frac{\partial \widetilde{\pi}_i}{\partial x_i} = G_i(x_i) - H(x_i + x_{-i}), \quad \text{where} \tag{4.45}$$

[6] For example, the Center for Disease Control and Prevention (CDC) purchases more than 50% of all routinely administered vaccines in the United States through the Vaccine Assistance Act (Section 317 of the Public Health Service Act, 1963) and the VFC (Vaccines For Children Act) program, which was established in 1994. To enforce minimum reliability standards, the CDC together with the US Food and Drug Administration (FDA) established current Good Manufacturing Practices (cGMPs) which required many of the vaccine manufacturers to renovate their facilities, see [24]. Many manufacturers institute qualification processes for which any potential supplier must compete to become part of the supplier base, see [19] for an example of such a qualification process prepared by Semiconductor companies such as Motorola, Infineon Technologies, Phillips and Texas Instruments. [34] describes the qualification processes in the data storage industry, and [31] those employed by Hitachi. Also, many firms require suppliers to comply with qualification processes such as ISO 9000.

$$G_i(x_i) = \frac{-\xi_i^{P\prime}(x_i)}{c - \xi_i^P(x_i)} + \frac{1}{x_i}, \quad \text{and} \quad \mathrm{H(R)} = -\left\{\log\left[\frac{\mathrm{T\,(R)}}{\mathrm{R}}\right]\right\}' \tag{4.46}$$

Thus, the marginal profit increase of a firm due to a marginal increase in his reliability level, depends on the competitors' strategic choices, only via their sum, x_{-i}. The dependence is captured by the function $H\left(\sum_{j=1}^{N} x_j\right) = -\partial \log\left(Y_E^*/\sum_{j=1}^{N} x_j\right) / \partial x_i$, the marginal *decrease* in the logarithm of the expected effective supply per industry-wide reliability level, due to an increase in firm i's reliability.

[16] has shown that the competition model represents a so-called (log-) supermodular game. This guarantees that a Nash equilibrium exists and permits a full characterization of the set of equilibria. To this end, consider a starting point where all suppliers operate at their minimum reliability levels $\{\underline{x}_i\}$. Let $\underline{R} \stackrel{\text{def}}{=} \sum_{i=1}^{N} \underline{x}_i$ denote the aggregate minimum reliability of the suppliers. Let $\underline{S}$ denote the set of suppliers who would be worse off by making marginal improvements to this minimum level, i.e., by (4.45), $\underline{S} \stackrel{\text{def}}{=} \left\{i : \frac{\partial \log \pi_i}{\partial \underline{x}_i}(x) = G_i(\underline{x}_i) - H(\underline{R}) \leqslant 0\right\}$. Thus, each supplier i is characterized by an index $I_i \stackrel{\text{def}}{=} G_i(\underline{x}_i)$; note that this index value only depends on the supplier's own cost rate function, his minimum reliability level and the sales price. Without loss of generality, number the suppliers in increasing order of their index values, i.e., $I_1 \leqslant I_2 \leqslant \cdots \leqslant I_N$. With this numbering, $\underline{S} = \{1, \ldots, |\underline{S}|\}$ and $|\underline{S}|$ is the highest indexed supplier whose index value I_i is below $H(\underline{R})$, i.e., $|\underline{S}| = \max\{i : I_i \leqslant H(\underline{R})\}$. The following theorem was proven in [Theorem 4,16], the latter providing a more complete characterization of the equilibria.

Theorem 7 (Characterization of the Set of Equilibria)

1. *The competition game is (log-)supermodular and has at least one equilibrium. The set of equilibria is a lattice; in particular, there exists a component wise smallest equilibrium* $\underline{\mathbf{x}}^*$ *and a component wise largest equilibrium* $\overline{\mathbf{x}}^*$. *Among all equilibria,* $\underline{\mathbf{x}}^*[\overline{\mathbf{x}}^*]$ *maximizes [minimizes] the expected profit for all suppliers and minimizes [maximizes] the expected cost for the buyer.*
2. *Assume* $\xi_i^P(\cdot)$ *is convex for all i. Let* $\underline{S}(\mathbf{x}^*) \stackrel{\text{def}}{=} \{i : x_i^* = \underline{x}_i\}$ *and* $S^+(\mathbf{x}^*) \stackrel{\text{def}}{=} \{i : x_i^* > \underline{x}_i\}$ *denote the set of suppliers which operate at and above their minimum reliability standards, respectively.*
   (2-i) *For every equilibrium* $\mathbf{x}^*$, *there* exists some $k^*(\mathbf{x}^*)(0 \leqslant k^* \leqslant |\underline{S}|)$ *such that* $\underline{S}(\mathbf{x}^*) = \{1, \ldots, k^*\}$ *and* $S^+(\mathbf{x}^*) = \{k^* + 1, \ldots, N\}$.
   (2-ii) *Any pair of equilibria* $\mathbf{x}^*$ *and* $\mathbf{x}^{**}$ *is completely ordered, i.e., either* $\mathbf{x}^* \leqslant \mathbf{x}^{**}$ or $\mathbf{x}^{**} \leqslant \mathbf{x}^*$. *Moreover, if* $\mathbf{x}^* \leqslant \mathbf{x}^{**}$, $k^*(\mathbf{x}^*) \geqslant k^*(\mathbf{x}^{**})$; *all suppliers are better off under* $\mathbf{x}^*$ *as compared to* $\mathbf{x}^{**}$, *while the buyer is worse off.*
3. *Assume Condition (C) applies:*
   Condition (C): $\xi_i^P(\cdot)$ *is convex and* $G_i^{-1} \circ H(\cdot)$ *is strictly concave for every i.*
   There exist at most $|\underline{S}| + 2$ *equilibria, as follows: for any* $1 \leqslant k \leqslant |\underline{S}|$, *there exists at most one equilibrium* $\mathbf{x}^*$ *such that* $k^*(\mathbf{x}^*) = k$. *In addition, there exist at most two interior equilibria* $\mathbf{x}^*(0)$ *and* $\overline{\mathbf{x}}^*(0)$.

Since the competition model is a supermodular game, both the smallest and largest equilibrium is obtained by applying a simple tatônnement scheme (with $\underline{\mathbf{x}}$ and $\bar{\mathbf{x}}$ as the starting point, respectively), in which, in each iteration, each firm determines his best response to the competitors' choices.

While the equilibrium set is always a lattice, the structure and properties of the equilibrium set depend on two factors. When the marginal procurement cost of each firm grows *convexly* with its reliability measure, the equilibrium set is restricted to a number of componentwise progressively larger equilibria. The restriction to a maximum of $|\underline{S}| + 2$ equilibria, as characterized in part (3) of the Theorem, arises when, in addition, the functions $G_i^{-1} \circ H(\cdot)$ are concave. The shape of the G_i-functions only depends on the shape of the procurement cost functions $\xi_i^P(\cdot)$ as well as the magnitude of the expected revenues per unit sold. In contrast, the shape of the $H(\cdot)$ functions is independent of the characteristics of the suppliers; it depends only on the parameters of the demand distribution as well as the maximum permitted shortfall probability. In general, the (additional) concavity property may be somewhat difficult to verify. However, Corollary 1 in [12] shows that this concavity property is guaranteed, when either (i) the cost function $\xi_i^P(\cdot)$ is convex and $G_i(\cdot)$ is concave or (ii) $\xi_i^P(\cdot)$ is linear.

The phenomenon of multiple equilibria is not just a theoretical possibility. We have encountered many instances with either two or three equilibria, even when all of the procurement cost functions are linear. Moreover, the equilibria are often far apart. Assume that firms dynamically adjust their choices before converging to an equilibrium, perhaps by iteratively selecting best responses to the choices made by their competitors. The adopted equilibrium is then critically dependent on the starting conditions of the industry. Since the game is supermodular, we know that the (component wise) *smallest* equilibrium is adopted when the firms start off at or close to the vector of *minimum* reliability standards x, while the (component wise) *largest* equilibrium arises when the firms start off at high levels of reliability close to the $\bar{\mathbf{x}}$-values; see [Theorem 4.3.2 and Theorem 4.3.4, 35]. Assuming the cost rate functions $c_i^P(\cdot)$ are convex, the different equilibria are progressively more beneficial to *all* of the suppliers, simultaneously, as we move from the largest equilibrium to smaller ones; conversely, the buyer is progressively worse off, since her total expense is given by cY_E^*, which by Lemma 1 is decreasing in $R^* = \sum_{i=1}^N x_i^*$.

The above observations have the following public policy implication: to ensure that the industry adopts a long-term equilibrium with relatively high reliability measures, it may pay to provide short-term incentives, via tax credits, subsidies or the like, for the firms to invest in reliability improvements, thus inducing a high performance equilibrium. Even if the incentives are eliminated after a while, firms are likely to readjust to a high performance equilibrium, given their starting conditions. In addition, an increase of the minimum reliability standards x may be used to induce a much larger impact on the industry's equilibrium behavior, as exemplified by Example 1 in [16].

The following Theorem shows that the smallest and the largest equilibrium, $\mathbf{x}^{*L}$ and $\mathbf{x}^{*H}$, is a monotone function of a number of the model parameters. In addition, under these equilibria, any new entrant to the industry, causes all firms to improve their reliability and the buyer to enjoy a cost reduction. As shown in Theorem 7(1), this pair of equilibria is especially important, since among all equilibria, *all* suppliers are best (worst) off under $\mathbf{x}^{*L}$ ($\mathbf{x}^{*H}$), while the opposite applies to the buyer.

Theorem 8 (Comparative Statics with Respect to the Equilibria)

1. *All equilibria depend on the parameters of the demand distribution only via its c.v.* γ_D.
2. $\mathbf{x}^{*L}$ *and* $\mathbf{x}^{*H}$ *are componentwise increasing in* c, γ_D *and the maximum permitted shortfall probability* α. *In particular, for fixed* $\sigma(\mu)$, *the two equilibria are component wise decreasing (increasing) in* $\mu(\sigma)$. *Under linear* $\xi_j^P(\cdot)$*-functions with* $\xi_j^P(x_j) = \xi_j x_j$, *these equilibria are decreasing in any of the marginal costs* $\{\xi_j\}$ *as well.*
3. *Assume* $\xi_i^{P\prime}(x_i | p_i^0)$ *is decreasing in* p_i^0.
   (3-i) $\mathbf{x}^{*L}$ *and* $\mathbf{x}^{*H}$ *are componentwise increasing in any of the firms' yield target* p_j^0, $j = 1, \dots, N$.
   (3-ii) *In both the smallest and largest equilibrium, a new entrant (firm* $N+1$*) causes all incumbent firms to increase their reliability measures, resulting in a decrease of the buyer's cost.*

One implication of the first monotonicity result is that the buyer "pays" for a lower unit price by having to cope with less reliable yield processes, at *all* of the suppliers. For example, in the vaccine supplier industry, the CDC is chartered to pay as little, for *established* vaccines, as it is able to negotiate. Indeed, Table 2 in [24] shows that the federally contracted prices are on average 40% lower than the catalog price which applies to the private sector sales. The National Vaccine Advisory Committee has identified this fact and the resulting reduced profit margins as one of the primary reasons why suppliers have left the industry. In the United States, the number of vaccine manufacturers has dropped from 26 in 1967 to a mere 6 in 2006. Indeed, it follows from part (3-ii) that the exit of many suppliers causes the equilibrium reliability choices to go down, by itself. However, not recognized in the committee's report is the fact that the highly reduced prices may well have eliminated incentives to improve yield reliabilities among those suppliers that chose to stay in the market. Thus, vaccine supplies may have become increasingly unreliable, not just because the number of suppliers decreased, but also because the federal contracts incentivized the remaining suppliers to adopt low levels of yield reliability, a phenomenon explained by part (2). In contrast, if *new* vaccines become covered by the VFC program, the CDC is required to purchase them at a price close to the supplier's catalog price. This policy has the unintended effect of incentivizing the industry to concentrate on new vaccines rather than to exploit the learning curve and improve the manufacturing processes for more established products.

Theorem 8 covers the comparative statics with respect to all of the model parameters, except for the minimum reliability standards $\underline{\mathbf{x}}$. Indeed, Example 2 in [16] shows that, for example, the largest equilibrium may fail to be monotone in this minimum reliability standard.

The following Theorem considers the special case where the model is symmetric, i.e., all suppliers have identical characteristics, and their cost rate function is convex. In this case, let $\xi^P(\cdot) \stackrel{\text{def}}{=} \xi_1^P(\cdot) = \cdots = \xi_N^P(\cdot)$ and $G(\cdot) \stackrel{\text{def}}{=} G_1(\cdot) = \cdots = G_N(\cdot)$. If $G(\underline{x}) > 0$, define $N^1(\underline{x}) \stackrel{\text{def}}{=} \min\{N \geqslant 2 : H(N\underline{x}) \leqslant G(\underline{x})\} < \infty$, since $\lim_{R\uparrow\infty} H(R) = 0$. Also, define $\underline{x}^0$ as the unique root of $G(\cdot)$. (Since $\xi^P(\cdot)$ is convex, $G(\cdot)$ is strictly decreasing, while $\lim_{x\downarrow 0} G(x) = \infty$ and $\lim_{x\uparrow\overline{x}} G(x) = -\infty$.)

Theorem 9 (Symmetric Case) *Assume identical suppliers, with convex cost rate function $\xi^P(\cdot)$.*

1. *Assume the minimum reliability standard $\underline{x} \geqslant \underline{x}^0$. The vector $\underline{\mathbf{x}}$ is the unique equilibrium, irrespective of the number of firms in the industry.*
2. *Assume $\underline{x} < \underline{x}^0$ and Condition (C). There exists a number of suppliers $N^0(\underline{x}) \leqslant N^1(\underline{x})$ such that*
   (2-i) *if $N > N^0(\underline{x})$, there exists a unique equilibrium $\mathbf{x}^*$ which is symmetric and interior and whose common component x^* is the larger (or unique) root of the characterization equation:*
$$G^{-1} \circ H(Nx) - x = 0 \tag{4.47}$$
   (2-ii) *if $N = N^0(\underline{x})$, the set of equilibria consists of one or two symmetric and interior equilibria, the common component of which is one of the (at most two) roots of the characteristic equation* (4.47).
   (2-iii) *if $N < N^0(\underline{x})$, the set of equilibria consists of x, possibly in conjunction with one or two symmetric and interior equilibria, the common component of which is one of the (at most two) roots of the characteristic equation* (4.47).
3. *Assume $\underline{x} < \underline{x}^0$ and Condition (C) applies. When $N > N^0(\underline{x})$, the unique equilibrium increases with every new entering supplier.*

Thus, when the minimum reliability standard $\underline{\mathbf{x}} \leqslant \underline{\mathbf{x}}^0$, a unique equilibrium is guaranteed, as long as the number of competitors is sufficiently *large*, and under this unique equilibrium, all firms exceed the minimum standard, and increase their reliability measure as the competition becomes fiercer, i.e., the number of competing suppliers grows. When $\underline{\mathbf{x}} > \underline{\mathbf{x}}^0$, the minimum reliability standard is set at a high enough level that $\underline{\mathbf{x}}$ arises as the unique equilibrium, irrespective of the number of firms in the industry.

We conclude this section with a discussion of what happens when condition (4.43) is violated, but

$$\sum_{j=1}^{N} \overline{x}_j > z_\alpha^2 \tag{4.48}$$

i.e., under some but not all reliability measure vectors $\mathbf{x}$, the buyer is incapable of meeting her service constraint. Under such vectors $\mathbf{x}$, no orders will be placed, resulting in zero profit for each supplier. It is easily verified that *no* (pure) equilibrium exists under which the buyer is serviced, if

$$\sum_{j=1}^{N} \overline{x}_j - \max_{1\leqslant j\leqslant N}(\overline{x}_j - \underline{x}_j) \leqslant z_\alpha^2 \tag{4.49}$$

(Let $\overline{x}_i - \underline{x}_i = \max_{1\leqslant j\leqslant N}(\overline{x}_j - \underline{x}_j) > 0$. Under any equilibrium $\mathbf{x}^*$ under which the buyer is serviced, $\sum_{j=1}^{N} x_j^* > z_\alpha^2$. If firm i decreases his reliability measure from x_i^* to $z_\alpha^2 + \epsilon - \sum_{j\neq i} x_j^* \geqslant z_\alpha^2 + \epsilon - \sum_{j\neq i} \overline{x}_j \geqslant \underline{x}_i + \epsilon$ by (4.49), the new total reliability value is $z_\alpha^2 + \epsilon$. It follows from (4.37) that as ϵ continues to decrease, the total order placed by the buyer goes to infinity, as does the order received by firm i, since his market share approaches $(z_\alpha^2 - \sum_{j\neq i} x_j^*)/z_\alpha^2 > \underline{x}_i/z_\alpha^2$. Finally, firm i's profit margin approaches $c - \xi_i^P(z_\alpha^2 - \sum_{j\neq i} x_j^*) > 0$. In other words, as $\epsilon \downarrow 0$, firm i's profit grows infinitely large, contradicting the assumption that $\mathbf{x}^*$ is an equilibrium.) Under (4.49), at least one of the suppliers is an *essential market maker*, in the sense that, irrespective of his competitors' choices, this firm is capable of creating an infeasible situation for the buyer.

The most complex situation arises in the intermediate case where (4.43) is violated, i.e. some reliability choices result in an infeasible solution, but no single firm is an essential market maker, i.e., $\sum_{j=1}^{N} \overline{x}_j - \max_{1\leqslant j\leqslant N}(\overline{x}_j - \underline{x}_j) > z_\alpha^2$. Assuming Condition (C) holds, the following is, however, a necessary and sufficient condition for a vector $\mathbf{x}^*$ to be an equilibrium: $(\partial\tilde{\pi}_i(\cdot, x_{-i}^*)/\partial x_i$ has, under (C), at most two roots, so that $\tilde{\pi}_i(\cdot, x_{-i}^*)$ has at most two local maxima on $[\underline{x}_i, \overline{x}_i]$; call x_i' the second local maximum of firm i, if any.)

1. $\mathbf{x}^*$ is a *local* Nash equilibrium, i.e., every firm's choice is a *local* maximum of his profit function,
2. $\tilde{\pi}_i(x_i', x_{-i}^*) \leqslant \tilde{\pi}_i(\mathbf{x}^*)$ for all $i = 1, \ldots, N$, and
3. $\sum_{j=1}^{N} x_j^* - \max_{1\leqslant j\leqslant N}(x_j^* - \underline{x}_j) > z_\alpha^2$.

To verify the sufficiency, note that under (3), no individual firm can create an infeasible situation by deviating. Moreover, by (1), x_i^* is a *local* maximum and by (2), the only other possible local maximum has an inferior profit value. The necessity of each of the parts (1), (2) and (3) is immediate.

4.5 Conclusions

In this chapter, we have surveyed recent work aimed at characterizing procurement strategies under the simultaneous presence of demand and supply risks. We have considered settings where the inventory manager has access to an arbitrary set of competing suppliers with different cost and yield characteristics.

In characterizing the procurement strategies, we have focused on the following four fundamental questions: (i) how many suppliers to maintain and diversify one's purchase orders amongst; (ii) how to select the desired number of suppliers from the set of potential suppliers; (iii) how to adjust one's inventory strategy to account for the supply risk, in particular how *total* purchase quantities should be set in the simultaneous presence of supply and demand risks; and (iv) how aggregate orders are to be split among the selected suppliers and whether the tradeoffs between reliability and cost differentials among the suppliers can be captured in terms of simple allocation rules. Our chapter has also focused on various comparative statics questions, i.e., on the effect various model primitives have on each of the above four questions. We have given separate treatment to the case where shortfalls are controlled by the specification of service constraints (see the SCM model of Sect. 4.2), and those where direct and indirect cost consequences of these shortfalls are added to the cost objective (the TCM model of Sect 4.3). As explained, in the presence of multiple less than fully reliable suppliers, these two modeling approaches exhibit many parallel results but also a number of important qualitative differences, not observed in standard inventory models (see Sect 4.3.3). Section 4.4 has surveyed how the above characterizations of the optimal procurement strategies can be used by the competing suppliers to select their yield distributions, in an attempt to maximize equilibrium market shares or profits. These analyses have been based on game-theoretical models.

It is important to investigate how the above results can be generalized, along several important dimensions: first, we have focused on the so-called stochastically proportional yield model, where the number of effective units y^e resulting from any given order of size y is obtained by multiplying the order size with a given random yield factor X. Some settings require a different or more general relationship between y^e, y and X, i.e., $y^e = Q(y, X)$ for a general function $Q(\cdot, \cdot)$. This approach was taken, for example by [7] in a single period TCM-type model. Indeed, these authors show that the characterization of the optimal set of retained suppliers as *consecutive* in the effective cost rates, carries over to a broad class of yield models $Q(\cdot, \cdot)$. It is of interest to explore how various *other* structural results obtained under the stochastically proportional yield model can be generalized to a broader class of yield models $Q(\cdot, \cdot)$.

The above planning models have also ignored any rapid recourse options after observation of the actual realizations of yields and possibly demands. More specifically, we have assumed that the recourse options are restricted to adjustments of future order sizes at future, regularly planned replenishment epochs. However, buyers may have access to more immediate recourse options. It is important to develop an understanding of how and when these recourse options should be used.

Finally, existing supplier competition models, as surveyed in Sect., 4.4, focus on settings where the suppliers compete in terms of their yield characteristics, under a given (common) wholesale price. Future work should generalize this to settings where the suppliers are differentiated in terms of price differences, either exogenously specified or endogenously determined as part of the competition model.

References

1. Anupindi R, Akella R (1993) Diversification under supply uncertainty. Manag Sci 39(8):944–963
2. Babich V, Burnetas AN, Ritchen PH (2007) Competition and diversification effects in supply chains with supplier default risk. Manuf Serv Oper Manag 9(2):123–146
3. Baker H, Ehrhardt R (1995) A dynamic inventory model with random replenishment quantities. Omega 23:109–116
4. Boyaci T, Gallego G (2001) Minimizing holding and ordering costs subject to a bound on backorders is as easy as solving a single backorder cost model. Oper Res Lett 29(4):187–192
5. Choi TY, Hartley JL (1996) An exploration of supplier selection practices across the supply chain. J Oper Manag 14:333–343
6. Cornuejols G, Fisher ML, Nemhauser GL (1977) Location of bank accounts to optimize float: an analytic study of exact and approximate algorithms. Manag Sci 23(8):789–810
7. Dada M, Petruzzi NC, Schwarz LB (2007) A newsvendor's procurement problem when suppliers are unreliable. Manuf Serv Oper Manag 9(1):9–32
8. Denardo E, Tang C (1992) Linear control of a Markov production system. Oper Res 40:259–278
9. Department of Energy (2005) Country analysis briefs. 12 January
10. Ehrhardt R, Taube L (1987) An inventory model with random replenishment quantities. Int J Prod Res 25:1795–1804
11. Federgruen A, Groenevelt H (1986) The greedy procedure for resource allocation problems: necessary and sufficient conditions for optimality. Oper Res 34(6):909–918
12. Federgruen A, Yang N (2007) Quality competition under yield uncertainty. Working paper
13. Federgruen A, Yang N (2008) Selecting a portfolio of suppliers under demand and supply risks. Oper Res 56(4):916–936
14. Federgruen A, Yang N (2008) Multi-period procurement strategies in the presence of unreliable suppliers (The unabridged version). Working paper
15. Federgruen A, Yang N (2009) Optimal supply diversification under general supply risks. Oper Res 57(6):1451–1468
16. Federgruen A, Yang N (2009) Competition under generalized attraction models: applications to quality competition under yield uncertainty. Manag Sci 55(12):2028–2043
17. Federgruen A, Yang N (2010) Procurement strategies with unreliable suppliers. Oper Res (forthcoming)
18. Federgruen A, Yang N (2010) Infinite horizon strategies for replenishments systems with a general pool of suppliers. Working paper
19. Gerling WH, Preussger A, Wulfert FW (2002) Reliability qualification of semiconductor devices based on physics-of-failure and risk and opportunity assessment. Qual Reliab Eng Int 18:81–98
20. Heinrich J (2001) Flu vaccine: supply problems heighten need to ensure access for high-risk people. GAO-01-624, 15 May, Washington, DC
21. Heinrich J (2001) Flu vaccine: steps are needed to better prepare for possible future shortages. GAO-01-786T. 30 May, Washington DC
22. Heinrich J (2001) Flu vaccine: recent supply shortages underscore ongoing challenges. GAO-05-177T. 18 November, Washington DC
23. Hill T (2000) Manufacturing Strategy. 3rd edn edn. McGraw-Hill/Irwin, MA
24. Klein JO, Myers MG (2006) Strengthening the supply of routinely administrered vaccines in the United States: problems and proposed solutions. Clin Infect Dis 42(3):S97–S103
25. Longitudes (2004) Mapping the new world of synchronized commerce. Harvard Business School Publishing, Cambridge, MA
26. Müller A, Stoyan D (2002) Comparison methods for stochastic models and risks. Wiley, NY
27. Nemhauser GL, Wolsey LA (1978) Best algorithms for approximating the maximum of a submodular set function. Math Oper Res 3(3):177–188
28. Nemhauser GL, Wolsey LA (1988) Integer and combinatorial optimization. Wiley, NY

29. Nemhauser GL, Wolsey LA, Fisher ML (1978) An analysis of approximations for maximizing submodular set functions—I. Math Program 14:265–294
30. New York Times (2004) With few suppliers, shortage of flu vaccine was long in the making. 17 October
31. Özer Ö, Uncu O, Cheng D (2007) An integrated framework to optimize time-to-market and production decisions for hitachi global storage technologies, Inc. Working paper
32. Swaminathan JM, Shanthikumar JG (1999) Supplier diversification: effect of discrete demand. Oper Res Lett 24(5):213–221
33. Tang C (1990) The impact of uncertainty on a production line. Manag Sci 36:1518–1531
34. Terwiesch C, Bohn RE, Chea KS (2001) International product transfer and production ramp up: a case study from the data storage industry. R D Manag 31:435–451
35. Topkis DM (1998) Supermodularity and complementarity. Princeton University Press, Princeton
36. Yano CA, Lee HL (1995) Lot sizing with random yields: a review. Oper Res 43(2):311–334
37. Zipkin PH (2000) Foundations of inventory management. McGraw-Hill Book Co, New York NY

Chapter 5
Inventory Strategies to Manage Supply Disruptions

Zümbül Atan and Lawrence V. Snyder

5.1 Introduction

Disruptions in supply chains occur routinely—both large ones, due to natural disasters, labor strikes, or terrorist attacks, and small ones, due to machine breakdowns, supplier stockouts, or quality problems (to name a few examples). Companies whose supply processes are affected by disruptions may experience delays in transportation and dysfunction in some of their facilities, which may result in inventory shortages. Although firms can take measures to prevent them, some disruptions are inevitable. Hence, in order to avoid the drastic impact of these disruptions, firms need to protect against them. There are multiple tactics that companies can choose from for managing the risk of disruptions. One of the most common tactics is to use inventory to buffer against the additional uncertainty. The main concern in inventory management problems is to find the optimal replenishment policy that tells when, from whom and how much to order.

The optimal management of inventory systems subject to supply disruptions may require an increase in inventory levels beyond those that would be required in a disruption-free environment. This extra inventory incurs extra holding costs, and therefore it may not be desirable by managers, especially since disruptions are often considered rare events. On the other hand, the increase in cost from proactively stocking extra inventory is often dwarfed by the cost that would result from a disruption that strikes an unprotected system. Therefore, there is a trade-off between the cost resulting from disruptions and the cost resulting from the protection. Where a firm falls on this trade-off—i.e.,whether it is beneficial for the firm to stock a lot of extra

Z. Atan
Department of Industrial Engineering and Innovation Sciences,
Eindhoven University of Technology, Eindhoven, The Netherlands

L.V. Snyder (✉)
Department of Industrial and Systems Engineering, Lehigh University, Bethlehem PA, USA
e-mail: larry.snyder@Lehigh.EDU

H. Gurnani et al. (eds.), *Supply Chain Disruptions: Theory and Practice of Managing Risk*, DOI: 10.1007/978-0-85729-778-5_5,

inventory or only a little to protect against disruptions—depends in large part on the "profile" of the disruptions. It has been shown that inventory is a more attractive disruption–mitigation strategy if disruptions tend to be frequent but short, while other strategies (such as supply redundancy) are more useful if disruptions tend to be rare but long [30, 34].

The use of inventory as a buffer against demand uncertainty has been studied for decades. In principle, there is no difference between using inventory to protect against supply uncertainty and using it to protect against demand uncertainty. Therefore, one might wonder whether classical models for demand uncertainty can be used to solve the problems of companies facing disruptions. The short answer is "no"—the optimal strategies can be quite different under the two types of uncertainty (see e.g., [29]).

When a location is subject to supply disruptions, it can choose to order more from a single supplier or it can choose to manage its inventories by having more than one supplier. The same is true for multi-echelon inventory systems. For both systems, inventory optimization is critical to achieve minimum expected costs and maximum customer service levels.

In this chapter, we summarize the inventory models proposed in the literature for single- and multi-echelon systems subject to supply disruptions. Our aim is not to provide a comprehensive review of these models, but rather to present some of the basic models, including some of their mathematical details, in an effort to demonstrate the direction that the field has taken thus far and to stimulate future research. For other reviews, we refer to the reader to [32], which provides a more comprehensive review of the literature on inventory management (and other topics) with supply disruptions, and [35], which provides an excellent overview of the literature on supply disruptions.

We present models for continuous-review systems in Sect. 5.2, discussing the economic order quantity (EOQ) model subject to disruptions, and its extensions, in detail. In addition, we discuss the effect of supply disruptions on inventory models in manufacturing environments. In Sect. 5.3, we change our focus to periodic-review models, including a discussion of the optimality of base-stock and (s, S) policies. In Sect. 5.4, we discuss models for multi-echelon inventory systems subject to supply disruptions. Finally, in Sect. 5.5, we present a summary and some suggestions for future research directions.

5.2 Continuous-Review Models

5.2.1 The EOQ Model with Disruptions (EOQD)

Consider a single-location, single-item inventory system that faces deterministic and continuous demand with rate d units per year. (Although this system may represent any location in a supply chain—factory, warehouse, etc—for convenience, we will generally refer to it as a "retailer.") Assume that there is a fixed order cost K per order and a holding cost h per unit per year. This is the classical EOQ model.

Now also suppose that the supplier becomes unavailable (i.e., is disrupted) at random points in time, and for a random duration no orders can be placed. We refer

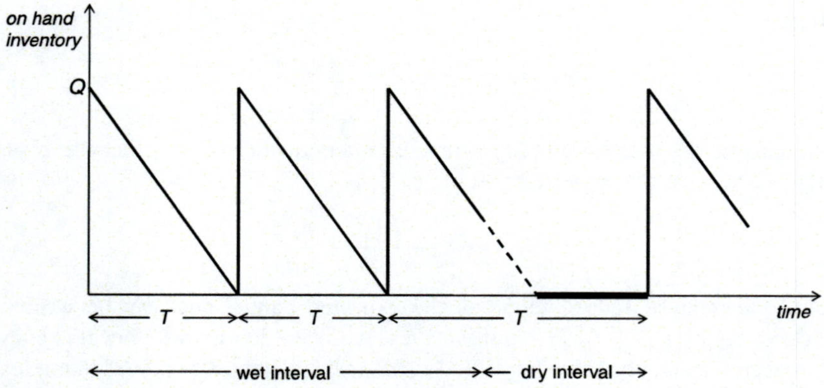

Fig. 5.1 Inventory curve for EOQD model

to the intervals during which the supplier is disrupted as "dry" intervals and intervals during which the supplier functions normally as "wet" intervals.[1] If the retailer runs out of inventory when its supplier is in a dry interval, the customer demands occurring before the start of the next wet interval are lost, with lost sales incurring a cost of p per unit. We will assume that the duration of dry and wet intervals is exponentially distributed with rates μ and λ, respectively. Therefore, the disruption process constitutes a two-state continuous-time Markov chain (CTMC). Although other distributions are possible, the exponential distribution is commonly employed in the literature since it provides mathematical tractability and is often a reasonable model of real disruptions.

The inventory curve for this problem is pictured in Fig. 5.1. Note that, since excess demands are lost, the inventory level is never negative.

This problem has come to be known as the economic order quantity with disruptions (EOQD). The EOQD was first introduced by [21]. Using the renewal reward theorem, the authors derive an expression for the expected annual cost and prove its convexity for exponentially distributed dry and wet intervals. [4] points out that the model in [21] is erroneous in two respects: first, it implicitly assumes that stockouts occur during every dry interval (which need not happen if the disruption begins and ends while the retailer still has inventory), and second, it treats the lost sales cost as though it is incurred per unit per year, rather than simply per unit. Our analysis below is based on the corrected model presented by [4]. We first derive an expression for the expected annual cost, and then investigate properties of the optimal order quantity, Q^*.

Define a cycle T as the time between receipts of successive orders (a random variable). Then the expected cycle length is

$$E[T] = \frac{Q}{d} + \frac{\beta}{\mu},$$

[1] These terms are common in the literature, as are others, such as off/on and down/up.

where

$$\beta = \frac{\lambda}{\lambda+\mu}\left(1 - e^{-(\lambda+\mu)\frac{Q}{d}}\right) \tag{5.1}$$

is the probability that the supplier is in a dry interval when the inventory level at the retailer hits zero. The expected cost per cycle is

$$E[C] = K + \frac{hQ^2}{2d} + \frac{dp\beta}{\mu}.$$

Using the renewal reward theorem, the expected annual cost can be written as $C(Q) = E[C]/E[T]$. *C(Q)* is quasiconvex, but is not known whether it is convex. Moreover, it cannot be solved in closed-form. Instead, it must be solved numerically, although this can be done efficiently since line search techniques can be applied to quasiconvex functions. [4] demonstrates numerically that the optimal order quantity is nondecreasing in *K*, *p* and *d*. These results are consistent with the classical EOQ model. In addition, the optimal order quantity is nondecreasing in the availability ratio λ/μ, which implies that the retailer orders more when its supplier is disrupted more frequently and/or for longer intervals.

Snyder [31] introduces a simple method that approximates the cost function by a convex function. This approximation yields a closed-form expression for the optimal order quantity, which allows insights that cannot be obtained from numerical solutions. In addition, a closed-form solution is useful since one can directly embed it into more complex models. (For example, see the discussion at the end of Sect. 5.2.2.) In particular, [31] proposes approximating *C(Q)* by replacing[2] β with $\beta' = \lambda/(\lambda+\mu)$. This is the steady-state probability that the supplier is in a dry interval (whereas β accounts for recent history and is therefore not a steady-state probability). Therefore, β' is a good approximation to β when the system reaches steady-state reasonably quickly, which happens, for instance, when the order cycle time is relatively long compared to the durations of wet and dry intervals. Under this approximation, the order quantity to minimize *C(Q)* is

$$Q^* = \sqrt{\frac{2Kd}{h} + a^2 + b} - a$$

where

$$a = \frac{\beta' d}{\mu} \quad \text{and} \quad b = \sqrt{\frac{2d^2 p\beta'}{h\mu}}.$$

Given that the optimal order quantity of the classical EOQ model is $\sqrt{\frac{2Kd}{h}}$, it is straightforward to conclude that Q^* is larger. The same relation holds for the optimal costs.

[2] More accurately, [31] uses $\beta' = r\lambda/(\lambda+\mu)$ for a constant r, but here we consider the special case of $r = 1$ since it is simpler and has a more natural interpretation.

Depending on the system parameters, the difference between the EOQ solution and Q^* can be quite large. Hence, using the classical model when there is supply uncertainty can be very costly. The cost of ignoring disruptions is a decreasing function of λ and an increasing function of μ. Therefore, the retailer should not ignore the possibility of supply disruptions when the disruption rate is high and the recovery rate is low, and to avoid high lost-sales costs, the retailer should hold more inventory than it would in the absence of supply disruptions.

5.2.2 EOQD with Disruptions at the Retailer

Consider a retailer that faces random disruptions both internally and externally. (In contrast, the EOQD considers only external disruptions, to the retailer's supplier.) An internal disruption causes all inventory to be destroyed, and the retailer cannot place a new order until the disruption is over. External disruptions, in contrast, affect the retailer only if its inventory level is zero and it attempts to place an order with the supplier. In this case, the retailer must wait until the supplier has recovered to place an order, just as in the EOQD. Hence, the retailer holds its order until both internal and external disruptions are over. It cannot satisfy customer demands either when it is disrupted or when it is waiting for a supplier disruption to end. Given that the retailer follows an EOQ-type model with these disruption processes, our objective, as in Sect. 5.2.1, is to determine the optimal order quantity, Q^*. This problem is studied by [24].

The expected cost function, which is the sum of the ordering, holding and shortage costs, can be obtained using the renewal reward theorem and is quasiconvex in Q. Therefore, like the EOQD, Q^* can be found using any method for solving single-dimensional unconstrained quasiconvex optimization problems, such as bisection or golden section search. However, also like the EOQD, one cannot derive a closed-form solution for Q^*. Instead, [24] proposes an effective approximation for the average cost function that uses a similar idea as that of [31]. They derive an approximate but closed-form expression for Q^* by replacing one exponential term in the objective function with zero and another with its second-order Taylor-series expansion.

Given that the optimal order quantity (from the approximate model) is $\hat{Q}$ and the optimal order quantity of the EOQ model is Q_E, Qi et al. demonstrate the following numerically:

- One can think of the quantity $\hat{Q} - Q_E$ as the "safety stock" that the retailer holds to protect against disruptions. This quantity tends to be large when the supplier is often unavailable, and it tends to be small, or even negative, when the retailer is often disrupted. The reason is that when the supplier is unavailable, the retailer needs to hold safety stock to avoid high-shortage costs, whereas when the retailer itself is disrupted, it has to keep less inventory to avoid the risk of inventory loss caused by disruptions.
- $\hat{Q}$ is small when the retailer is disrupted very often or the supplier has high availability. Under these conditions, the retailer loses less inventory and reduces its

ordering and holding costs by ordering smaller amounts. Hence, the cost benefit of using $\hat{Q}$ instead of Q_E is significant.

- The retailer's availability has a more significant impact on the fill rate than the supplier's availability does. This implies that uncertainty in the part of the supply chain closer to the customers has more negative effects than uncertainty farther upstream. (We discuss a similar finding, in a different modeling context, in Sect. 5.4.)
- The retailer has little ability to buffer against the supplier's disruptions when it is also disrupted very often. Hence, the supplier's availability influences the average cost most dramatically when the retailer is often unavailable.

This model, and the EOQD, is simple to solve from an optimization perspective since their objective functions are quasiconvex. It is therefore natural to ask why we need approximations to these models. One answer is that closed-form solutions can often be used to derive insights, like the ones given above, that cannot be derived from the exact model. Another is that closed-form solutions can be embedded into other models much more readily. For example, [25] presents a joint location–inventory model with disruptions in which the inventory cost is expressed in the objective function using the closed-form approximation of [24]. If the exact cost function were used instead, the resulting model would have additional decision variables (for the order quantities) and would be highly nonlinear. Moreover, under the approximate model, the cost of the optimal solution is a concave function of the demand (a fact that could not be proven for the exact model), and this property allows [25] to apply an existing algorithm that is quite effective.

5.2.3 Disruption Models with Non-Zero Reorder Points

A zero-inventory ordering (ZIO) policy, in which a replenishment order is placed only when the inventory reaches level zero, is optimal for the classical EOQ model. However, in the EOQD it may be optimal to order when the inventory level is strictly positive in case the supplier will be disrupted when the inventory level reaches zero. More generally, positive inventories are used as a buffer against uncertainties in both supply and demand processes, so it is reasonable to expect non-zero reorder points to be optimal in the EOQD.[3] In this section, we discuss continuous-review inventory policies with non-zero reorder points and supply disruptions.

Consider a retailer that orders each time its inventory level drops to r. If its supplier is in a wet interval, it places an order for Q units. This replenishment occurs instantaneously and the retailer's inventory level increases to $Q + r$. On the other hand, if the supplier is in a dry interval when the retailer's inventory level reaches r, the retailer must wait until the supplier becomes available and it orders enough to bring inventories up to $Q + r$ units. (Actually, this policy is not technically a (Q, r) policy, since the order quantity may not equal exactly Q at each order. More accurately, it is an (s, S) policy with $s = r$ and $S = Q + r$.)

[3] There are conditions under which a ZIO policy is actually optimal, even if there are uncertainties in the system. However, this requires a sufficiently efficient design [5].

As in most stochastic inventory problems, the key trade-off here is between holding and stockout costs: if we choose a large r, we protect against disruptions and reduce stockouts but incur higher holding costs, whereas if we choose a small r, we reduce the holding cost but increase the stockout risk.

Next, we develop an expression for the average cost as a function of Q and r and present an approximate solution. The analysis in this section is due to [22], which introduces the exact model, and [14], which proposes the approximation.

We use most of the same assumptions as in Sect. 5.2.1. The only difference is that here we assume that unsatisfied demands are backordered, incurring a cost of p for each backordered unit per unit time. The inventory level curve for this model is pictured in Fig. 5.2. Defining a cycle as the time between two consecutive times at which the inventory level is increased to $Q + r$ and using the renewal reward theorem, the average cost function can be written as follows:

$$C(Q,r) = \frac{K + h\frac{Q^2}{2d} + hQr + \beta d C(r)}{\frac{Q}{d} + \frac{\beta}{\mu}},$$

where $C(r)$ is defined as

$$C(r) = \frac{h(\mu r - 1) + e^{-\mu r}(p\mu + h)}{\mu^2}$$

and β is as defined in (5.1). The values of Q and r that minimize $C(Q,r)$ can be found numerically. Alternately, we can use the same logic as in Sect. 5.2.1 to find approximate values, approximating β by $\beta' = \lambda/(\lambda + \mu)$ and rewriting $C(Q,r)$ accordingly. Let $C'(Q,r)$ be the resulting (approximate) cost function, and let Q^* and r^* be the values that minimize $C'(Q,r)$. Setting the partial derivatives of C with respect to Q and r to 0 and solving the resulting equations simultaneously, we obtain

$$Q^* = \frac{d}{\mu}(1 - \beta') + \sqrt{\frac{2Kd}{h} + \frac{d^2}{\mu^2}(1 - \beta')^2}$$

$$r^* = -\frac{1}{\mu}\ln\left[\frac{h}{\beta'(p\mu + h)}\left(1 + \sqrt{\frac{2K\mu^2}{dh} + (1 - \beta')}\right)\right]$$

If $r^* < 0$, then we must replace r^* by 0; then the minimum cost occurs at $(Q^*(0), 0)$, where $Q^*(0)$ is the optimal order quantity if $r = 0$. In this case, a ZIO policy ($r^* = 0$) is optimal.

This approximation is most accurate when β' is close to β, i.e. when the supplier is disrupted relatively frequently and/or can recover relatively quickly. Computational results suggest that the average percentage difference between the costs of the exact and approximate solutions usually is less than 1%. Numerical results suggest that the average cost savings from using an (Q, r) policy instead of a ZIO policy is 8.5%,

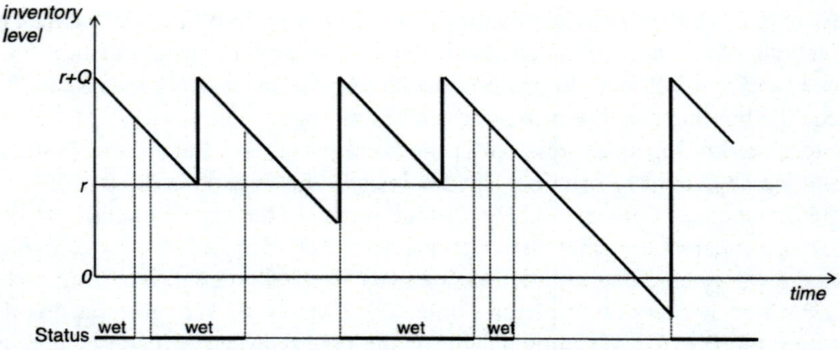

Fig. 5.2 Inventory curve for (Q, r) model with disruptions

and that the most significant savings occurs in the most adverse situations, in which the supplier is the most unreliable and recovers the slowest.

The (Q, r) model discussed in this section can be extended in multiple ways by considering random demand processes, random leadtimes, and different disruption and recovery processes. It is possible to find exact expressions for the average cost functions for these problems, but solving for the optimal system parameters is difficult, if not impossible. Most of the literature relies on either numerical solutions or approximations; see, for example, [1, 13, 15–20]. The overall conclusions of these models are the same:

- (Q, r) polices are better than ZIO policies when suppliers are unreliable.
- Ignoring disruptions when choosing policy parameters may cause high operating costs, especially when disruptions are long and stockouts are costly.
- The average cost and stockout risk increase with the leadtime variability, part of which is due to the supply disruptions.

5.2.4 Supply Disruptions in Manufacturing Environments

Although much of the literature tends to focus on rare, catastrophic disruptions, more minor disruptions are quite common in manufacturing environments, stemming from machine breakdowns or the halting of a production process due to maintenance requirements. Moreover, as noted in Sect. 5.1, these frequent-but-short disruptions can be most easily managed using inventory. Therefore, economic lot sizing and safety stock decisions need to consider these types of disruptions.

In this section, we discuss an economic manufacturing quantity (EMQ) model for an unreliable manufacturing environment. (Our analysis is based on the models by [9] and [10].) As in the classical EMQ model, the demand is deterministic and constant with rate d units per day, and production is similarly continuous, with finite rate η units per day. Inventory accumulates gradually during the production interval and is then depleted until the inventory level reaches zero, at which point production begins again. However, in this system, random breakdowns occur; if these

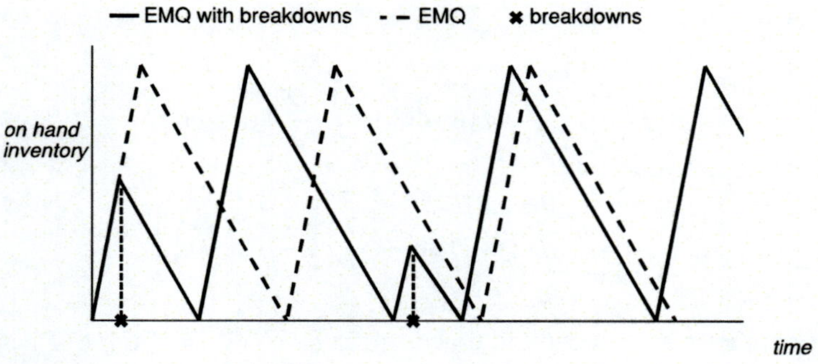

Fig. 5.3 Inventory curve for EMQ model with machine breakdowns

occur during production intervals, then production stops, corrective maintenance is performed, and the next production interval begins when the inventory level reaches zero. The on-hand inventory curve for this problem is pictured in Fig. 5.3.

Two types of maintenance occur in this system. Corrective maintenance is carried out after a breakdown, whereas regular maintenance occurs at the end of each production interval. The same initial working conditions are obtained after each maintenance action, of either type. Corrective maintenance incurs both a setup cost K and a maintenance cost M, whereas regular maintenance incurs only the former cost. Neither type of maintenance is assumed to be time consuming.

As in the EOQ model, a linear holding cost of h is charged per unit per day. Given that the system is wet, the time-to-breakdown is W days, where W is a random variable with density function $f(w)$ and cumulative distribution function $F(w)$. Note that the wet–dry process continues to evolve at all times, regardless of whether the system is in production mode or not. We aim to find the optimal lot size, Q^*, so that the long-run average cost per unit time is minimized.

As in the EOQD, we use the renewal reward theorem to obtain an expression for the average cost function. We define a cycle as the time between starts of successive production runs. The expected cycle cost is

$$\begin{aligned} E[C] &= \int_0^{Q/\eta} \left[K + M + \frac{1}{2} h(\eta - d) \frac{\eta}{d} w^2 \right] f(w)\,\mathrm{d}w \\ &\quad + \int_{Q/\eta}^{\infty} \left[K + \frac{1}{2} h(\eta - d) \frac{\eta - d}{\eta d} Q^2 \right] f(w)\,\mathrm{d}w \\ &= K + MF\left(\frac{Q}{\eta}\right) + \frac{1}{2} h(\eta - d) \frac{\eta}{d} \\ &\quad \times \left[\left(\frac{Q}{\eta}\right)^2 \left(1 - F\left(\frac{Q}{\eta}\right)\right) + \int_0^{Q/\eta} w^2 f(w)\,\mathrm{d}w \right] \end{aligned}$$

The expected cycle length is

$$E[T] = \int_0^{Q/\eta} \frac{\eta}{d} w f(w)\,\mathrm{d}w + \int_{Q/\eta}^{\infty} \frac{Q}{d} f(w)\,\mathrm{d}w$$

$$= \frac{\eta}{d} \int_0^{Q/\eta} t f(w)\,\mathrm{d}w + \frac{Q}{d}\left[1 - F\left(\frac{Q}{\eta}\right)\right]$$

The long-term average cost per unit time is given by $C(Q) = E[C]/E[T]$ and can be written as

$$C(Q) = \frac{K + MF(\frac{Q}{\eta}) + \frac{1}{2}h(\eta - d)\frac{\eta}{d}\left[\left(\frac{Q}{\eta}\right)^2\left(1 - F(\frac{Q}{\eta})\right) + \int_0^{Q/\eta} w^2 f(w)\,\mathrm{d}w\right]}{\frac{\eta}{d}\int_0^{Q/\eta} w f(w)\,\mathrm{d}w + (Q/d)(1 - F(Q/\eta))}$$

Although it is possible to solve for Q^*, the lot size minimizing *C(Q)*, numerically for any type of disruption distribution, fairly simple formulas can be derived for the exponential distribution. We omit the mathematical details but summarize the main insights as follows:

- The average corrective maintenance cost is independent of the lot size Q. Hence, the value of *M* does not play a role in determining Q^*.
- When the system approaches perfect reliability, the optimal lot size approaches the ordinary EMQ, with

$$Q^* \to \sqrt{\frac{2Kd\eta}{h(\eta - d)}}.$$

- The optimal lot size and the optimal objective value are increasing functions of the disruption rate. Hence, combined with the previous property, we can conclude that at optimality, a manufacturing system subject to frequent machine breakdowns should operate with larger lot sizes, and this costs the system more than the disruption-free system.
- Using the classical EMQ lot size when the system is subject to machine breakdowns results in an average cost increase that is guaranteed to be no more than 2% greater than the optimal average cost, i.e. $C(EMQ)/C(Q^*) \leq 1.02$. The reason for this surprisingly small cost ratio is that even though Q^* may be more than twice as large as EMQ, the difference in the average actual lot size achieved is much smaller, due to breakdowns. (In contrast, recall from Sect. 5.2.1 that the analogous ratio in the EOQD setting may be arbitrarily large, at least under the approximation by [31].)

In this model, we assume that machine breakdowns interrupt the production of the current lot and that a new lot starts when inventory is depleted. This is optimal if the cost of resumption equals the ordinary setup cost, K. However, if the cost of resuming the production run after a disruption is less than K, it may be cheaper to resume production after a disruption rather than to let inventory drop back to zero before starting a new run. Therefore, the decision to abort or resume can depend on the amount of on-hand inventory and on the setup and resumption costs. In fact, the optimal lot size under this assumption approaches Q^* when the resumption cost approaches the setup cost. Moreover, the ratio of the average cost per unit time in the original problem to that in the modified problem, with resumption, is bounded by the ratio of the setup cost to the resumption cost. In addition, if a threshold-like policy is used to determine the starting inventory levels for a new manufacturing period, significant cost savings can be achieved.

Up to this point, we have assumed that corrective maintenance times are negligible. If repairing failed machines is time consuming (as is often the case in reality), then safety stocks are required to ensure smooth deliveries when machines are being repaired. Therefore, during each production run, a certain fraction of the items produced can be diverted into the safety stock, while the rest of the items are used to meet regular customer demands. Assuming that safety and cycle stocks are maintained separately and that safety stocks are used only when a machine breakdown occurs, models to specify service levels and lot sizes can be developed, and it can be shown that the optimal lot size and expected safety stock increase with the disruption rate, required service level, demand rate, and setup and repair times.

The models and results presented in this section can be used in production scheduling, resource allocation, and capacity planning decisions. There is a clear trade-off between the overall investment in increasing the maintenance level and the resulting savings in safety stocks and repair costs, and these analyses provide guidelines for production managers who consider improvements in corrective and preventive maintenance activities. In general, slight overinvestments in machine maintenance are significantly less expensive than similarly sized underinvestments. Hence, investing in maintenance activities with the objective of avoiding costly machine breakdowns is an important operating strategy in unreliable manufacturing environments.

5.3 Periodic-Review Models

5.3.1 The Base-Stock Problem with Disruptions

In this section, we examine optimal inventory policies for a retailer who uses a periodic-review base-stock policy and is subject to supply disruptions and deterministic demand. The problem under consideration is similar to the multi-period newsboy model with inventory carryover and backordered demands, except that instead of random demand we have supply disruptions.

The most common way to model the disruption process for periodic-review systems is using a two-state discrete-time Markov chain (DTMC). Let I_t denote the state of the supply process in period t, with $I_t = 1$ denoting a wet period and $I_t = 0$ denoting a dry period. Transition probabilities are given by

$$\alpha = P(I_t = 0 | I_{t-1} = 1)$$
$$\beta = P(I_t = 1 | I_{t-1} = 0),$$

where α is called the *disruption probability* and β is called the *recovery probability*. The lengths of wet and dry intervals are therefore geometrically distributed. (Note the analogy to the two-state continuous-time Markov chain used in Sect. 5.2.1.) We will use this model throughout Sect. 5.3.

In this subsection, it will be mathematically convenient to work instead with a more granular, infinite-state DTMC whose states are numbered $0, 1, 2, \ldots$, in which state 0 represents being in a wet interval and state $n \geq 1$ represents being in the nth period of a dry interval. Let θ_t be the random variable representing the state in period t. Let π_n be the steady-state probability of being in state $n \geq 0$ and let $F(n) = \sum_{i=0}^{n} \pi_i$ be its cdf; that is, *F(n)* is the probability of being in a wet interval or a dry interval that has lasted n periods or fewer. (This approach is also more general, allowing, for example, recovery probabilities that depend on the current length of the disruption; see, e.g., [34].)

In this subsection, we assume that there are no setup costs for orders placed, but we relax this assumption in Sect. 5.2. In both models, we assume that the retailer does not know the state of the supplier when it attempts to place an order. If the supplier is in a dry interval when the retailer orders, the order is ignored. Actually, this assumption is inconsequential in this model, since the model is equivalent to one in which the retailer knows the state of the supplier and only orders during wet intervals. However, in the next model (Sect. 5.3.2), we assume that the retailer pays a setup cost for attempting an order, whether or not the order is successful, so that model is not equivalent to one in which the retailer knows the supplier's state.

A base-stock policy is optimal for this system [33]. In each period, the sequence of events is as follows:

1. The retailer observes the current inventory level.
2. The retailer attempts to place an order to bring its inventory level up to the base-stock level S. If the supplier is in a wet interval, the order is received immediately. Otherwise, the order is ignored.
3. Demand is observed and is subtracted from the inventory level. We assume that the demand is d units in every period. Demands are met from stock to the extent possible, and unsatisfied demands are backordered.
4. On-hand inventory and backorders incur a holding cost of h and penalty cost of p per item per period, respectively.

The decision variable for this model is the base-stock level, S, and the objective is to minimize the sum of the expected holding and backordering cost per period. We can write the expected cost per period as

$$C(S) = \sum_{n=0}^{\infty} \pi_n \left[h\left(S - (n+1)d\right)^+ + p\left((n+1)d - S\right)^+ \right],$$

since, in the nth period of a disruption ($n \geq 0$), $n+1$ periods' worth of demand have occurred since the most recent replenishment.

It can be shown that S^*, the S that minimizes $C(S)$, is an integer multiple of d. Hence, difference equations (rather than derivatives) must be used to determine it. In particular, let $\Delta C(S) = C(S+d) - C(S)$. Then S^* is the smallest S such that $\Delta C(S) \geq 0$, since $C(S)$ is a convex function of S, i.e. $\Delta^2 C(S) \geq 0$. One can show that

$$\Delta C(S) = (h+p)F\left(\frac{S}{d} - 1\right) - p,$$

and therefore $S^* = kd$, where k is the smallest integer such that

$$F(k-1) \geq \frac{p}{p+h}.$$

Let $F^{-1}(\cdot)$ be the inverse cdf of the disruption process. For a given $0 \leq r \leq 1$, if there is no n such that $F(n) = r$, then we use the convention that $F^{-1}(r)$ equals the smallest n such that $F(n) \geq r$. Then $F^{-1}(r)$ is well defined and integer-valued for all r. This allows us to write the optimal base-stock level of a periodic-review system with deterministic demand and supply disruptions as

$$S^* = d + dF^{-1}\left(\frac{p}{p+h}\right) \tag{5.2}$$

It is well known that, for a system with demand uncertainty (but no disruptions), if the lead time is deterministic and the leadtime-demand is normally distributed demand with mean μ and standard deviation σ, then the optimal base-stock level is

$$S^* = \mu + \sigma\Phi^{-1}\left(\frac{p}{p+h}\right), \tag{5.3}$$

where $\Phi^{-1}(\cdot)$ is the inverse standard normal cumulative distribution function. Here, μ is the cycle stock used to satisfy the expected demand, while $\sigma\Phi^{-1}\left(p/(p+h)\right)$ is the safety stock used to protect against the demand uncertainty. Note that S^* in (5.2) has a nearly identical structure: d is the cycle stock (used to satisfy the current period's demand) and $dF^{-1}\left(p/(p+h)\right)$ is the safety stock (used to protect against the supply uncertainty). Whereas in (5.3), the safety stock level indicates that the system should protect against $\Phi^{-1}\left(p/(p+h)\right)$ standard deviations' worth of demand uncertainty, in (5.2), the safety stock level indicates that the system should protect against $F^{-1}\left(p/(p+h)\right)$ periods' worth of disruptions. In both cases, the system incurs stockouts when the random variable exceeds the threshold specified by the safety-stock level.

Assuming that the disruptions arise from a two-state DTMC with disruption probability α and recovery probability β, i.e., geometrically distributed wet and dry intervals, one can show (see, e.g., [27]) that the optimal base-stock level is $S^* = kd$, where k is the smallest integer such that

$$1 - (1-\beta)^{k-1}\frac{\alpha}{\alpha+\beta} \geq \frac{p}{p+h}.$$

It is possible to write similar inequalities for different disruption processes as long as one has an expression for $F(n)$. From this inequality, the optimal base-stock levels can be calculated.

As with the continuous-review model of Sect. 5.2.1, the model in this section can be extended to include different assumptions such as non-stationary demands and disruption probabilities [11], partial disruptions [12], stochastic demand [27], and advanced warning of disruptions [30]. It is possible to obtain exact expressions for the optimal base-stock levels in some of these extensions, but in some cases they must be solved numerically or approximated. In general, these models agree about the necessity of holding more inventory as the level of uncertainties in the system increases.

5.3.2 Periodic-Review Problems with Setup Costs

In this section, we consider a finite-horizon, periodic-review problem with setup costs.[4] We argued in Sect. 5.3.1 that if there is no setup cost, a base-stock policy is optimal for infinite-horizon models. The same is true for finite-horizon problems, as well; see [11]. In this section, we argue that when there is a positive setup cost, the optimal policy is a state-dependent (s, S) policy in which the re-order level and the order-up-to level depend on the supplier's disruption status in the previous period. In addition to setup costs, there are inventory holding and backordering costs.

The sequence of events is identical to that in Sect. 5.3.1 except that we assume that the demand in period t, denoted D_t, is stochastic. The D_t are i.i.d. random variables with a continuous density function $r(\cdot)$. (In the no-setup-cost model of Sect. 5.3.1, we assumed the demand was deterministic because it allowed us to obtain simple expressions for the optimal base-stock level. In the model with setup costs, no such expressions are available, whether the demand is deterministic or stochastic. Therefore, we make the more general assumption and allow the demand to be stochastic.) As in the previous model, excess demands are backordered. The planning horizon contains T periods. Periods are numbered in reverse order; therefore, in period t there are t periods remaining until the end of the planning horizon.

The retailer is subject to disruptions. Let I_t, $t = 1, \ldots, T$ denote the state of the supply process, with $I_t = 1$ denoting a wet period and $I_t = 0$ denoting a dry

[4] The analysis in this section is due to [23].

period. We assume that $\{I_t; t = 1, \dots, T\}$ is a two-state Markov chain with transition probabilities

$$P(I_{t-1} = 0 | I_t = 1) = \alpha$$
$$P(I_{t-1} = 1 | I_t = 0) = \beta,$$

as in the previous section. For notational convenience, we define π_{ij} as the transition probability from state i to state j, that is, $\pi_{10} = \alpha, \pi_{11} = 1 - \alpha, \pi_{01} = \beta$, and $\pi_{00} = 1 - \beta$.

We assume that there are two types of setup costs. A cost of K is incurred whenever an order is placed (whether or not it is filled), and an additional cost of K_1 is incurred only when an order is filled. The unit ordering cost, unit backordering cost per period and unit holding cost per period are given by c, p and h, respectively. The discount factor is given by $0 < \gamma \leq 1$. (In the previous section, we assumed $\gamma = 1$.)

We represent the initial inventory at the beginning of period t by x_t and the order quantity placed in period t by u_t. Then we can write the initial inventory in period $t - 1$ as

$$x_{t-1} = x_t + I_t u_t - D_t, \qquad t = 1, \dots, T$$

Next, we derive the total expected cost. Given that $I_t = i$, the expected cost of an order of size u placed in the next period, $t - 1$, is

$$O(u) = \begin{cases} K + \pi_{i1} K_1 + \pi_{i1} c u & \text{if } u > 0 \\ 0 & \text{otherwise} \end{cases} \tag{5.4}$$

The expected one-period holding and backorder cost when the system was in state i in the previous period and we place an order of size u in the current period is

$$E^i_{I,D}[h(x + Iu - D)^+ + p(D - x - Iu)^-],$$

where the expectation is taken over I and D, the current period's supply state and demand, respectively. The superscript i indicates that the expectation is conditioned on the fact that the supply state in the previous period was i. We can rewrite this cost as

$$J^i(y, x) \equiv \pi_{i1} L(y) + \pi_{i0} L(x), \tag{5.5}$$

where $y \equiv x + u$ and

$$L(y) \equiv h \int_0^y (y - z) r(z) dz + p \int_y^\infty (z - y) r(z)\, dz \tag{5.6}$$

is the standard newsboy cost function.

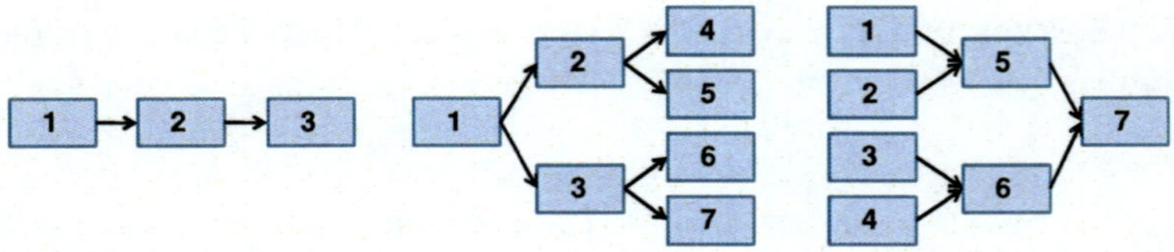

Fig. 5.4 Examples of (a) a serial system, (b) a distribution system, (c) an assembly system

We will analyze the total expected cost using a dynamic programing approach, similar to the approach taken in classical inventory models with no disruptions. To that end, let $f_t^i(x)$ be the optimal expected discounted cost for periods t through the end of the horizon, given that the system was in state i in period $t+1$. Then

$$f_t^i(x) = \min_{y \geq x} \left\{ E_I^i[O(y-x)] + J^i(y,x) + \gamma E_{I,D}^i[f_{t-1}^I(x + Iu - D)] \right\}.$$

It can be proved that $f_t^i(x)$ is $(K + \pi_{i1} K_1)$-convex in x. This implies that the optimal policy for period t is an (s_t^i, S_t^i)-policy, with $S_t^i > s_t^i$. Under this policy, if $x \leq s_t^i$, then the retailer orders up to S_t^i, and if $x > s_t^i$, it does not order. Note that the optimal parameters depend on i, the supply state in the previous period. In fact, it can be shown that the optimal S_t^i is independent of i. Hence, the optimal policy is an (s^i, S) policy. Unfortunately, calculating the control parameters is difficult, and numerical optimization or approximations are required.

5.4 Disruptions in Multi-Echelon Inventory Systems

In this section, we analyze two multi-echelon inventory systems subject to disruptions. Three "archetype" network topologies—serial, distribution, and assembly systems—are illustrated in Fig. 5.4. In a serial system, each node has at most one successor and predecessor; in a distribution system, each node has at most one predecessor; and in an assembly system, each node has at most one successor.

We discuss serial systems (in Sect. 5.4.1) and distribution systems (in Sect. 5.4.2). We do not discuss assembly systems in detail. However, it is worth noting here that, although assembly systems with no disruptions can be transformed into equivalent serial systems and solved using algorithms for such systems [26], no such equivalence is possible when disruptions are present [7]. Hence, results obtained from the analysis of serial systems subject to supply disruptions cannot be applied directly to unreliable assembly systems.

5.4.1 Serial Systems

In this section, we consider disruptions in a serial system. For ease of exposition, we consider only a two-echelon system, but these results can be extended to serial systems with more than two echelons. In our system, node 2 faces customer demands and orders from node 1, which in turn orders from an external supplier with infinite supply. We consider a discrete-time, infinite-horizon model in which demands across periods are independent and identically distributed. Demand is given by a general non-negative distribution. The replenishment lead times are constant. There are linear holding costs at both nodes and a linear backordering cost at node 2. There are no setup costs. This is the well-known problem introduced by [6].

We introduce supply disruptions to both nodes of this system. Disruptions at node 2 prevent node 2 from ordering from node 1, while disruptions at node 1 prevent node 1 from ordering from the external supplier. Both nodes may continue to satisfy demands from on-hand inventory during a disruption. The disruption process at each node follows a two-state DTMC, as in Sect. 5.3; the process at node j ($j = 1, 2$) has disruption probability α_j and recovery probability β_j.[5] Following the notation in Sect. 5.3.1, we let ϕ^j represent the state of node j, with $\phi^j = 0$ indicating that node j is not disrupted and $\phi^j = n$ indicating that node j is in its nth disrupted period.

Define $G(Q^1, \bar{x}^1, \bar{x}^2, \phi^1, \phi^2)$ as the optimal discounted cost for the entire (infinite) horizon given that, at the start of the horizon, there are Q^1 items in transit to node 1, the local net inventory levels at nodes 1 and 2 are $\bar{x}^1$ and $\bar{x}^2$, and the states at the two nodes are ϕ^1 and ϕ^2. Atan et al. [3] shows that $G(Q^1, \bar{x}^1, \bar{x}^2, \phi^1, \phi^2)$ can be written as the sum of the optimal objective values of two optimization problems. Specifically, we have

$$G(Q^1, \bar{x}^1, \bar{x}^2, \phi^1, \phi^2) = G^1(Q^1, \bar{x}^1, \phi^1, \phi^2) + G^2(\bar{x}^2, \phi^2),$$

where $G^1(Q^1, \bar{x}^1, \phi^1, \phi^2)$ and $G^2(\bar{x}^2, \phi^2)$ are the optimal objective values of the problems that find the optimal inventory policies at nodes 1 and 2, respectively. In fact, $G^2(\bar{x}^2, \phi^2)$ corresponds to the minimization of a convex function of $\bar{x}^2$, and this convexity implies that a state-dependent base-stock policy is optimal for node 2, in which the optimal base-stock level depends on the state of node 2; in particular, it equals $-\infty$ (implying no order should be placed) if $\phi^2 > 0$. Once the base-stock level of node 2 is fixed, $G^2(\bar{x}^2, \phi^2)$ becomes a constant and the problem reduces to the optimization problem required to obtain $G^1(Q^1, \bar{x}^1, \phi^1, \phi^2)$. It can be proven that in this optimization problem, the objective function is convex with respect to $\bar{x}^1$. Therefore, a base-stock policy is optimal for node 1, as well. Moreover, the optimal base-stock level depends not only on the state of the disruption at node 1 but also on the state of the disruption at node 2. Hence, we can express the optimal base-stock levels at nodes 1 and 2 as $S_1^*(\phi^1, \phi^2)$ and $S_2^*(\phi^2)$, respectively. Note that $S_2^*(\phi^2) = -\infty$ if $\phi^2 > 0$ and $S_1^*(\phi^1, \phi^2) = -\infty$ if $\phi^1 > 0$, but $S_1^*(0, \phi^2)$ may be greater than $-\infty$, even if $\phi^2 > 0$, and may be different for different values of ϕ^2.

As mentioned above, it is possible to extend these results to serial systems with more than two locations. The dependence structure among the base-stock levels and disruption states implies that the base-stock levels may be optimized sequentially, starting from the downstream node, as in the Clark–Scarf algorithm (for serial systems

[5] The results presented below also hold for a more general disruption process in which the recovery probability depends on the current length of the disruption.

without disruptions). Therefore, the base-stock optimization problem is much simpler than if all base-stock levels had to be optimized simultaneously.

The following results apply to serial systems with two or more echelons and with disruptions:

- In the system with Bernoulli disruptions, that is, the disruption probability at node j does not depend on the disruption state in the previous period, i.e., $\alpha_j = 1 - \beta_j$, an echelon base-stock policy is optimal. The base-stock levels do not depend on any state of the inventory system.
- In the same system but with disruptions that are governed by a DTMC (that is, the disruption probability depends on the disruption state in the previous period), the base-stock level of a node depends on the state of the disruption process at the node itself and at all the remaining downstream locations.
- The state-dependent base-stock levels are monotonically increasing in the number of disrupted periods. Hence, when a node's downstream nodes are experiencing a long disruption, it is cost effective for the node to keep more inventory.

5.4.2 Distribution Systems

The previous section suggests that the disruptions in one part of the supply chain can affect the inventory optimization decisions made in other parts of the supply chain, depending on the location and the nature of disruptions. In this section, we consider a two-echelon distribution system, known as a one-warehouse, multiple-retailer (OWMR) system, and analyze how supply disruptions affect the optimal inventory levels. As before, we assume deterministic demand and zero lead time in order to obtain a tractable model that can serve as a foundation for future, more complex models.

We consider a locally controlled one-warehouse, N-retailer system. Each location monitors only its own inventory level. Retailers observe their customer demands and place orders with the warehouse. The warehouse observes the orders from each retailer and places its own order with an outside supplier, which is assumed to have infinite capacity. The inventory levels are reviewed periodically and base-stock policies are used for replenishment.[6] We assume that all retailers are identical, with demand d per period, holding cost h_r per unit per period, and stockout cost p_r per unit per period. (Excess demands are backordered.) A holding cost of h_0 per unit per period is incurred at the warehouse.

We consider the possibility of having disruptions in the supply processes of the warehouse, the retailers, or both. Disruptions at the retailers occur simultaneously at all retailers. Let $\pi_{i,j}$ be the probability that the warehouse is in the ith consecutive period of a disruption and the retailers are in the jth consecutive period of a disruption.

---

[6] The optimal stocking and allocation polices for distribution systems with random demand and disruption-free supply systems are unknown, as are the optimal policies for the system under consideration here.

We make the simplifying (and usually realistic) assumption that, if both echelons are subject to disruptions, then disruptions at the two echelons never overlap, that is, $\pi_{i,j} = 0$ if i and j are both nonzero.

5.4.2.1 Optimization

In this section, we want to find the optimal base-stock levels for the warehouse, S_0^*, and the retailers, S_r^*, so that the expected system cost is minimized. The expected cost, $C(S_0, S_r)$, can be written as the sum of three functions: the expected cost when both supply systems are nondisrupted, the expected cost when only the supply system of the warehouse is disrupted, and the expected cost when only the supply systems of the retailers are disrupted. This decomposition facilitates the analysis.

Closed-form solutions are available for the special cases in which either the retailers or the warehouse are disruption-free, and we first consider those two special cases. First suppose that the retailers are disruption-free, that is, $\pi_{i,j} = 0$ for all $j \geq 1$. In this case, we can replace $\pi_{i,j}$ with π_i, which is identical to the disruption process defined in Sect. 5.3.1. The optimal base-stock levels depend on the relative magnitude of the warehouse and retailer holding costs. In particular, if $h_r \geq h_0$, then

$$\begin{aligned} S_0^* &= kNd \\ S_r^* &= d \end{aligned} \tag{5.7a}$$

where k is the smallest integer such that $(h_0 + p_r)F(k-1) + p_r\pi_k - p_r \geq 0$ and $F(i) = \sum_{n=0}^{i} \pi_n$ is the cdf of the disruption pmf, π_i. If, instead, $h_0 \geq h_r$, then

$$\begin{aligned} S_0^* &= Nd \\ S_r^* &= \max\left\{dF^{-1}\left(\frac{p_r}{p_r + h_r}\right), d\right\} \end{aligned} \tag{5.7b}$$

where, as in Sect. 5.3.1, we use the convention that $F^{-1}(r)$ equals the smallest n such that $F(n) \geq r$.

Next that suppose the warehouse is disruption-free, that is, $\pi_{i,j} = 0$ for all $i \geq 1$. Similar to the previous case, we can replace $\pi_{i,j}$ with π_j, with cdf *F(i)*, and solve for this system optimally:

$$\begin{aligned} S_0^* &= Nd \\ S_r^* &= d\left(F^{-1}\left(\frac{p_r}{p_r + h_r}\right) + 1\right) \end{aligned} \tag{5.8}$$

If the supply processes were all disruption-free, all locations would maintain only their cycle stock, that is, enough inventory to meet the current period's demand. In particular, the warehouse and retailer base-stock levels would be *Nd* and *d*, respectively. The solutions given in (5.7)–(5.8) suggest that unreliable distribution systems require more inventory than disruption-free ones. This extra inventory constitutes safety stock, which is used as a precaution against disruptions. When the warehouse can be disrupted, although both the warehouse and the retailers are affected by the disruptions, only one of them (the one with the lower holding cost) holds the safety stock. When the retailers are subject to disruptions, the warehouse is not affected by the disruptions, and only the retailers hold safety stock. Of course, in all cases, each location holds at least its cycle stock, i.e., one period's demand. Note that, when the retailers can be disrupted, the location of the safety stock does not depend on the relative magnitude of the holding costs. This is because, even if the warehouse has a smaller holding cost, it does not benefit the system to hold inventory there as these inventories cannot be sent to the retailers during disruptions.

The cases with disruptions at only one echelon are easy to solve since the expected cost function behaves nicely as a result of the random variable representing the disruptions being reduced from bi-variate to uni-variate. However, the same is not true when disruptions occur at both echelons. For this case, [2] develops a heuristic procedure to find the optimal base-stock levels. This heuristic is easy to implement, and for all the instances tested, it found the optimal solution. Relaxing the identical-retailer assumption also results in a less tractable cost function, and [2] develops a heuristic for this problem, too. We summarize the results and conclusions drawn from these heuristics as follows:

- Both locations hold more inventory when their supply systems are subject to disruptions than they do when the overall system is disruption-free.
- Let α_i and β_i [α_r and β_r] be the disruption and recovery probability at the warehouse [the retailers]. The effects of these parameters on the optimal base-stock levels and the corresponding expected cost can be summarized as follows:
  - S_0^* increases with α_0 and decreases with β_0. S_r^* is not affected significantly by either parameter. The optimal expected cost decreases as β_0 increases but does not change significantly with α_0.
  - S_r^* increases with α_r and decreases with β_r. Counterintuitively, S_0^* decreases with *both* α_r and β_r. It decreases with β_r due to the reduced need for safety stock and decreases with α_r since, as retailer disruptions become more likely, warehouse inventory becomes a greater liability since it incurs costs but provides no value during a disruption.
- At all locations, it is more beneficial to focus on reducing the duration of disruptions (increasing β) than reducing their probability of occurrence (decreasing α).
- Ignoring supply disruptions close to the customers (that is, assuming incorrectly that $\alpha_r = 0$) is more costly than ignoring disruptions upstream.

5.4.2.2 Centralization and Decentralization in OWMR Systems

Consider again an OWMR system subject to disruptions. We consider two scenarios. The first is a *centralized* system in which inventories are stocked at the warehouse only, and the second is a *decentralized* system in which inventories are stocked at the retailers only. Supply disruptions affect only the locations keeping inventory, and therefore the other echelon effectively disappears. We have similar settings as in the previous sections: customer demands are deterministic, lead times are zero and disruptions follow a random process governed by the probability mass function π_i, where i is the number of consecutive periods during which a given stage has been disrupted. Effectively, the decentralized system behaves like N copies of the single-location model in Sect. 5.3.1 and the centralized system behaves like a single copy with N times the demand.

The question is, which echelon should hold the inventory? Intuition suggests that it is preferable to hold inventory at the retailers since disruptions at the warehouse would affect the whole system, whereas retailer disruptions each affect only a part of the system. In this section we will confirm this intuition. Note that, under demand uncertainty, the *risk-pooling effect* [8] dictates that it is preferable to have a single centralized inventory location rather than N separate locations. Under disruptions, then, the opposite result is true. This phenomenon is known as the *risk-diversification effect* [28, 29]. Both effects describe the effects of centralization or decentralization of inventory on the holding and stockout costs only; they ignore the costs of transportation, facility location, and other logistics costs in order to isolate the effects on inventory management.

Let C_D and C_C be the single-period costs in the decentralized and centralized systems, respectively; these are random variables. Let $E[\cdot]$ and $V[\cdot]$ denote expectation and variance, respectively. Let S_D^* and S_C^* be the base-stock levels (at the retailers and the warehouse, respectively) that minimize $E[C_D]$ and $E[C_C]$, respectively. Then one can prove the following relations[7]:

$$S_C^* = N S_D^* \tag{5.9}$$

$$E[C_C^*] = E[C_D^*] \tag{5.10}$$

$$V[C_C^*] = NV[C_D^*] \tag{5.11}$$

Equation 5.9 suggests that the total inventories are equal for the centralized and decentralized systems—either N retailers each order up to S or one warehouse orders up to NS. The other two equations confirm that the decentralized system is preferable to the centralized one, but not because it has a smaller expected cost. Indeed, the expected costs of the two systems are equal (by (5.10)), but the variance is N times larger in the centralized system (by (5.11)). Since most decision-makers are risk-averse, the decentralized system is preferable. This is the risk-diversification effect. It occurs because a given retailer is affected by disruptions the same percentage of periods in both systems, but in the centralized system, disruptions are lower-frequency, higher-impact (and therefore higher-variance) events than in the decentralized system.

[7] These results are due to [28].

Under the risk-diversification effect, the inventory, cost mean, and cost variance are equal, equal, and greater, respectively, in the centralized system than in the decentralized system. It is worth mentioning that, under the risk-pooling effect, the reverse is true, that is, the inventory, cost mean, and cost variance are less, less, and equal, respectively, in the centralized system. (Smaller inventory and cost mean is proven by [8]; equal variances is proven by [28]).

Under demand uncertainty, one wants to maintain fewer inventory locations; under disruptions, one wants to maintain more. A natural question is, if both demand and supply uncertainty are present, which system is preferable? Numerical results suggest that, if the decision-maker uses a mean–variance objective to assess the two systems, then a decentralized system is almost always preferable, that is, risk diversification almost always trumps risk pooling. Exceptions occur when the service level (newsboy fractile) is very small, the decision-maker is very risk neutral, and/or the system is very reliable.

5.5 Conclusions

In this chapter, we studied the use of inventory to mitigate disruptions. This requires holding more inventory than would be necessary if there were no risk of disruption. On the other hand, this extra inventory provides a valuable buffer against the additional uncertainty introduced by disruptions. The amount of extra inventory required depends on the severity of the disruptions; in general, it is an increasing function of the disruption probability and a decreasing function of the recovery probability.

The optimal replenishment policy in a system subject to disruptions depends on the cost structure, demand process, and other factors. It is often quite difficult to determine and prove the form of the optimal policy, and therefore many researchers (and practitioners) choose a policy type and then try to optimize the parameters of that policy. This optimization, too, is sometimes easy, if closed-form solutions or convex objectives are available, and sometimes difficult, due to nonconvexities. In the latter case, approximations are sometimes available. Optimization is generally significantly more difficult for multi-echelon systems, for which heuristics are often used.

Inventory is only one of many strategies that firms may use to mitigate disruptions. Other strategies are categorized by [34] and are discussed in other chapters of this book. The most appropriate strategy for a given system depends both on the nature of disruptions and on the objectives of the firm. In general, inventory is a more attractive strategy for frequent–short disruptions than for rare–long ones, for which other mitigation strategies such as dual sourcing become preferable.

There are many promising avenues for future research on the subject of inventory models with supply disruptions. One such avenue involves more general and more realistic disruption processes. Most of the existing disruption models assume a particular stationary disruption process, such as Markovian disruptions, and optimize based on this assumption. However, in reality, disruption risks are often non-Markovian and nonstationary. For example, hurricanes are more likely in the summer than in other seasons. Therefore, it is more appropriate to consider more general disruption models, such as nonstationary ones, and as a result to propose inventory policies that incorporate these features of the disruptions into the decision-making process. Although there are a few models in the literature that consider a more general disruption process than the ones discussed in this chapter, there is still a need for inventory policies that can handle more realistic disruption processes.

One criticism that is frequently raised is that, since disruptions are infrequent, it can be hard to estimate the parameters of the disruption process, and therefore any model relying on them may be inaccurate. Therefore, another important avenue for future research is the development of models that are robust with respect to data errors, especially in the disruption parameters. A related question is how to update these estimates based on recent historical data using Bayesian or other methods.

As in the case of demand uncertainty, the analysis of multi-echelon inventory systems subject to supply disruptions is much harder than the analysis of single-location ones. The models discussed in this chapter provide some basic understanding of the effects of supply disruptions on serial systems and distribution systems, but they rely on several simplifying assumptions. There is a need for studies that consider multi-echelon systems with more general demand and supply disruption processes, as well as more difficult network topologies such as assembly systems and non-tree networks.

References

1. Arreola-Risa A, DeCroix G (1998) Inventory management under random supply disruption and partial backorders. Nav Res Logist 45:687–703
2. Atan Z, Snyder L (2011) Disruptions in one-warehouse multiple-retailer systems. Working paper, Working paper, P.C. Rossin College of Engineering and Applied Sciences, Lehigh University, Bethlehem, PA
3. Atan Z, Rong Y, Snyder L (2011) Optimality of base-stock policies for serial supply chains. Working paper, Working paper, P.C. Rossin College of Engineering and Applied Sciences, Lehigh University, Bethlehem, PA Lehigh University
4. Berk E, Arreola-Risa A (1994) Note on future supply uncertainty in EOQ models. Nav Res Logist 41:129–132
5. Bielecki T, Kumar P (1988) Optimality of zero-inventory policies for unreliable manufacturing systems. Oper Res 36(4):532–541
6. Clark AJ, Scarf H (1960) Optimal policies for a multi-echelon inventory problem. Manag Sci 6(4):475–490
7. DeCroix GA (2010) Inventory management for an assembly system subject to component supplier disruptions. Working paper, Wisconsin School of Business, University of Wisconsin-Madison, Madison, WI
8. Eppen GD (1979) Effects of centralization on expected costs in a multi-location newsboy problem. Manag Sci 25(5):498–501
9. Groenevelt H, Pintelon L, Seidmann A (1992) Production lot sizing with machine breakdowns. Manag Sci 38(1):104–123

10. Groenevelt H, Pintelon L, Seidmann A (1992) Production batching with machine breakdowns and safety stocks. Oper Res 40(5):959–971
11. Güllü R, Önol E, Erkip N (1997) Analysis of a deterministic demand production/inventory system under nonstationary supply uncertainty. IIE Trans 29:703–709
12. Güllü R, Önol E, Erkip N (1999) Analysis of an inventory system under supply uncertainty. Int J Prod Econ 59:377–385
13. Gupta D (1996) The (Q, r) inventory system with an unreliable supplier. INFOR 34(2):59–76
14. Heimann D, Waage F (2007) A closed-form approximation solution for an inventory model with supply disruptions and non-ZIO reorder policy. J Syst Cyber Inform 5(4):1–12
15. Mohebbi E (2003) Supply interruptions in a lost-sales inventory system with random lead time. Comput Oper Res 30:411–426
16. Mohebbi E (2004) A replenishment model for the supply-uncertainty problem. Int J Prod Econ 87(1):25–37
17. Mohebbi E, Hao D (2006) When suppliers availability affects the replenishment lead time an extension of the supply-interruption problem. Eur J Oper Res 175:992–1008
18. Mohebbi E, Hao D (2008) An inventory model with non-resuming randomly interruptible lead time. Int J Prod Econ 114:755–768
19. Moinzadeh K, Aggarwal P (1997) Analysis of a production/inventory system subject to random disruptions. Manag Sci 43(11):1577–1588
20. Parlar M (1997) Continuous-review inventory problem with random supply interruptions. Eur J Oper Res 99:366–385
21. Parlar M, Berkin D (1991) Future supply uncertainty in EOQ models. Nav Res Logist 38:107–121
22. Parlar M, Perry D (1996) Inventory models of future supply uncertainty with single and multiple suppliers. Nav Res Logist 43:191–210
23. Parlar M, Wang Y, Gerchak Y (1995) A periodic review inventory model with Markovian supply availability. Int J Prod Econ 42:131–136
24. Qi L, Shen ZJM, Snyder LV (2009) A continuous-review inventory model with disruptions at both supplier and retailer. Prod Oper Manag 18(5):516–532
25. Qi L, Shen ZJM, Snyder LV (2010) The effect of supply disruptions on supply chain network design. Transp Sci 44(2):274-289
26. Rosling K (1989) Optimal inventory policies for assembly systems under random demands. Oper Res 37(4):565–579
27. Schmitt A, Snyder L, Shen ZJM (2010) Inventory systems with stochastic demand and supply: properties and approximations. Eur J Oper Res 206(2):313-328
28. Schmitt A, Snyder L, Shen ZJM (2011) Centralization versus decentralization: risk pooling, risk diversication, and supply uncertainty in a one-warehouse multiple-retailer system. Working paper, P.C. Rossin College of Engineering and Applied Sciences, Lehigh University, Bethlehem, PA
29. Snyder L, Shen ZJM (2006) Supply and demand uncertainty in multi-echelon supply chains. Working paper, P.C. Rossin College of Engineering and Applied Sciences, Lehigh University, Bethlehem, PA
30. Snyder L, Tomlin B (2008) Inventory management with advanced warning of disruptions. Working paper, P.C. Rossin College of Engineering and Applied Sciences, Lehigh University, Bethlehem, PA
31. Snyder LV (2011) A tight approximation for a continuous-review inventory model with supplier disruptions. Working paper, P.C. Rossin College of Engineering and Applied Sciences, Lehigh University, Bethlehem, PA
32. Snyder LV, Atan Z, Rong Y et al (2010) Supply chain disruptions: a review. Working paper, P.C. Rossin College of Engineering and Applied Sciences, Lehigh University, Bethlehem, PA

33. Song JS, Zipkin P (1996) Inventory control with information about supply conditions. Manag Sci 42(10):1409–1419
34. Tomlin B (2006) On the value of mitigation and contingency strategies for managing supply chain disruption risks. Manag Sci 52(5):639–657
35. Vakharia AJ, Yenipazarli A (2008) Managing supply chain disruptions. Found Trends Technol Inf Oper Manag 2(4):243–325

Chapter 6
Manufacturer Competition and Subsidies to Suppliers

Adam A. Wadecki, Volodymyr Babich and Owen Q. Wu

6.1 Introduction

Dealing with risky suppliers is a part of everyday business for manufacturers. For example, the domestic automotive supply industry has faced numerous hardships as some of its largest firms have flirted with bankruptcy, or have been subsumed by Chapter 11 over the past few years. Nearly 30% of the pre-existing North American automotive supply base had filed for bankruptcy by the end of 2008. Half are predicted to file for bankruptcy before the end of 2010 [1]. Despite a $5 billion cash injection from the federal government in early 2009, auto suppliers continue to struggle. With the US economy rebounding more slowly than first expected, the short-term outlook for the entire auto industry is bleak. "Bottomed out" auto sales have not yet begun to rebound significantly and nearly two-thirds of Tier 1 suppliers remain financially distressed [5].

If a supplier defaults, its operations may temporarily or permanently cease, crippling downstream manufacturers and starving them of necessary production

A. A. Wadecki
Industrial and Operations Engineering, University of Michigan,
1205 Beal Ave., Ann Arbor,
MI 48109, USA
e-mail: awadecki@umich.edu

V. Babich (✉)
McDonough School of Business, Georgetown University,
37th and O Streets, NW 546 Rafik B Hariri Building,
Washington, DC 20057, USA
e-mail: vob2@georgetown.edu

O. Q. Wu
Ross School of Business, University of Michigan,
701 Tappan St., Ann Arbor,
MI 48109, USA
e-mail: owenwu@umich.edu

H. Gurnani et al. (eds.), *Supply Chain Disruptions: Theory and Practice of Managing Risk*, DOI: 10.1007/978-0-85729-778-5_6,

inputs. In some cases, manufacturers source products from a single supplier. This is especially true of high-technology items where suppliers may hold patents on the products they produce. When a supplier defaults, the manufacturer's operations are in jeopardy. Considexr the recent hardships faced by American Axle & Manufacturing (AAM). In September 2009, AAM received $110 million in cash and a $100 million loan from General Motors (GM) to keep the supplier out of bankruptcy [6]. These payments also prevented shutdowns at numerous GM facilities that were dependent on AAM-sourced parts. Ironically, AAM was once a part of GM's Saginaw Steering Division before being sold to a group of investors in 1994. GM also experienced a similar situation in December 2008 when Cadence Corporation, a supplier of interior components for GM vehicles, filed for bankruptcy. This bankruptcy caused Cadence to shut down its operations, delaying production of GM's 2010 Chevrolet Camaro.

Financially distressed suppliers pose significant operational risks to manufacturers. Manufacturers may be able to switch suppliers in the event of a default, but in an environment where nearly all suppliers are financially distressed, this is not beneficial. Supplier diversification is expensive for manufacturers when procuring non-commodity products. The only remaining option for manufacturers is subsidies.

Research suggests that publicly-traded firms suffering from supply disruptions experience abnormal stock returns that are roughly 40% lower than their the academics [7]. Despite the continual emphasis on supply chain robustness both within the academics and industry, a majority of supply chain studies and practices have focused on increasing supply chain efficiency rather than mitigating disruptions. Our paper addresses this gap.

In order to mitigate risks arising from supplier financial distress, manufacturers may elect to provide subsidies to their suppliers. These subsidies can take the form of cash, agreements for future contracts, or targeted supplier development. By providing suppliers with subsidies, manufacturers reduce the risk of supplier default, thus increasing the reliability of their supply chain.

In our model, manufacturers can select from two sourcing options. They can procure goods from a dedicated supplier who does not serve other manufacturers (a *dedicated supplier*), or from a supplier who is shared by multiple manufacturers (a *shared supplier*). For example, in 1996, Delphi (then a division of GM Corporation) served as a dedicated supplier for GM, and derived 83% of its revenues from its parent company. In 2007, however, 63% of Delphi Corporation's revenues came from non-GM customers.

Manufacturers can also participate in two downstream retail market environments: a *competitive* (or *oligopolistic*) and a *non-competitive* (or *monopolistic*) environment. Over the past 40 years, the North American automotive marketplace has been transformed from an oligopoly dominated by four manufacturers (GM Corporation, Ford Motor Company, and Chrysler Corporation, and American Motors) to a fiercely competitive environment with a plethora of players. Increased competition has commensurately decreased profits for automakers, forcing them to pay significant attention to their material procurement decisions in attempting to control costs.

We consider four supply chain structures (see Fig. 6.1): (1) monopolistic manufacturers with dedicated suppliers, (2) oligopolistic manufacturers with dedicated suppliers, (3) oligopolistic manufacturers with a shared supplier, and (4) monopolistic manufacturers with a shared supplier.

Procuring goods from shared suppliers permits manufacturers to share subsidy costs (thus *cross-subsidizing* the supplier), reducing the burden of each manufacturer of ensuring that its supplier is financially viable. On the other hand, by using a dedicated supplier, manufacturers have more direct control over their supplier's reliability and can potentially exploit potential monopoly power should their competitor's supplier falter. In this chapter we study (1) the cross-subsidy benefit to a manufacturer and how this benefit depends on manufacturer competition, (2) the benefit to a supplier from working with several manufacturers and how this benefit depends on manufacturer competition, and (3) the consumer surplus and quantities released to the market and how these quantities depend on the choice of a supply chain structure.

Our analysis shows that when the market size of each monopolistic manufacturer is the same as that of competing manufacturers, suppliers receive less subsidies when manufacturers compete than if they are monopolists. Less competition among manufacturers leads to higher subsidies provided to suppliers, more reliable suppliers, and greater benefits to consumers. If the combined market size of monopolistic manufacturers is equal to that of competing manufacturers, consumers may prefer competing or non-competing manufacturers depending on supplier reliability.

We also find that manufacturers' subsidy costs are less when manufacturers share suppliers, irrespective of whether or not manufacturers compete in a retail market. Interestingly, in the scenario where manufacturers do not compete, the total amount of subsidies received by a shared supplier is greater than the payment received by each dedicated supplier. However, in the scenarios where manufacturers compete, manufacturers face a tradeoff between using a shared supplier and dedicated suppliers. By sharing a supplier, manufacturers enjoy decreased subsidy costs because they reap the benefits of competitor-provided subsidies. On the other hand, by using dedicated suppliers, manufacturers may become monopolists when their competitor's supplier defaults. In this scenario, whether a shared or dedicated supplier receives a greater subsidy depends on the difference between monopolistic and oligopolistic manufacturers' profits. When this difference is large, dedicated suppliers receive greater subsidies and are more reliable; if this difference is insignificant, a shared supplier receives greater subsidies and is more reliable.

6.2 Literature Review

Our paper contributes to two streams of the operations management literature: supply risk and manufacturer-level competition. We examine both issues in a novel framework that quantifies the optimal subsidy decisions of manufacturers.

The literature on supply risk uncertainty is surveyed in [14]. Silver authored an early paper on this topic using the economic order quantity (EOQ) framework [11]. His paper considered yield uncertainties in which the standard deviation of received goods was proportional to lot size, and also when it was not proportional to lot size. [4] uses Silver's framework and the EOQ model to jointly determine yield variability and lot sizes when yield variability can be decreased (e.g. improved) through investment. In special cases of the investment function, they derive closed-form solutions for the optimal investment and lot size levels. Gerchak and Parlar's idea of reducing yield variability through investment is very similar to a central theme in our paper: manufacturers can reduce yield variability by providing financial subsidies to suppliers.

More recent papers [3, 12] have modeled supply chain disruptions in light of strategic competition among manufacturers, which is represented by Cournot competition in their models.

Deo and Corbett examine the impact of yield uncertainty on manufacturer-level production and entry into a retail marketplace. They use a two-stage model in which, during the first stage, firms decide whether or not they will enter a retail market model, and, during the second stage, each firm selects the target production quantity of goods. They also examine the effects of yield uncertainty on consumers as measured through the consumer surplus. The authors find that yield uncertainty decreases competition at the manufacturer level and also decreases the expected consumer surplus.

Tang and Kouvelis examine the benefits of supplier diversification for competing manufacturers. They consider a two-stage model in which the suppliers' output is affected by proportional random yield similar to Deo and Corbett. In the first stage, manufacturers engage in a sourcing strategy game, while in the second stage, manufacturers compete in the Cournot sense. The authors find that manufacturers should never choose to use the same supplier and that increasing correlation between suppliers' yields decreases manufacturer-level profits.

Our analysis differs from each of the aforementioned papers in several ways: (1) we use a different model of yield uncertainty that is based on the supplier's financial state; (2) we assume manufacturers can directly affect the supplier's financial state through subsidies; and (3) we focus on the optimal manufacturer subsidy decisions in both competitive and non-competitive manufacturer environments.

Babich [2] employs financial models of bankruptcy similar to that in our model. He solves an N-period optimization problem, examining both the optimal order quantities and financial subsidies of a manufacturer dealing with a single, risky supplier. He asserts that the supplier's ability to deliver goods is increasing in its state of financial health, defined as the ratio of the supplier's assets to its liabilities. Our analysis differs from Babich's work in that we first seek to quantify the optimal subsidy decisions of manufacturers participating in a competitive retail market. We modify Babich's model of financial health in a multi-manufacturer setting, and analyze the optimal subsidy decisions of such firms when they procure goods from both dedicated and shared suppliers. We also examine the impact of manufacturer-level competition on subsidies and the effect of yield uncertainty on the consumer surplus.

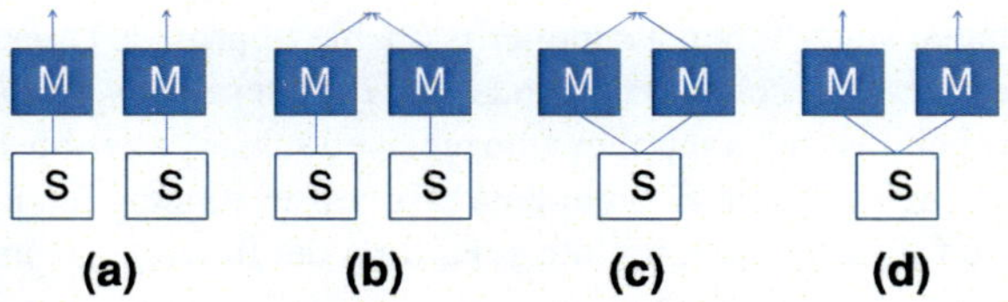

Fig. 6.1 Supply chain structures. *Shaded boxes* labeled with an *M* represent manufacturers. *Unshaded boxes* labeled with an *S* represent suppliers. Retail competition is denoted by two intersecting *arrows*. **a**, **d** Non-competitive retail market scenarios, **b**, **c** competitive scenarios

6.3 Model

We model supply chain interactions as two-stage games of complete information. Each of these stages represents a subgame in our model. In the first stage, manufacturers simultaneously select subsidies, θ. These subsidies represent a promise contingent on the supplier being financially viable by the second stage. Suppliers' capacity is also realized in this first stage. In the second stage, manufacturers release goods, z, to the downstream market if their supplier is financially sound and able to deliver products. Our assumption that the production decision occurs after the uncertainty about the supplier's financial status is resolved is based on current automotive industry practices. Intense competition among Tier 1 manufacturers has afforded much power to manufacturers in sourcing goods. In fact, according to Chrysler Group LLC's Senior Vice President of Purchasing and Supplier Quality, Dan Knott, current contract terms allow manufacturers to "drop a [currently-contracted] supplier because 'I didn't like the way you look' " [13]. In other words, if a manufacturer senses a supplier will be unable to deliver goods, current contract terms generally allow the manufacturer to terminate the relationship without significant repercussions.

We find the subgame perfect equilibrium by backward induction. Recall that our model is used to analyze manufacturers' decisions in each of four supply chain structures that differ along two dimensions: competition among manufacturers and the use of a dedicated or shared supplier (see Fig. 6.1).

We assume that the total amount of subsidies θ provided by manufacturers is always non-negative and improves supplier reliability by elevating the supplier's financial state. Manufacturers must always reimburse suppliers for their total production costs. Subsidies, then, describe any contributions provided by manufacturers to suppliers in excess of production costs. These subsidies could take the form of promises for future contracts, loans, or cash. Let θ_i denote the subsidies received by the supplier from manufacturer i. In the case where two manufacturers share a common supplier, that supplier receives $\theta_1 + \theta_2$. In the case where two manufacturers use dedicated suppliers, each supplier $i \in \{1, 2\}$ receives θ_i from its dedicated manufacturer.

In each of our four supply chain structures, a supplier is able to provide sufficient capacity to fully satisfy manufacturers' orders or no capacity at all, depending on the supplier's financial state. When a supplier is unable to provide capacity, we assume that its manufacturer(s) receive(s) no goods, and cannot sell any products in the retail market. The probability that a supplier who received subsidy θ is able to deliver goods is $p(\theta)$. For each unit of θ that a manufacturer promises its supplier, it must pay $\delta(\theta)$.

We make the following assumptions regarding the functional forms of $p(\theta)$ and $\delta(\theta)$.

Assumption 1 Function p(θ) is increasing and log-concave in θ.

Assumption 2 Function $\delta(\theta)$ is increasing and convex in θ.

Assumption 1 holds for many different probability distributions, including the normal and exponential distributions. Assumption 2 is intuitive. We offer the following lemma that results directly from Assumption 1:

Lemma 1 *The quantity* $\frac{p'(\theta)}{p(\theta)}$ *is decreasing in* θ.

Proof Since $p(\theta)$ is log-concave by Assumption 1, the first-order derivative of $\log[p(\theta)\,]$ is decreasing. That is, $\frac{d}{d\theta}[log\ p(\theta)] = \frac{p'(\theta)}{p(\theta)}$ is decreasing. □

Function $p(\cdot)$ can take many forms. For example, a structural model of a supplier's bankruptcy similar to [10] yields

$$p(\theta) = Pr[I(T) - L + \theta \geq 0], \tag{6.1}$$

where $I\ (T)$ is a random variable representing the supplier's net income by time T and L represents the supplier's financial obligations. Earnings, E, follow a Brownian motion process:

$$dE(t) = \mu dt + \sigma dW(t). \tag{6.2}$$

In this equation, μ and σ are the drift and diffusion coefficients of the Brownian motion process. Bankruptcy occurs when earnings fall below liabilities by an exogenous level at the end of the period, time T. In our model, we specify this barrier as 0. However, this quantity can be changed without affecting the qualitative results of our analysis. This interpretation of $p(\theta)$ allows $p(\cdot)$ to be increasing and log-concave as required by Assumption 1.

A reduced-form model of default similar to [9] yields

$$p(\theta) = Pr[\tau(\theta) > T] = e^{-\lambda(\theta)T}, \tag{6.3}$$

where $\tau(\theta)$ is random variable representing the arrival event of a Poisson process with rate $\lambda(\theta)$. This alternative interpretation of $p(\theta)$ also allows $p(\cdot)$ to be increasing and log-concave as required by Assumption 1 as long as $\lambda(\theta)$ is decreasing and convex in θ.

We define the manufacturer's optimal second-stage subgame profit by π using appropriate superscripts and subscripts where necessary in both monopolistic and oligopolistic models. In monopolistic models, π^1 represents a manufacturer's optimal expected profit if its supplier did not declare bankruptcy by stage 2; π^0 represents a manufacturer's profit if the supplier files for bankruptcy before stage 2. We assume $\pi^0 = 0$. In oligopolistic models, we will use two superscripts to indicate the delivery status of suppliers. Subscripts will denote to which manufacturer the equilibrium profit pertains. For instance, π_1^{11} is manufacturer 1's expected equilibrium profit when suppliers of both manufacturers are in sound financial state by stage 2. In this case, manufacturers engage in oligopolistic competition in the second-stage subgame and π_i^{11} is manufacturer i's equilibrium profit. π_1^{10} is profit of manufacturer 1 when its supplier survived stage 1 and its competitor's (manufacturer 2's) supplier did not survive stage 1. Hence, manufacturer 1 becomes a monopolist in the market in this scenario. π_1^{01} is the profit of manufacturer 1 when its supplier did not survive stage 1, but its competitor's supplier did survive stage 1. In this circumstance, manufacturer 1 has no goods to sell and its competitor is a monopolist. We will assume that $\pi_1^{01} = 0$ and $\pi_2^{10} = 0$. π_1^{00} is manufacturer 1's profit when both suppliers declared bankruptcy. We will assume that $\pi_i^{00} = 0$ for $i \in \{1, 2\}$.

Additionally, we make the following assumption regarding manufacturer profits.

Assumption 3 Manufacturers' expected profits are positive in equilibrium for all supply chain scenarios.

We do not need to specify how profits are derived for most of our analysis. However, a convenient illustration is the retail market model where prices are determined by a linear inverse demand function (as in Cournot competition)

$$P(z) = d - z. \tag{6.4}$$

In (6.4), d is the market size parameter and z is the quantity of goods released to the retail market. For this illustrative model we will assume that manufacturers have a constant marginal cost of production c. We would like to emphasize that the Cournot competition model is used for illustration and that most of our results hold for more general retail models.

We next analyze the models for the supply chain structures shown in Fig. 6.1. We begin with an analysis of a benchmark case consisting of two monopolistic manufacturers and two dedicated suppliers. This scenario is presented in Fig. 6.1a.

6.3.1 Benchmark Case: Monopolistic Manufacturers with Dedicated Suppliers

In the benchmark case, there is no strategic interaction among manufacturers. In the second-stage subgame, manufacturers are monopolists earning expected profits π^1 or π^0 contingent on their supplier's status. For example, for linear demand model (6.4), if the supplier is able to deliver goods, the manufacturer chooses an order quantity z to maximize its expected profit $\Pi(z)$

$$\pi^1 = \max_z\{\Pi(z) = zP(z) - cz\}. \tag{6.5}$$

P(z) is the inverse demand function given by (6.4).

If the supplier did not default, the equilibrium quantity of goods released to the market by the manufacturer in the second-stage subgame is given by

$$z^* = \arg\max_z \Pi(z) = \frac{(d-c)}{2} \tag{6.6}$$

for inverse demand (6.4). When the supplier does default, the manufacturer has no goods to sell in the retail market and $z^* = 0$.

For inverse demand (6.4)

$$\pi^1 = \frac{(d-c)^2}{4}. \tag{6.7}$$

Turning now to the first-stage subgame, manufacturers select the amount of subsidies to provide to their dedicated suppliers according to the following optimization problem:

$$\max_{\theta \geq 0} p(\theta)[\pi^1 - \delta(\theta)]. \tag{6.8}$$

Note that manufacturer's profit and subsidy costs are contingent on the supplier being financially viable by the second stage as we have assumed subsidies represent promises for future contracts. If the supplier is not available by the second stage, these future contracts need not be awarded: the manufacturer will not pay them when the supplier defaults.

The first order condition of (6.8) is given by

$$p'(\theta)[\pi^1 - \delta(\theta)] - p(\theta)\delta'(\theta) = 0, \tag{6.9}$$

or, equivalently

$$\frac{p'(\theta)}{p(\theta)} = \frac{\delta'(\theta)}{\pi^1 - \delta(\theta)}. \tag{6.10}$$

We now offer the following lemma related to this optimization problem.

Lemma 2 *The optimization problem in* (6.8) *is log-concave.*

Proof Since $p(\theta)$ is log-concave by Assumption 1, we must demonstrate $\pi^1 - \delta(\theta)$ is log-concave. Define

$$f(\theta) \stackrel{\text{def}}{=} \log[\pi^1 - \delta(\theta)]. \tag{6.11}$$

The function $f(\theta)$ is a composition of an increasing concave and concave function. Therefore, $f(\theta)$ is itself concave and $\pi^1 - \delta(\theta)$ is a log-concave function. □

As (6.8) is log-concave, the first order condition represents a sufficient condition for finding optimal subsidy levels.

The following proposition details the manner in which subsidies vary with the manufacturer's optimal profit in our second-stage subgame.

Proposition 1 *The optimal amount of manufacturer-provided subsidies θ^* is increasing in π^1.*

Proof Taking the cross partial derivative of (6.8), with respect to π^1 and θ yields

$$\frac{\partial^2}{\partial \pi^1 \partial \theta}\{p(\theta)[\pi^1 - \delta(\theta)]\} = p'(\theta) > 0. \tag{6.12}$$

Therefore, (6.8) is supermodular in (θ, π^1) as we have assumed $p(\cdot)$ is an increasing function. Because (6.8) is supermodular in (θ, π^1), the optimal amount of manufacturer-provided subsidies θ^* is increasing in π^1 as shown in [8] Chap. 8.□

For demand model (6.4), Lemma 1 leads to the following corollary.

Corollary 1 *The optimal amount of manufacturer-provided subsidies θ^* is increasing in market size d and decreasing in marginal cost of production c, assuming $d > c$.*

Proof The assumption of linear demand (6.4), and a monopoly environment for the manufacturer means that manufacturers will each sell the monopoly quantity, (d-c)/2, of goods in the downstream retail market. This means that the optimal manufacturer's profit when the supplier is able to deliver goods is

$$\pi^1 = \frac{(d-c)^2}{4}. \tag{6.13}$$

We have already shown θ^* is increasing in π^1. Therefore, with (6.13), θ^* is increasing in d and decreasing in c. □

6.3.2 Oligopolistic Manufacturers with Dedicated Suppliers

We now analyze the effect of competition on the subsidies by comparing the benchmark case with the competitive scenario shown in Figure 6.1(b). Manufacturers engage in competition by supplying z_i^{jk} to the retail market, where subscript i is used to distinguish between manufacturers ($i \in \{1, 2\}$) and superscripts j and k denote the delivery status of each manufacturer's supplier, similar to the explanation of π_i^{jk} presented earlier in Sect. 3.

The following proposition describes the second-stage equilibrium quantities and profits for demand model (6.4).

Proposition 2 *For demand model (6.4), the equilibrium order quantities for manufacturers 1 and 2 in the second-stage game are given by*

$$
(z_1^{jk}, z_2^{jk}) = \begin{cases} \left(\frac{d-c}{3}, \frac{d-c}{3}\right); & \textit{Suppliers 1 and 2 did not default } (j=1, k=1); \\ \left(\frac{d-c}{2}, 0\right); & \textit{Supplier 1 did not default while Supplier 2 defaulted } (j=1,\ k=0); \\ \left(0, \frac{d-c}{2}\right); & \textit{Supplier 1 defaulted while Supplier 2 did not default } (j=0,\ k=1); \\ (0,0); & \textit{Suppliers 1 and 2 defaulted } (j=0,\ k=0). \end{cases} \tag{6.14}
$$

The equilibrium profits for manufacturers 1 and 2 are

$$
(\pi_1^{jk}, \pi_2^{jk}) = \begin{cases} \left(\frac{(d-c)^2}{9}, \frac{(d-c)^2}{9}\right); & \textit{Suppliers 1 and 2 did not default } (j=1,\ k=1); \\ \left(\frac{(d-c)^2}{4}, 0\right); & \textit{Supplier 1 did not default while Supplier 2 defaulted } (j=1,\ k=0); \\ \left(0, \frac{(d-c)^2}{4}\right); & \textit{Supplier 1 defaulted while Supplier 2 did not default } (j=0,\ k=1); \\ (0,0); & \textit{Suppliers 1 and 2 defaulted } (j=0,\ k=0). \end{cases} \tag{6.15}
$$

Proof When manufacturer 1 is unable to supply goods to the retail market, its equilibrium order quantity is necessarily 0. When manufacturer 1 can supply goods to the retail market while manufacturer 2 cannot, manufacturer 1 acts as a monopolist and releases the monopoly quantity to the downstream market. This result is similar to that discussed in the previous section. When both manufacturers can supply goods to the retail market, manufacturer 1 releases the standard oligopoly quantity. These equilibrium quantities yield the optimal profits shown above through substitution into (6.4). □

In the first period, manufacturers choose an appropriate amount of non-negative subsidies to provide to their supplier. Manufacturer 1's expected profit is

$$
p(\theta_1)\{p(\theta_2)\pi_1^{11} + [1-p(\theta_2)]\pi_1^{10} - \delta(\theta_1)\} = p(\theta_1)[\pi_1^{10} - p(\theta_2)(\pi_1^{10} - \pi_1^{11}) - \delta(\theta_1)]. \tag{6.16}
$$

Similar expressions apply for manufacturer 2's profit.

Manufacturer 1's best response function r_1 for the first-stage subgame is

$$
r_1(\theta_2) = \arg\max_{\theta_1 \geq 0} p(\theta_1)[\pi_1^{10} - p(\theta_2)(\pi_1^{10} - \pi_1^{11}) - \delta(\theta_1)]. \tag{6.17}
$$

The first-order condition of (6.16) for manufacturer 1 is

$$
p'(\theta_1)[\pi_1^{10} - p(\theta_2)(\pi_1^{10} - \pi_1^{11}) - \delta(\theta_1)] - p(\theta_1)\delta'(\theta_1) = 0, \tag{6.18}
$$

or equivalently,

$$
\frac{p'(\theta_1)}{p(\theta_1)} = \frac{\delta'(\theta_1)}{\pi_1^{10} - p(\theta_2)(\pi_1^{10} - \pi_1^{11}) - \delta(\theta_1)}. \tag{6.19}
$$

Assuming $\pi_1^{10} = \pi_2^{01}, \pi_1^{00} = \pi_2^{00}$ and $\pi_1^{11} = \pi_2^{11}$ (a "symmetric equilibrium"), $\theta = \theta_1 = \theta_2$ and

$$\frac{p'(\theta)}{p(\theta)} = \frac{\delta'(\theta)}{\pi_1^{10} - p(\theta)(\pi_1^{10} - \pi_1^{11}) - \delta(\theta)}. \tag{6.20}$$

As the following lemma shows, (6.16) is log-concave in θ_1.

Lemma 3 *Manufacturer 1's expected profit* (6.16) *is log-concave in* θ_1.

Proof Assumption 1 states $p(\cdot)$ is log-concave in its argument. Cost function $\delta(\cdot)$ is convex by Assumption 2. Hence, (6.16) is log-concave as it is a product of a log-concave function, $p(\theta_1)$, with a positive concave function: $\pi_1^{10} - p(\theta_2)(\pi_1^{10} - \pi_1^{11}) - \delta(\theta_1)$ is concave in θ_1. The result is log-concave function (6.16). □

Similarly, manufacturer 2's profit is log-concave in θ_2. Lemma 4 yields the existence of a Nash Equilibrium in the first-stage game between manufacturers.

Lemma 4 *There exists a Nash Equilibrium in the oligopolistic manufacturers with dedicated suppliers scenario.*

Proof Manufacturers' action spaces are compact and convex when their suppliers are available. Additionally, all components of (6.16) are continuous, therefore, their payoff functions are continuous. We have also shown in Lemma 3 that payoff functions are log-concave. As log-concavity implies quasi-concavity, there exists at least one pure strategy Nash Equilibrium in the oligopolistic manufacturers with dedicated suppliers scenario. □

6.3.2.1 Insights on Manufacturer Competition

Having presented an analysis of supply chain structures with and without competition, we can examine the effect of competition on expected quantities of goods released to the retail market, subsidies, manufacturer profits, and the consumer surplus. We introduce subscripts "*m*" and "*o*" to denote the monopolistic and oligopolistic manufacturers, respectively, and the superscript "*d*" to denote a dedicated supplier scenario.

Proposition 3 examines the effect of competition on subsidies.

Proposition 3 *If* $\pi^1 = \pi_1^{10}$, *then, as long as* $\pi_1^{10} > \pi_1^{11}$, *the level of subsidies provided by manufacturers to suppliers will be higher when manufacturers are monopolists than when they are oligopolists* ($\theta_m^d > \theta_o^d$) *in dedicated supplier scenarios.*

Proof Rewriting (6.10) and (6.20), we compare solutions of the following two equations:

$$\text{Monopolistic Model:} \quad \pi^1 = \frac{\delta'(\theta_m^d)p(\theta_m^d)}{p'(\theta_m^d)} + \delta(\theta_m^d) \tag{6.21}$$

$$\text{Oligopolistic Model:} \quad \pi_1^{10} = \frac{\delta'(\theta_o^d)p(\theta_o^d)}{p'(\theta_o^d)} + \delta(\theta_o^d) + p(\theta_o^d)(\pi_1^{10} - \pi_1^{11}) \tag{6.22}$$

Lemma 1 states that the quantity $\frac{p'(\cdot)}{p(\cdot)}$ is decreasing when it is governed by Assumption 1. Therefore, the first expression in the right-hand side of both equations is increasing in θ. Also, using Assumption 2, $\delta(\cdot)$ is increasing in its argument. Given $\pi^1 = \pi_1^{10}$, θ_o^d must be lower than θ_m^d as long as $\pi_1^{10} - \pi_1^{11} > 0$ in order for the right-hand side of (6.21) and (6.22) to remain the same. □

We call $\pi_1^{10} - \pi_1^{11}$ the *competition intensity* as it represents the loss of profits experienced by manufacturers in competitive settings over non-competitive settings. The condition $\pi^1 = \pi_1^{10}$ (i.e., the profit of the manufacturer in the monopolistic model when its supplier delivers goods equals the profit of the manufacturer in the oligopolistic model when its supplier delivers goods while its competitor's supplier does not) in model (6.4) with linear inverse demand is the same as assuming that the market size d is the same in the monopolistic and oligopolistic models.

We see from (6.22) that subsidies decrease with higher competition intensity (lower π_1^{11}, holding π_1^{10} constant) in the oligopoly model. Intuitively, these subsidies decrease as competition intensity increases because the expected profits of manufacturers decrease. Hence, manufacturers will reduce the amount of subsidies provided to suppliers to curb subsidy costs.

Let us now assume that the market size increases in the oligopoly model that is, $\pi_1^{10} > \pi^1$. This market size increase could be due to synergies (e.g. combined advertising) realized from multiple manufacturers producing and selling goods. This effect is fairly common within the automotive industry when new products are introduced. For instance, as more and more vehicle manufacturers added hybrid vehicles to their lineup, demand increased considerably. Hybrid vehicle sales rose from 210 thousand units in 2005 to 324 thousand in 2008. Over the same time period, the number of vehicle nameplates offering hybrid vehicles expanded from 8 to 15.

If $\pi_1^{10} > \pi^1$, the size of the manufacturer-provided subsidies in the oligopoly model may be higher than the monopoly model. Referring to (6.21) and (6.22), we see that when $\pi_1^{10} > \pi^1$, the left-hand side of (6.22) is greater than the left-hand side of (6.21), as the right-hand side of both equations is an increasing function of θ.

Turning now to analyze manufacturers' optimal profits, we offer the following proposition.

Proposition 4 *If* $\pi^1 = \pi_1^{10}$, *expected profits are lower for manufacturers in the competitive setting as compared to the non-competitive setting for symmetric manufacturers.*

Proof Expected profits in the monopoly setting are given by

$$p(\theta_m^d)[\pi^1 - \delta(\theta_m^d)], \tag{6.23}$$

and by the following expression in the oligopoly setting

$$p(\theta_o^d)[\pi_1^{10} - p(\theta_o^d)(\pi_1^{10} - \pi_1^{11}) - \delta(\theta_o^d)] = p(\theta_o^d)[\pi_1^{10} - \delta(\theta_o^d)] - p(\theta_o^d)^2(\pi_1^{10} - \pi_1^{11}). \tag{6.24}$$

When $\pi^1 = \pi_1^{10}$, and because θ_m^d maximizes (6.8), we have

$$p(\theta_m^d)[\pi^1 - \delta(\theta_m^d)] \geq p(\theta_o^d)[\pi_1^{10} - \delta(\theta_o^d)]. \tag{6.25}$$

Hence, $p(\theta_m^d)[\pi^1 - \delta(\theta_m^d)] > p(\theta_o^d)[\pi_1^{10} - \delta(\theta_o^d)] - p(\theta_o^d)^2(\pi_1^{10} - \pi_1^{11})$, in other words, expected profits are lower in the oligopoly setting as compared to the monopoly setting. □

The previous proposition demonstrates that, not surprisingly, manufacturers prefer operating in a monopolistic environment, all other things equal.

Total quantities of goods released to *all* retail markets in the monopolistic manufacturers with dedicated suppliers scenario (Q_m^d) are

$$Q_m^d = \begin{cases} 0, & \text{w.p. } [(1 - p(\theta_m^d)]^2, \\ z^1, & \text{w.p. } 2p(\theta_m^d)[1 - p(\theta_m^d)], \\ 2z^1, & \text{w.p. } p(\theta_m^d)^2. \end{cases} \tag{6.26}$$

In the oligopolistic manufacturers with dedicated suppliers scenario, the quantities released to retail markets, Q_o^d, are

$$Q_o^d = \begin{cases} 0, & \text{w.p. } [1 - p(\theta_o^d)]^2, \\ z^{10}, & \text{w.p. } p(\theta_o^d)[1 - p(\theta_o^d)], \\ z^{01}, & \text{w.p. } p(\theta_o^d)[1 - p(\theta_o^d)], \\ 2z^{11}, & \text{w.p. } p(\theta_o^d)^2. \end{cases} \tag{6.27}$$

The expected total quantity of goods released to retail markets is then

$$EQ_m^d = 2p(\theta_m^d)z^1, \tag{6.28}$$

$$EQ_o^d = p(\theta_o^d)[2z^{11}p(\theta_o^d) - (z^{01} + z^{10})(p(\theta_o^d) - 1)]. \tag{6.29}$$

If $z^{10} = z^{01} = z^1$, (6.29) becomes

$$EQ_o^d = 2p(\theta_o^d)[z^1 - p(\theta_o^d)(z^1 - z^{11})]. \tag{6.30}$$

As discussed in Proposition 3, $p(\theta_m^d) > p(\theta_o^d)$ under reasonable assumptions. Under these assumptions, $EQ_m^d > EQ_o^d$ as long as $z^1 > z^{11}$, which is the case in the Cournot model.

While our discussion up to this point has focused on the manufacturer and its decisions, it is also important to consider the benefits consumers can reap as a consequence of each supply chain structure. We measure benefits to consumers using the consumer surplus. Lemma 5 discusses this quantity in the benchmark scenario.

Lemma 5 *For demand model* (6.4), *the expected consumer surplus (CS) in the monopolistic manufacturers with dedicated suppliers scenario is*

$$E(CS_m^d) = p(\theta_m^d) \cdot \frac{(d - c)^2}{4}. \tag{6.31}$$

Proof Using the inverse demand function specified in (6.4), the equilibrium price of goods in the retail market is

$$P(z^*) = d - \frac{d-c}{2} = \frac{d+c}{2}, \tag{6.32}$$

when the supplier is not in default. Using the standard formula for consumer surplus with linear demand and taking into account the fact that manufacturers serve two retail markets yields

$$CS_m^d(z^*) = 2 \cdot \frac{1}{2} \cdot \frac{d-c}{2} \cdot \left(d - \frac{d+c}{2}\right) \tag{6.33}$$

$$= 2\frac{(d-c)^2}{8}. \tag{6.34}$$

When the supplier is bankrupt, the manufacturer cannot release goods to the market, hence the consumer surplus is 0 as no goods are available for consumers to buy. Therefore, the expected consumer surplus, $E(CS_m^d)$ is given by (6.31). □

We now examine in Lemma 6 the consumer surplus in the oligopolistic manufacturers with dedicated suppliers scenario.

Lemma 6 *The expected consumer surplus in the oligopolistic manufacturers with dedicated suppliers scenario,* $E(CS_o^d)$*, assuming a symmetric equilibrium and perfectly correlated supplier asset shocks, is given by*

$$E(CS_o^d) = p(\theta_o^d)^2 \cdot \frac{2(d-c)^2}{9} + p(\theta_o^d)[1 - p(\theta_o^d)] \cdot \frac{(d-c)^2}{4}. \tag{6.35}$$

Proof Using the inverse demand function specified in (6.4), we find that the equilibrium price of goods in the retail market for the oligopolistic manufacturers with dedicated suppliers scenario is

$$P(z^*) = d - \frac{2(d-c)}{3} = \frac{d+2c}{3}, \tag{6.36}$$

when both suppliers are not bankrupt. Using the standard formula for consumer surplus with linear demand yields

$$CS_o^d = \frac{1}{2} \cdot \frac{2(d-c)}{3} \cdot \left(d - \frac{d+2c}{3}\right) \tag{6.37}$$

$$= \frac{2(d-c)^2}{9}. \tag{6.38}$$

When one supplier is bankrupt while the other is not, the consumer surplus is specified by an expression similar to CS_m^d (however, p_o^d is substituted for p_m^d) because only one manufacturer can sell goods in the retail market. The consumer surplus when both suppliers are bankrupt is zero due to the fact that no manufacturer can sell goods in the downstream retail market. The expected consumer surplus follows directly from these values and the probability they occur. □

It is important to note which scenario consumers prefer. We now compare the results shown in Propositions 5 and 6.

Proposition 5 *The consumer surplus is larger with manufacturers who are monopolists in the retail marketplace.*

Proof Rewriting expression (6.35) for $E(CS_o^d)$ yields

$$p(\theta_o^d)^2\left[\frac{2(d-c)^2}{9}-\frac{(d-c)^2}{4}\right]+p(\theta_o^d)\frac{(d-c)^2}{4} \tag{6.39}$$

$$=-p(\theta_o^d)^2\cdot\frac{(d-c)^2}{36}+p(\theta_o^d)\frac{(d-c)^2}{4}. \tag{6.40}$$

As we have shown previously, for any given level of expected quantities released to the market, $p(\theta_m^d) > p(\theta_o^d)$. Therefore, comparing (6.40) with the expression for $E(CS_o^d)$ given in (6.31), $E(CS_m^d) > E(CS_o^d)$ and consumers are better off if manufacturers are monopolists in retail markets. □

The result shown in Proposition 5 arises from the fact that $p(\theta_m^d) > p(\theta_o^d)$. Because of this relationship, suppliers will be more reliable when manufacturers are monopolists, and, in expectation, consumers will reap greater benefits because manufacturers will be more likely to release goods to the retail market in this scenario.

We wish to point out that our previous discussion of expected quantities released to retail markets and the consumer surplus examined these quantities across *all* retail markets. In other words, when manufacturers operate as monopolists, each addressed its own market, whereas when manufacturers operate as oligopolists, they addressed one market in total. If, instead, we normalize the combined size of the retail markets in the monopolistic manufacturers with dedicated suppliers scenario to equal that of the single market in the oligopolistic manufacturers with dedicated suppliers scenario, we have a different expression for Q_m^d:

$$Q_m^d=\begin{cases}0, & \text{w.p. } [(1-p(\theta_m^d)]^2,\\ \frac{z^1}{2}, & \text{w.p. } 2p(\theta_m^d)[1-p(\theta_m^d)],\\ z^1, & \text{w.p. } p(\theta_m^d)^2.\end{cases} \tag{6.41}$$

The expected consumer surplus $E(CS_m^d)$ would also differ from (6.31):

$$E(CS_m^d)=p(\theta_m^d)\cdot\frac{(d-c)^2}{8}. \tag{6.42}$$

Conventional wisdom would suggest that manufacturer-level competition would benefit consumers and increase the consumer surplus. However, we find through Proposition 5 that manufacturer-level competition leads to decreased subsidies, causing suppliers to become less reliable. Therefore, the net effect of competition on the consumer surplus with normalized market sizes is dependent on the reliability of suppliers, inclusive of subsidies. For instance, when suppliers for competing manufacturers are highly reliable, $E(CS_o^d)$ will be greater than $E(CS_m^d)$, but when suppliers for competing manufacturers are not reliable, it is the case that $E(CS_m^d)$ will exceed $E(CS_o^d)$.

6.3.3 Monopolistic Manufacturers with a Shared Supplier

Having analyzed the effects of manufacturer-level competition, we now examine the effect of using a shared supplier on the manufacturers' decisions when manufacturers do not compete. Specifically, we will analyze the supply chain structure in Figure 6.1(d) against the benchmark case in Figure 6.1(a).

Because manufacturers do not compete, the equilibrium quantity of goods released to the market is the same as in the benchmark model. If its supplier does not default, a manufacturer releases z^* to the market. For linear demand model (6.4), $z^* = \frac{(d-c)}{2}$ when the supplier does not default, and $z^* = 0$ when the shared supplier defaults. Definitions of π^1 and π^0 are given in our explanation of the benchmark model.

In the first-stage subgame, manufacturer 1 solves

$$\max_{\theta_1 \geq 0} p(\theta_1 + \theta_2)[\pi^1 - \delta(\theta_1)], \tag{6.43}$$

where $p(\cdot)$ is now a function of θ_1 and θ_2 because the supplier receives subsidies from both manufacturers. The first-order condition of (6.43) is given by

$$p'(\theta_1 + \theta_2)[\pi^1 - \delta(\theta_1)] - p(\theta_1 + \theta_2)\delta'(\theta_1) = 0, \tag{6.44}$$

or equivalently,

$$\frac{p'(\theta_1 + \theta_2)}{p(\theta_1 + \theta_2)} = \frac{\delta'(\theta_1)}{\pi^1 - \delta(\theta_1)}. \tag{6.45}$$

In a symmetric equilibrium, (6.45) simplifies to

$$\frac{p'(2\theta_m^s)}{p(2\theta_m^s)} = \frac{\delta'(\theta_m^s)}{\pi^1 - \delta(\theta_m^s)}, \tag{6.46}$$

where the subscript "*m*" denotes monopolistic manufacturers and the superscript "*s*" denotes a shared supplier scenario. Similar to Lemma 2, (6.43) is log-concave. Therefore, the first-order condition represents a sufficient condition for finding optimal subsidy levels. The results from Proposition 1 and Corollary 1 also hold in this supply chain scenario.

6.3.3.1 Insights on Cross-Subsidies

We now analyze the impact of cross-subsidies on the expected quantities of goods released to the retail market, subsidies, manufacturer profits and the consumer surplus.

The following proposition compares the amount of subsidies provided by each manufacturer in the monopolistic manufacturers with a shared supplier scenario to those in the monopolistic manufacturers with dedicated suppliers scenario.

Proposition 6 *The optimal amount of subsidies provided by each manufacturer in the monopolistic manufacturers with a shared supplier scenario,* θ_m^s *is less than the optimal amount of subsidies in the monopolistic manufacturers with dedicated suppliers scenario,* θ_m^d.

Proof Comparing (6.46) and (6.10), one can see that the right-hand sides of both expressions are identical. The only difference between these expressions is that θ has been replaced by 2θ in the left-hand side of (6.46). We also know $\frac{p'(\cdot)}{p(\cdot)}$ is decreasing by Lemma 1, and $\frac{\delta'(\theta)}{\pi^1-\delta(\theta)}$ is increasing as

$$\frac{\partial}{\partial\theta}\left[\frac{\delta'(\theta)}{\pi^1-\delta(\theta)}\right]=\frac{\delta''(\theta)[\pi^1-\delta(\theta)]+[\delta(\theta)^2]}{[\pi^1-\delta(\theta)]^2}>0. \tag{6.47}$$

Therefore, θ_m^s must be less than θ_m^d in order for first-order condition (6.46) to hold.□

If, instead of examining the subsidies *provided* by each manufacturer, we examine the subsidies *received* by each supplier, we must take into account the fact that a shared supplier receives subsidies from two manufacturers. In other words, if each manufacturer that uses a shared supplier provides θ_m^s, the supplier receives $2\theta_m^s$. Proposition 7 compares the amount of subsidies received by shared and dedicated suppliers when manufacturers are monopolists.

Proposition 7 *The amount of subsidies received by a shared supplier,* $2\theta_m^s$, *is greater than the amount of subsidies received by a dedicated supplier* θ_m^d.

Proof The first-order condition in the shared supplier scenario is

$$\frac{p'(2\theta_m^s)}{p(2\theta_m^s)}=\frac{\delta'(\theta_m^s)}{\pi^1-\delta(\theta_m^s)}, \tag{6.48}$$

and, for the dedicated suppliers scenario

$$\frac{p'(\theta_m^d)}{p(\theta_m^d)}=\frac{\delta'(\theta_m^d)}{\pi^1-\delta(\theta_m^d)}. \tag{6.49}$$

Because $\frac{p'(\cdot)}{p(\cdot)}$ is decreasing by Lemma 1 and $\frac{\delta'.}{\pi^1-\delta(\cdot)}>0$, it follows that $\theta_m^d<2\theta_m^s$.

One can see that (1) manufacturers each provide less subsidies to a shared supplier than they do to a dedicated one, and (2) the total subsidies received by a shared supplier is greater than the total subsidies received by a dedicated supplier. □

Proposition 8 discusses the implications of cross-subsidies on manufacturer-level profits.

Proposition 8 *Manufacturer profits in monopolistic settings are higher when manufacturers share a supplier than if they used dedicated suppliers.*

Proof Manufacturer-level profits in the monopolistic manufacturers with dedicated suppliers and monopolistic manufacturers with a shared supplier models are respectively,

$$p(\theta_m^d)[\pi^1 - \delta(\theta_m^d)], \tag{6.50}$$

$$p(2\theta_m^s)[\pi^1 - \delta(\theta_m^s)]. \tag{6.51}$$

From Propositions 6 and 7,

$$2\theta_m^s > \theta_m^d, \tag{6.52}$$

$$\theta_m^s < \theta_m^d. \tag{6.53}$$

Using this information together with Assumptions 1 and 2 yields

$$p(2\theta_m^s) > p(\theta_m^d), \tag{6.54}$$

$$\delta(\theta_m^s) < \delta(\theta_m^d). \tag{6.55}$$

In other words, supplier reliability increases when manufacturers use a shared supplier in non-competitive settings while decreasing their costs of subsidies. Therefore, manufacturer-level profits in monopolistic settings are higher when manufacturers use a shared supplier. □

In the monopolistic manufacturers with a shared supplier scenario, the quantities released to *all* retail markets, Q_m^s, are

$$Q_m^s = \begin{cases} 0, & \text{w.p.} \quad 1 - p(2\theta_m^s), \\ 2z^1, & \text{w.p.} \quad p(2\theta_m^s), \end{cases} \tag{6.56}$$

while the quantities released to *all* retail markets in the monopolistic manufacturers with dedicated suppliers scenario, Q_m^d, is specified by (6.26).

The expected quantities released to the market are then:

$$EQ_m^s = 2z^1 p(2\theta_m^s), \tag{6.57}$$

$$EQ_m^d = 2z^1 p(\theta_m^d). \tag{6.58}$$

As $2\theta_m^s > \theta_m^d$, $EQ_m^s > EQ_m^d$.

With respect to the preferences of consumers, we examine the consumer surplus in the monopolistic manufacturers with a shared supplier scenario and compare it to the consumer surplus in the monopolistic manufacturers with dedicated suppliers scenario.

Lemma 7 *The expected consumer surplus in the monopolistic manufacturers with a shared supplier scenario is given by*

$$E(CS_m^s) = p(2\theta_m^s) \cdot \frac{(d-c)^2}{4}. \tag{6.59}$$

Proof The expected consumer surplus in this scenario is identical to the expected consumer surplus in the monopolistic manufacturers with dedicated suppliers scenario except that the shared supplier receives subsidies from both manufacturers. These subsidies total $2\theta_m^s$. □

The expected consumer surplus in the monopolistic manufacturers with dedicated suppliers scenario is identical to that stated in Lemma 5. Comparing these two expressions, we have the following proposition.

Proposition 9 *Consumers are always better off when non-competing manufacturers use a shared supplier.*

Proof The proof results directly from (6.52) and Assumption 1. □

6.3.4 Combined Effects of Competition and Cross-Subsidies

In the previous sections, we saw that manufacturer-level competition decreases the amount of subsidies to suppliers while simultaneously decreasing manufacturers' profits. We have also seen that cross-subsidies make the total subsidy received by a supplier larger while increasing manufacturers' profits. In this section we answer the question of which of these effects will dominate the other when they are combined. We do so by incorporating an analysis of the supply chain structure shown in Fig. 6.1c.

The results in this section provide an interesting extension to an observation in [12] that competing manufacturers should never choose to share a supplier. We show that when cross-subsidy benefits are significant, a shared supplier is the preferred choice for manufacturers.

Within the oligopolistic manufacturers with dedicated suppliers scenario, recall that manufacturer 1 solved best response function (6.17) which yielded first-order condition (6.18).

The first-stage subgame optimization problem for manufacturers in the oligopolistic manufacturers with a shared supplier scenario is

$$\max_{\theta_1 \geq 0} p(\theta_1 + \theta_2)[\pi_1^{11} - \delta(\theta_1)]. \tag{6.60}$$

It is easy to see that this problem is again log-concave, and that the symmetric equilibrium in this scenario is given by

$$\frac{p'(2\theta)}{p(2\theta)} = \frac{\delta'(\theta)}{\pi_1^{11} - \delta(\theta)}. \tag{6.61}$$

We offer the following proposition that compares subsidies provided by competitive manufacturers when they use dedicated suppliers or a shared supplier.

Proposition 10 *The amount of subsidies provided by each manufacturer to its supplier in the oligopolistic manufacturers with a shared supplier scenario is less than the amount of subsidies provided by each manufacturer in the oligopolistic manufacturers with dedicated suppliers scenario when* $\pi_1^{10} > \pi_1^{11}$.

Proof Equations (6.20) and (6.61) determine the equilibrium amount of subsidies for dedicated suppliers and shared supplier settings, respectively, when manufacturers compete (reproduced here with distinguishing notation):

$$\frac{p'(\theta_o^d)}{p(\theta_o^d)} = \frac{\delta'(\theta_o^d)}{[1 - p(\theta_o^d)]\pi_1^{10} + p(\theta_o^d)\pi_1^{11} - \delta(\theta_o^d)}, \tag{6.62}$$

and

$$\frac{p'(2\theta_o^s)}{p(2\theta_o^s)} = \frac{\delta'(\theta_o^s)}{\pi_1^{11} - \delta(\theta_o^s)}. \tag{6.63}$$

Let $\bar{\theta}$ satisfy

$$\frac{p'(\bar{\theta})}{p(\bar{\theta})} = \frac{\delta'(\bar{\theta})}{\pi_1^{11} - \delta(\bar{\theta})}. \tag{6.64}$$

Comparing (6.62) and (6.64), we have $\theta_o^d > \bar{\theta}$, and comparing (6.63) and (6.64), we have $\bar{\theta} > \theta_o^s$. Therefore, $\theta_o^d > \theta_o^s$. □

In analyzing the amount of subsidies received by suppliers we must again compare the first-order conditions associated with the oligopolistic manufacturers with dedicated suppliers ("*od*") and oligopolistic manufacturers with a shared supplier ("*os*") scenarios. For the oligopolistic manufacturer with dedicated suppliers scenario ("*od*")

$$\frac{p'(\theta_o^d)}{p(\theta_o^d)} = \frac{\delta'(\theta_o^d)}{\pi_1^{10} - p(\theta_o^d)(\pi_1^{10} - \pi_1^{11}) - \delta(\theta_o^d)}, \tag{6.65}$$

and, for the shared supplier scenario ("*os*")

$$\frac{p'(\theta_o^s)}{p(\theta_o^s)} = \frac{\delta'\left(\frac{\theta_o^s}{2}\right)}{\pi^{11} - \delta\left(\frac{\theta_o^s}{2}\right)}. \tag{6.66}$$

We define

$$G(\theta_o^d) = \frac{\delta'(\theta_o^d)}{\pi_1^{10} - p(\theta_o^d)(\pi_1^{10} - \pi_1^{11}) - \delta(\theta_o^d)}, \tag{6.67}$$

$$H(\theta_o^s) = \frac{\delta'\left(\frac{\theta_o^s}{2}\right)}{\pi^{11} - \delta\left(\frac{\theta_o^s}{2}\right)}. \tag{6.68}$$

Examining (6.67) and (6.68), we can see that if $\pi_1^{10} \approx \pi_1^{11}$, total subsidies are lower in the oligopolistic manufacturers with dedicated suppliers scenario than the shared supplier scenario because the $\frac{\theta}{2}$ effect will dominate. However, when competition intensity $(\pi_1^{10} - \pi^{11})$ increases, subsidies to dedicated suppliers increase. In the extreme case, when $\pi_1^{10} >> \pi_1^{11}$, the amount of total subsidies in the oligopolistic manufacturers with dedicated suppliers scenario will exceed the amount of total subsidies in the shared supplier scenario.

We now compare the quantity of goods released to retail markets in the oligopolistic manufacturers with dedicated suppliers (Q_o^d) and oligopolistic manufacturers with a shared supplier (Q_o^s) scenarios. Q_o^d is specified by (6.27), while Q_o^s is specified by

$$Q_o^s = \begin{cases} 0, & \text{w.p. } 1 - p(2\theta_m^s), \\ 2z^{11}, & \text{w.p. } p(2\theta_m^s). \end{cases} \tag{6.69}$$

If $z^{10} = z^{01} = z^1$, expected quantities released to retail markets in both scenarios are

$$EQ_o^d = 2p(\theta_o^d)[z^1 - p(\theta_o^d)(z^1 - z^{11})], \tag{6.70}$$

$$EQ_o^s = 2z^{11}p(2\theta_o^s). \tag{6.71}$$

We know $z^{11} < z^1 - p(\theta_o^d)(z^1 - z^{11})$. Therefore, if $2\theta_o^s < \theta_o^d$, $EQ_o^d > EQ_o^s$.

However, we have previously seen that $2\theta_o^s > \theta_o^d$ when $\pi^{10} - \pi^{11} \approx 0$. Therefore, if competition is weak, the manufacturer using a shared supplier will, in expectation, release more goods to the retail market.

6.4 Conclusions

Intense competition in the automotive retail marketplace has forced manufacturers to carefully examine their cost structures. A major component of these costs is purchases from suppliers. Manufacturers strive to procure the highest quality goods at the lowest prices from reliable suppliers. They can elect to purchase goods from a dedicated supplier, or share a supplier with other manufacturers. Manufacturers' profitability is directly related to supplier reliability, as well as subsidies provided to suppliers. These subsidies can directly influence supplier reliability.

We have examined how optimal manufacturers' profits, subsidies provided to suppliers, and quantities released to the retail market vary with manufacturer-level competition and cross-subsidies by analyzing the four supply chain structures shown in Fig. 6.1.

We show that there exists a tradeoff between the benefits of cross-subsidies and benefits of reduced competition in making the decision to share a supplier. The use of shared or dedicated suppliers will largely depend on whether the effects of competition or cross-subsidies dominates.

On one hand, cross-subsidies achieved through the use of a common supplier allow manufacturers to share subsidy costs. We show that the amount of subsidies provided by each manufacturer is less when manufacturers share a supplier. However, because suppliers receive subsidies from two manufacturers, the total subsidy received can be larger for a shared supplier. Thus, cross-subsidies can improve supplier reliability.

On the other hand, the use of shared suppliers poses additional problems for competing manufacturers. When suppliers are shared, all manufacturers are affected by their supplier's status. Manufacturers cannot differentiate themselves from their peers by the availability of supplies, so retail competition intensifies. If, instead, suppliers are dedicated, manufacturers possess an option to capture a larger portion of the market if their competitor's supplier fails.

Our model is applicable to the behavior of firms within the US automotive industry. For example, GM procured many goods from internal, dedicated suppliers for a large majority of its corporate life. These internal suppliers were organized as divisions—largely run autonomously—and provided GM with everything from radios and engine control modules (Delco Division), air conditioning compressors (Harrison Division), engine electrical systems (Packard Division), fuel delivery systems (Rochester Products Division), headlamps (Guide Division) and car bodies (Fisher Body Division). While some components produced by these divisions were sold to other automakers, their primary responsibility was to fulfill GM's needs. However, in the late 1990s, GM elected to combine many of these organizations into a single division, which it renamed Delphi. GM spun Delphi off in 1997 through an initial public offering, thus creating an independent parts supplier. In divorcing itself from Delphi, GM was hoping to reap the benefits of cross-subsidies from other manufacturers.

After its initial public offering, Delphi Corporation moved to diversify its customer base. In 2007, the year before it filed for bankruptcy, 63% of Delphi's revenues came from non-GM customers, up from 17% in fiscal year 1996. Delphi increasingly became a "shared supplier." However, as Delphi continued to diversify its customer base, competition in the automotive industry increased. According to Ward's Automotive, sales of cars produced by US automakers in 2009 represented 45% of total US vehicle sales, compared with over 70% of vehicle sales in 1996. This decreased market share is largely due to increased competition within the American marketplace.

As competition in the US market intensified, subsidies to Delphi from manufacturers began to decrease. One could argue that Delphi's new customers elected to use the firm for the same reason as GM: namely, to reap the benefit of cross-subsidies. However, as our paper has demonstrated, under certain conditions, the amount of subsidies received by a shared supplier decreases. This decreased level of subsidies makes shared suppliers less reliable than dedicated ones, and is exemplified by the continuing financial struggles facing Delphi: from 1996 through (Delphi filed for bankruptcy 2007), Delphi's pre-tax operating income decreased from $1.3 to –$2.0 billion in spite of the company's diversification efforts. (Here, we cite pre-tax income to negate the impact of tax-loss carry forwards and non-recurring/extraordinary items.)

Delphi is not alone in its financial struggles. Its fate largely mirrors that of Visteon Corporation, a parts maker that was formed in 1997 and spun off from the Ford Motor Company in 2000. Visteon filed for Chapter 11 bankruptcy in May 2009. Sales to Ford accounted for 28% of Visteon's revenues in 2009 compared with 84% of its revenues in 2000, the year it went public. In spite of its attempts to become a profitable "shared supplier," Visteon's strategy did not allow it to avoid a bankruptcy that was arguably caused, in part, by the decreased level of subsidies it received from its customers.

References

1. A. T. Kearney (2009) More than half of the automotive supply base could go bankrupt in 2009; at least 1 million job losses expected. http://www.atkearney.com/index.php/News-media/more-than-half-of-the-automotive-supply-base-could-go-bankrupt-in-2009.html
2. Babich V (2010) Independence of capacity ordering and financial subsidies to risky suppliers. Manuf Serv Oper Manag, Published online before print February 16, 2010
3. Deo S, Corbett C (2009) Cournot competition under yield uncertainty: the case of the U.S. influenza market. Manuf Serv Oper Manag 11(4):563–576
4. Gerchak Y, Parlar M (1990) Yield randomness, cost tradeoffs and diversification in the EOQ model. Nav Res Logist 37(3):341–354
5. Harvey D (2009) Happier times ahead for auto suppliers? www.atkearney.com/index.php/Publications/happier-times-ahead-for-auto-suppliers.html
6. Haywood K (2009) GM helps axle maker avoid Chapter 11. Wall Str J 18 September
7. Hendricks K, Singhal V (1995) An empirical analysis of the effect of supply chain disruptions on long-run stock price performance and equity risk of the firm. Prod Oper Manag 14(1):35–52
8. Heyman D, Sobel MJ (2004) Stochastic models in operations research, vol 2, Dover
9. Jarrow R, Turnbull S (1995) Pricing derivatives on financial securities subject to credit risk. J Finance 50(1):53–85
10. Merton R (1974) On the pricing of corporate debt: the risk structure of interest rates. J Finance 29(2):449–470
11. Silver E (1976) Establishing the order quantity when the amount received is uncertain. INFOR 14(1):32–39
12. Tang S, Kouvelis P (2009) Supplier diversification strategies in the presence of yield uncertainty and buyer competition. Working paper, University of Miami
13. Treece J (2010) Auto suppliers to get revised contract terms Chrysler group. Crain's Detroit Business, 11 June
14. Yano C, Lee H (1995) Lot sizing with random yields: a review. Oper Res 43(2):311–334

Chapter 7
Supply Contracting Under Information Asymmetry and Delivery Performance Consideration

Fuqiang Zhang

7.1 Introduction

A global economy and rapidly-changing market conditions have greatly intensified industry competition. This, in turn, has led to an ever-increasing level of outsourcing/offshoring activities by firms in order to gain cost advantage and market share [35, 40]. According to the Department of Commerce [29], typical US manufacturers spend more than half of their revenue on goods and services obtained from external suppliers. As a result, supply management has become a significant issue for many companies that rely more on their suppliers for the delivery of components, products, and services. When sourcing from outside suppliers, a buyer should consider both price and non-price factors. One of the major non-price factors is the supplier's delivery performance. The benefit of fast, reliable deliveries from a supplier is quite clear from an operations management perspective. It enables the buyer to lower inventory and provide superior service. In other words, the more responsive the supplier is, the lower the buyer's operating cost (e.g., inventory holding cost plus penalty cost for backorders).

The importance of supplier delivery performance in procurement has been emphasized by both practitioners and academics [12, 52]. It has been reported that many firms rank price, quality, and delivery performance as the top three criteria for selecting and evaluating suppliers [20, 43]. For example, Sun Microsystems considers procurement cost and delivery performance the two most important dimensions when choosing suppliers [32]. With the help of recent advances in information technologies, there has been a drastic increase in the use of online auction as a procurement tool [51, 53, 63]. A challenge in procurement auction design for B2B exchanges is how to bring non-price factors into consideration, including supplier

F. Zhang (✉)
Olin Business School, Washington University,
St. Louis, MO 63130, USA
e-mail: fzhang22@wustl.edu

H. Gurnani et al. (eds.), *Supply Chain Disruptions: Theory and Practice of Managing Risk*, DOI: 10.1007/978-0-85729-778-5_7,

delivery performance [31]. These observations call for theoretical research that can generate useful managerial guidelines for procurement process design while taking delivery performance into account.

This chapter studies a buyer's supply contract design problem under delivery performance consideration. A supplier's delivery performance depends on the supplier's capacity (if the supplier makes to order) or inventory (if the supplier makes to stock), both of which are costly to invest. For most practical situations, the supplier is a self-interested, independent organization. Thus the buyer needs to offer appropriate incentives to induce sufficient capacity or inventory investment from the supplier. In addition, the supplier may have private information about the cost for providing fast delivery. For instance, the buyer does not know the supplier's exact capacity or production cost when contracting with the supplier. This means the buyer often faces cost uncertainty when designing the procurement contract.

The objective of this chapter is to identify efficient and easy-to-implement procurement mechanisms for the buyer. We consider scenarios where the supplier is either a make-to-order service provider or a make-to-stock manufacturer. For both scenarios, we propose some simple mechanisms and evaluate their performances along two dimensions. First, we compare the buyer's profit in the simple mechanisms with the profit in the optimal (profit-maximizing) mechanism. This sheds light on how efficient the simple mechanisms are from the buyer's standpoint. Second, we examine the supply chain's performance under the simple mechanisms by comparing it to the supply chain's optimal solution. This reveals whether the simple mechanisms are efficient from the entire chain's perspective.

The rest of the chapter is organized as follows. Section 7.2 reviews the related literature. Section 7.3 studies a make-to-order supplier and Sect. 7.4 considers a make-to-stock supplier. Section 7.5 discusses some contracting issues and we conclude with Sect. 7.6

7.2 Related Literature

The models and analyses in this chapter are based on the studies by Cachon and Zhang [16, 65, 67]. We outline the model settings and summarize the main results from these studies. Below we briefly review the representative papers in the literature (this is by no means an exhaustive list of all related papers).

This chapter is closely related to the supply chain coordination literature. Most of the studies in this literature focus on the design of coordination contracts for decentralized supply chains under complete information. A comprehensive survey is provided by Cachon [14]. A few of the most relevant papers include Cachon and Zipkin [18], Ray et al. [55] for serial inventory systems, Caldentey and Wein [19] for production-inventory systems, Chen [22] for distribution systems, and Bernstein and DeCroix [5], Zhang [66] for assembly systems.

Complete information is not a realistic assumption for most decentralized supply chains. This has inspired a stream of research that studies supply chain contracting under asymmetric information. Corbett [26], Corbett and de Groote [27], Corbett et al. [28], Ha [38] are some of the representative studies that consider asymmetric cost information, whereas Burnetas et al. [13], Cachon and Lariviere [15], Özer and Wei [50] consider asymmetric demand information in a supply chain setting. Recently, Yang et al. [64] analyzes a manufacturer's sourcing strategy when the supplier possesses private information on supply disruption. Gümüş [37] studies the effect of supply guarantees when a buyer procures from two suppliers with different price and reliability characteristics. This chapter also falls into the category of supply contracting under asymmetric information, but with quite different model settings.

There is an extensive literature on procurement in economics. It focuses primarily on two issues. The first is how to select a cost-efficient supplier from a potential supplier pool, and the second is how to induce the selected supplier to invest in R&D and other improvement efforts. These two issues have been addressed by the auction theory and the theory of incentives. Surveys of this literature can be found in Klemperer [45], Laffont and Martimort [48]. The procurement papers in the economics literature are different from this chapter because they do not take operation factors such as delivery performance into consideration.

We study a multi-attribute procurement problem because the buyer cares about both price and delivery performance. A few papers also study multi-attribute procurement. Branco [10], Che [21] study a multi-dimensional auction in which price and quality are the two attributes in procurement. Chen et al. [24] studies procurement auction design for a third-party auctioneer with a focus on the price and transportation cost aspects. Multi-attribute multi-round auctions have been studied in Beil and Wein [3], where the buyer can learn about the suppliers in each round. Kostamis et al. [46] considers a procurement problem where both price and non-price factors are used for contract award decisions and the buyer may release the information about the suppliers' non-price factors.

Besides incentive contracts, a buyer may motivate suppliers to improve delivery performance through various competition mechanisms. In particular, the buyer may adopt a multi-sourcing strategy and allocate business to suppliers based on their past performance; see [4, 17, 36, 39] in the operations management literature for a few examples.

There are papers that study a buyer's procurement or replenishment strategies given exogenous suppliers characteristics (such as delivery lead time and price). Examples include [1, 33, 34, 49, 54]. A review of this literature can be found in Elmaghraby [30]. In this chapter, the suppliers' delivery performance is endogenous and can be influenced by the buyer's procurement strategy. There is also a growing literature on supply risk management, which analyzes mitigation strategies for buyers to manage various supply risks (e.g., price fluctuations and supply disruption). Reviews of this literature can be found in Tang [56], Tomlin [61].

The importance of simplicity in contract design has been emphasized both in the economics and operations management literatures. Bhattacharyya and Lafontaine [7], Holmstrom and Milgrom [41] point out that most real-world incentive schemes seem to take less extreme forms than the sophisticated policies predicted by economic theory. See [9, 25], and the references therein for more discussion of the performance of simple incentive contracts. In recent years, there has been a growing literature in operations management that investigates the performance of simple procurement contracts, including Kayiş [44], Taylor [57], Tunca [62]. Bolandifar et al. [8] studies the supply contract design problem for a newsvendor who faces uncertain demand for a single selling season. They demonstrate that a simple wholesale price could be theoretically optimal even under information asymmetry about the supplier's capacity cost. These papers do not consider suppliers' delivery performance and therefore generate quite different insights.

7.3 Contracting with Make-to-Order Supplier

This section focuses on the scenario where the buyer sources an input from a make-to-order supplier. For example, in the contract manufacturing industry, many firms assemble highly customized components for their customers on a make-to-order basis [11, 60]. In this case, the supplier does not hold inventory and thus can be modeled as a queueing system. An alternative interpretation is that the buyer procures a certain service from the supplier. The scenario where the supplier can hold inventory to improve delivery performance will be studied in Sect. 7.4

7.3.1 Basic Model

A buyer (e.g., a manufacturer or a retailer) procures a product or service from a supplier (e.g., a contract manufacturer or service provider) to satisfy consumer demand. Throughout the chapter we will use "she" for the buyer whereas "he" for the supplier. Consumer demand follows a Poisson arrival process with rate λ. The buyer uses a base-stock policy to manage her inventory. Let s be the base-stock level at the buyer. The buyer has to incur a cost rate h for holding each unit of the product. In the basic model we assume that h is fixed and independent of the procurement price the buyer pays the supplier. Sect. 7.3.4.2 relaxes this assumption and demonstrates that the qualitative insight will remain unchanged. Unmet demand at the buyer can be backlogged and there is a unit penalty cost rate b for backlogged demand.

The supplier adopts a make-to-order production strategy and does not hold inventory. Let μ be the capacity (production or service rate) chosen by the supplier. For a given μ, the production time is exponentially distributed with mean $1/\mu$. There is a constant unit cost rate c for maintaining a certain capacity level, which is a random draw from a distribution F (density f) on a support $[\bar{c}, \underline{c}]$. We follow the literature to assume that F is log-concave, i.e., $F(c)/f(c)$ is an increasing function. This condition holds for most commonly used distribution functions [2]. The distribution F represents the cost uncertainty for delivering the product. Such an uncertainty may be associated with research and development (R&D), production yield, and other random factors. The supplier observes the realization of c, whereas the buyer does not. However, we assume that the buyer has an unbiased belief about the distribution F.

The buyer is assumed to act as a Stackelberg leader in the contracting process. This applies to situations where the buyer is a major player in the industry and has an advantageous bargaining position. Both firms are risk-neutral and aim at maximizing the expected payoff function per unit of time. In addition, we assume that the supplier has a zero opportunity cost (the best outside alternative yields zero profit). Including a positive opportunity cost does not change the essence of the problem. The chronology of events is as follows: First, the supplier's capacity cost c is realized and only observed by the supplier; second, the buyer offers a take-it-or-leave-it contract to the supplier; if the contract is accepted, then the supplier invests in capacity μ and the procurement price R is determined accordingly; the buyer then chooses the base-stock level s; finally, the system runs over a sufficiently long horizon. In this chapter we use superscript "o" to denote the optimal mechanism for the buyer and "$*$" denote the optimal solution for the supply chain. Let E be the expectation operation, and let a prime denote the derivative of a function of a single variable. Define $(x)^+ = \max(0, x)$ and $(x)^- = \max(0, -x)$.

The above model setup is based on Cachon and Zhang [16] and Zhang [65]. Some modifications have been made on the notations (to maintain consistency with the notations in Sect. 7.4). For example, here we use c for the capacity cost, b for the backorder cost, C for the firms' cost functions, and subscripts 1 and 2 to denote the buyer (stage 1) and the supplier (stage 2), respectively. More details can be found in Cachon and Zhang [16], which provides a comprehensive treatment of the problem.

Next we derive the supply chain's optimal solution as preliminary analysis. Suppose the supplier has a capacity level μ and the buyer adopts a base-stock level s. Let N be the number of outstanding orders at the supplier in steady state, which follows a geometric distribution. Then the buyer's operating cost (inventory and backorder costs) is given by

$$C(\mu, s) = E[h(s-N)^+ + b(N-s)^-] = h\left(s - \frac{\lambda}{\mu - \lambda}\right) + (h+b)\left(\frac{\lambda}{\mu}\right)^s \frac{\lambda}{\mu - \lambda}. \tag{7.1}$$

To derive a closed-form expression for the optimal s, let us assume s is large enough so that we can treat it as a continuous variable. The buyer's (or the supply chain's) cost-minimizing base-stock level can be shown to be

$$s^*(\mu) = -\ln\left(\left(\frac{h}{h+b}\right)\left(\frac{\mu/\lambda - 1}{\ln(\mu/\lambda)}\right)\right) \Big/ \ln(\mu/\lambda). \tag{7.2}$$

Plugging $s^*(\mu)$ into the buyer's operating cost function gives

$$C(\mu) = C(\mu, s^*(\mu)) = h\left[\frac{1 - \ln\left(\left(\frac{h}{h+b}\right)\left(\frac{\mu/\lambda - 1}{\ln \mu/\lambda}\right)\right)}{\ln \mu/\lambda} - \frac{1}{\mu/\lambda - 1}\right]. \tag{7.3}$$

Then the supply chain total cost is $C_{sc}(c, \mu) = c\mu + C(\mu)$, where the subscript sc stands for the supply chain.

Theorem 1 *The buyer's operating cost $C(\mu)$ and the supply chain total cost $C_{sc}(c, \mu)$ are convex in $\mu \geq \lambda$.*

The proofs can be found in Cachon and Zhang [16], Zhang [67] and therefore are omitted in this chapter. The above theorem implies that the supply chain's optimal capacity $\mu^*(c)$ must satisfy

$$C'(\mu^*) = -c. \tag{7.4}$$

The cost functions $C(\mu)$ and $C_{sc}(c, \mu)$ are quite complex, and there is no closed-form expression for the supply chain's optimal solution. Cachon and Zhang [16] offers an approximation of these cost functions. Notice the exponential distribution is the continuous counterpart of the geometric distribution. Hence we may use an exponential distribution with mean $E(N)$ to approximate the geometric distribution for N. This approximation can be justified in a heavy-traffic queuing system [19]. It tends to underestimate the average delivery lead time; however, Cachon and Zhang [16] also demonstrates that this approximation is quite accurate when the system utilization is reasonably high. In the rest of the analysis, we will use ^ to denote the variables under the approximation. Specifically, the supply chain's optimal solution is given by

$$\hat{\mu}^*(c) = \lambda + \sqrt{\alpha/c} \text{ and } \hat{s}^*(c) = \sqrt{c\alpha/h^2}, \tag{7.5}$$

where $\alpha = h\lambda \ln((h + b)/h)$ is a constant. We will adopt this approximation for later analysis.

7.3.2 Buyer's Procurement Mechanisms

We present the buyer's procurement contracts in this section. (We will use the words "mechanism" and "contract" exchangeably in this chapter). Three contracts are considered: optimal mechanism (OM), late-fee mechanism (LF), and lead-time mechanism (LT). The performances of these contracts will then be compared in the next section.

7.3.2.1 Optimal Mechanism

The buyer needs to design a procurement mechanism to offer to the supplier. The mechanism is a mapping from the supplier's information space to the space of all possible action and payment schedules. Based on the revelation principle [48], there exists an optimal mechanism that is both direct (i.e., the supplier's information space is identical to his private cost values) and truth-inducing (i.e., it is in the supplier's best interest to announce the true cost). Thus, without losing generality, we will restrict our attention to direct, truth-inducing mechanisms when searching for the optimal one.

The buyer's optimal mechanism consists of a pair of functions $\{\mu(\cdot), R(\cdot)\}$: If the supplier announces a cost x (which may not necessarily equal the true cost c), then the supplier will build a capacity $\mu(x)$ and receives a unit price $R(x)$ from the buyer. The optimal mechanism must satisfy two constraints. The first is the incentive compatibility (IC) constraint, i.e., the supplier will truthfully announce his cost under the optimal mechanism:

$$c = \arg\max_x \pi_2(x) = R(x)\lambda - c\mu(x), \tag{7.6}$$

where π_2 is the supplier's profit function. The second is the individual rationality (IR) constraint, i.e., all supplier types will participate:

$$\pi_2(c) \geq 0 \text{ for all } c \in [\bar{c}, \underline{c}]. \tag{7.7}$$

An implicit assumption is that even the least efficient supplier will earn a non-negative profit that is, the buyer is willing to contract with all supplier types. The possibility of excluding certain supplier types will be discussed in Sect. 7.5

The buyer's mechanism design problem can be now written as

$$\begin{aligned} &\min_{\mu(\cdot), R(\cdot)} \int_{\underline{c}}^{\bar{c}} (R(x)\lambda + C(\mu(x))) f(x)dx \\ &\text{s.t. } (7.6), (7.7) \end{aligned} \tag{7.8}$$

where $R(x)\lambda + C(\mu(x))$ is the buyer's total cost function for a given supplier cost x. The following theorem characterizes the buyer's optimal procurement mechanism.

Theorem 2 *The buyer's optimal mechanism* $\{\mu^o(x), R^o(x)\}$ *(i.e., the solution to (6)) is characterized by*

$$C'(\mu^o(x)) = -x - F(x)/f(x) \tag{7.9}$$

$$R^o(x)\lambda = x\mu^o(x) + \int_x^{\bar{c}} \mu^o(y)dy. \tag{7.10}$$

Recall $C(\mu)$ is the buyer's optimal operating cost function for a given μ. From (7.4) and (7.8), we see that there is $\mu^o(x) \leq \mu^*(x)$. This means that the optimal mechanism induces a capacity level lower than the optimal solution of the supply chain. Generally, we do not have closed-form expression for the optimal functions $\mu^o(x)$, $R^o(x)$ and their associated total cost for the buyer. However, later we will use numerical method to evaluate the optimal mechanism.

7.3.2.2 Lead-Time Mechanism

The optimal mechanism (OM) can maximize the buyer's expected payoff function, but it is quite complex and may be difficult to implement in practice. Cachon and Zhang [16] also considers two suboptimal, but simpler mechanisms for the buyer. The first simple mechanism is called a lead-time mechanism (LT): The buyer posts a target (average) delivery lead time and sets a unit price that is, the lead-time mechanism consists of two parameters, μ^t and R^t, the supplier's required capacity and the buyer's price per unit, respectively (the superscript t denotes the lead time). Note there is a one-to-one relationship between the supplier's capacity μ^t and the average lead time $(\mu^t - \lambda)^{-1}$. The practice of specifying a target delivery performance has been widely used in service outsourcing [47] and online procurement auctions [53].

The supplier's expected profit is $\pi_2(c) = \lambda R^t - c\mu^t$ under the lead-time mechanism. Hence the optimal unit price must be $R^t(\mu^t) = \bar{c}\mu^t/\lambda$ to ensure participation. Then the buyer's total cost can be written as

$$C_1^t(\mu^t) = C\left(\mu^t\right) + \lambda R^t\left(\mu^t\right) = C\left(\mu^t\right) + \bar{c}\mu^t,$$

which is the supply chain's cost with the highest-capacity cost, $C_{sc}(\bar{c}, \mu^t)$. Thus the buyer will choose the supply chain's optimal capacity at cost $\bar{c}$: $\mu^t = \mu^*(\bar{c})$. Accordingly, the buyer pays the supplier $R^t(\mu^*(\bar{c}))$ to ensure participation. In fact, under the lead-time mechanism, the buyer sells the supply chain to the supplier and charges a price that is equal to the supply chain's optimal profit with the highest cost $\bar{c}$.

7.3.2.3 Late-Fee Mechanism

The second simple mechanism studied by Cachon and Zhang [16] is called a late-fee mechanism (LF): The buyer pays the supplier a unit price R^f and meanwhile charges the supplier η^f for each outstanding order per unit time. The superscript f stands for late fee. This mechanism is quite intuitive, and has been observed in practice [6]. For transparent analysis, we take advantage of the exponential approximation for N as described in Sect. 7.3.1. Under this approximation, recall $\hat{\mu}(c) = \lambda + \sqrt{\alpha/c}$ minimizes the supply chain's total cost $\hat{C}_{sc}$.

In the late-fee mechanism, the supplier is free to choose his capacity to maximize his own profit. It is straightforward to show that the supplier's optimal capacity depends on the late fee η^f:

$$\mu^f(c) = \lambda + \sqrt{\eta^f\lambda/c}.$$

Suppose we choose the late fee that equalizes $\mu^f(c)$ and $\hat{\mu}(c)$ (i.e., the late fee will induce a capacity that minimize the supply chain's total cost under the approximation). This requires

$$\eta^f = \alpha/\lambda. \tag{7.11}$$

Clearly, this may not be the buyer's optimal late fee she would like to charge. Nevertheless, it will be shown that this late fee yields excellent results for the buyer.

We need to check the individual rationality constraint. To ensure participation, the buyer must pay a unit price R^f such that

$$\pi_2(\bar{c}) = R^f\lambda - \bar{c}\hat{\mu}(\bar{c}) - \eta^f\left(\frac{\lambda}{\hat{\mu}(\bar{c}) - \lambda}\right) = 0,$$

which gives

$$R^f = \bar{c} + 2\sqrt{\alpha\bar{c}}/\lambda. \qquad (7.12)$$

It is worth noting that with the late fee, a base-stock policy may no longer be optimal for the buyer. This is because the buyer might have incentives to manipulate the ordering policy to take advantage of the late fee charged to the supplier. Here we assume the buyer is able to credibly commit to a base-stock policy. Alternative solution has been discussed in Cachon and Zhang [16] when the buyer is unable to make such a commitment. Assuming that the base-stock policy is used by the buyer, then the optimal base-stock level should be $s^*(\hat{\mu}(c))$ (since N does not depend on s).

7.3.3 Comparison of Mechanisms

The simple mechanisms are intuitive and easy to implement. In addition, they do not require the supplier's information as an input to determine the contract parameters. This is a desirable property for two reasons. First, in many situations the supplier may be reluctant to reveal his true efficiency level. Second, under the simple mechanisms, the contract parameters can be determined even before the supplier's cost is realized (this gives more flexibility in contract negotiation). Despite these merits of the simple mechanisms, they do not yield the optimal profit for the buyer. We investigate the performance of the simpler contracts (LF and LT) relative to the optimal mechanism (OM) in this section. This will help us quantify the value of using a more complex procurement contract in practice. We first compare the procurement contracts analytically by making some simplifying assumptions, then we use an extensive numerical study to confirm the generality of the insight from the analytical comparison.

7.3.3.1 Analytical Comparison

We compare the optimal mechanism and the late-fee mechanism analytically. The comparison of the optimal mechanism and the lead-time mechanism is similar and therefore omitted. The analysis with the actual cost functions is challenging, so we use the exponential approximation introduced in Sect. 7.3.1. Also, we focus on uniform distribution F for the supplier's cost c. Recall $[\underline{c}, \bar{c}]$ is the support of the supplier's cost distribution. Let $\bar{c} = \theta(1+\delta)$ and $\underline{c} = \theta(1-\delta)$, where θ is the mean cost and δ measures the cost variation.

With uniform distribution, there are $F(c) = \left(c - \underline{c}\right)/2\delta\theta$ $f(c) = 1/2\delta\theta$, and $F(c)/f(c) = c - \underline{c}$. The optimal mechanism in Theorem 2 satisfies

$$C'(\mu^o) = -c - F(c)/f(c) = -2c + \underline{c}. \tag{7.13}$$

By replacing $C'(\mu^o)$ with $\hat{C}'(\mu^o)$ in (7.13), we have

$$\hat{\mu}^o(c) = \lambda + \sqrt{\frac{\alpha}{2c - \underline{c}}}.$$

The buyer's operating cost can be then written as

$$\hat{C}(c) = \hat{C}(\hat{\mu}(c), c) = \sqrt{\alpha\left(2c - \underline{c}\right)}.$$

From Theorem 2, with the optimal contract the buyer's total cost is

$$\hat{C}_1^o(c) = \hat{C}(c) + c\hat{\mu}^o(c) + \int_c^{\bar{c}} \hat{\mu}(y)dy = \bar{c}\lambda + \sqrt{\alpha}\left(\frac{c}{\sqrt{2c - \underline{c}}} + \sqrt{2\bar{c} - \underline{c}}\right).$$

It can be readily shown that in the late-fee mechanism, the buyer pays the supplier $R^f\lambda = \bar{c}\lambda + 2\sqrt{\alpha\bar{c}}$, incurs an operating cost $\hat{C}(\mu^f(c)) = \sqrt{\alpha c}$, and collects total late fee $\sqrt{\alpha c}$. Thus the buyer's total cost is

$$\hat{C}_1^f = R^f\lambda = \bar{c}\lambda + 2\sqrt{\alpha\bar{c}}.$$

Note, the buyer's total cost is independent of c, which implies that with the late-fee mechanism the buyer is unable to extract any rents from efficient supplier types.

Now we are ready to compare the optimal and late-fee mechanisms. Notice that $\hat{C}_1^o(c)$ is increasing in c and $\hat{C}_1^o(\underline{c}) < \hat{C}_1^f < \hat{C}_1^o(\bar{c})$. Therefore, the buyer's expected cost with the optimal mechanism, $E[\hat{C}_1^o(c)]$, is approximately equal to the buyer's expected cost with the late-fee mechanism, $\hat{C}_1^f$, if $\hat{C}_1^o(c)$ is relatively flat. This is manifested by considering the following ratio:

$$\frac{\hat{C}_1^o(\bar{c})}{\hat{C}_1^o(\underline{c})} = \frac{\bar{c}\lambda + \sqrt{\alpha}\left(\frac{\bar{c}}{\sqrt{2\bar{c}-\underline{c}}} + \sqrt{2\bar{c} - \underline{c}}\right)}{\bar{c}\lambda + \sqrt{\alpha}\left(\sqrt{\underline{c}} + \sqrt{2\bar{c} - \underline{c}}\right)} < \frac{2 + 4\delta}{\sqrt{(1+3\delta)(1-\delta)} + (1+3\delta)}, \tag{7.14}$$

where the inequality follows because $\bar{c} > \sqrt{\underline{c}}\sqrt{2\bar{c} - \underline{c}}$. The right-hand side of (7.14) equals 1.025 and 1.075 with $\delta = 0.2$ and $\delta = 0.4$, respectively. So even if the supplier's cost can vary up to 40% around its mean ($\delta = 0.4$) and the demand rate is extremely small, in the optimal mechanism the buyer's total cost with the highest-cost supplier is no more than 7.5% higher than that with the lowest-cost supplier.

This suggests that the cost function is indeed flat, i.e., $\hat{C}_1^f$ in that case cannot be more than 7.5% higher than $E[\hat{C}_1^o(c)]$.

The above analysis suggests that the buyer's total cost is relatively insensitive to the supplier's capacity cost with the optimal mechanism that is, asymmetric information conveys substantial protection to a supplier: An efficient supplier is able to keep essentially all the benefit from having a low cost, even when the buyer adopts the complex, optimal mechanism. This implies that the optimal mechanism is not very effective in extracting the efficiency rents from the supplier. As a result, the late-fee mechanism performs very well even if it extracts no rents at all.

Finally, it is worth pointing out that the above observation is not because the supply chain's cost function is flat. With the supply chain's optimal cost we obtain the following ratio:

$$\frac{\hat{C}_{sc}^*(\bar{c})}{\hat{C}_{sc}^*(\underline{c})} = \frac{(1+\delta)\,\lambda + 2\sqrt{\alpha(1+\delta)}}{(1-\delta)\,\lambda + 2\sqrt{\alpha(1-\delta)}} > \sqrt{\frac{1+\delta}{1-\delta}}. \tag{7.15}$$

The value of the ratio is 1.22 and 1.53 for $\delta = 0.2$ and $\delta = 0.4$, respectively, that is, the supply chain's cost with the least efficient supplier could be much higher than that with the most efficient supplier (53% higher when $\delta = 0.4$).

7.3.3.2 Numerical Study

To confirm the results from the analytical comparison, Cachon and Zhang [16] presents a comprehensive numerical study with the following design: $h = 1$, $\lambda \in \{0.1, 1, 10, 100\}$, $b \in \{3, 40, 200\}$, c follows either a uniform or a normal distribution on the interval $[\underline{c}, \bar{c}]$, where $\underline{c} = \theta(1-\delta)$ and $\bar{c} = \theta(1+\delta)$, $\theta \in \{0.5, 5, 50, 200\}$ and $\delta \in \{0.05, 0.1, 0.2\}$. The value of δ measures the magnitude of cost uncertainty. For instance, $\delta = 0.05$ corresponds to reasonably small uncertainty in supplier's cost (within 5% of forecast) and the scenarios with $\delta = 0.20$ represent high uncertainty. With normal cost distribution, the mean is set to be θ and the standard deviation $\delta\theta/4$. The value of h is fixed because the buyer's total cost depends only on the relative magnitude of the parameters c, b, and h. The backorder cost b is allowed to range from a low value of three times h to a high value of two hundred times h. The mean capacity costs range from $\theta = 0.5$ to $\theta = 200$, which corresponds to very low and very high utilizations, respectively. There are totally 144 scenarios in this numerical study for each cost distribution.

The results from the numerical study are summarized in Table 7.1. The percentiles of the buyer's percentage cost increase in each simple mechanism relative to the optimal mechanism (OM) are listed. For example, with the lead-time mechanism (LT), the 90th percentile of the buyer's percentage cost increase relative to the optimal mechanism is only 0.24% that is, if the buyer adopts the lead-time mechanism rather than the optimal one, the percentage cost increase is less than 0.24% for 90 percent of the scenarios. Among all tested scenarios, the buyer's maximum percentage cost increase is 3.53% (with the LF mechanism and normal cost distribution).

Table 7.1 Buyer's percentage cost increase relative to the optimal mechanism (with a single supplier)

		Average (%)	90th percentile (%)	Maximum (%)
Uniform cost distribution	LT	0.09	0.24	0.56
	LF	0.20	0.19	2.85
Normal cost distribution	LT	0.50	1.20	2.39
	LF	0.60	1.38	3.53

Table 7.2 Supply chain's percentage cost increase relative to the supply chain optimal solution, i.e., the supply chain inefficiency or the value of renegotiation (with uniform cost distribution)

		Average (%)	90th percentile (%)	Maximum (%)
Single supplier	OM	0.08	0.23	0.51
	LT	0.10	0.30	0.67
	LF	0.25	0.36	3.53
$n = 2$ suppliers	OM	0.05	0.14	0.30
	LT	0.06	0.18	0.41
	LF	0.25	0.39	3.78

Overall, Table 7.1 shows that the lead-time and late-fee mechanisms perform quite well, for both uniform and normal cost uncertainties.

Why does the optimal mechanism perform poorly in extracting the supplier's information rent in this model setting? An intuitive explanation is as follows. Note that the market demand is exogenously given (and so is the supply chain's revenue). Inducing truth telling by using a cost-contingent service-level can reduce the supply chain's operating cost for satisfying the market demand; however, it is the buyer who is responsible for such an operating cost. Hence the supplier has little incentive to share the true cost information unless he will keep essentially all the benefits from a low-cost realization.

Cachon and Zhang [16] also considers the firms' incentives to renegotiate the contract in such a problem setting. In particular, they compare the supply chain performance under these mechanisms to the supply chain's centralized optimal performance using the same scenarios above. The result is given in Table 7.2 (the case with two suppliers is also presented; see Sect. 7.3.4.1 for the analysis of multiple competing suppliers). Interestingly, the numerical study indicates that these mechanisms are generally associated with very small supply chain inefficiency, i.e., there is a negligible renegotiation value for the supply chain firms. This also suggests that supply chain coordination can be nearly achieved even when the buyer is self-interested and chooses her own procurement mechanism under information asymmetry.

There are a couple of additional observations in the more detailed numerical analysis in Cachon and Zhang [16]. First, it has been found that η^f is nearly optimal from the buyer's standpoint (recall η^f is derived by equating the induced capacity with the supply chain's optimal capacity), and therefore it is not necessary to search for the true optimal late fee. However, this approximation causes the LT mechanism to perform slightly better than the LF mechanism as shown in Table 7.1. Second, they also test extremely large cost uncertainty with $\delta = 0.40$ and obtain similar results. This confirms the robustness of the results in Tables 7.1 and 7.2

7.3.4 Extensions

7.3.4.1 Multiple Competing Suppliers

We first extend the basic model to include multiple competing suppliers. In this case, the buyer may select the most efficient supplier through competitive bidding. Suppose there are $n \geq 2$ suppliers. Each supplier's cost c^i is a random draw from a common distribution F (density f). Let $\mathbf{c} = (c^1, \cdots, c^n)$ be the cost vector of the suppliers. Similar to the single-supplier case, the buyer's optimal mechanism consists of a menu of contracts, $\{q^i(\cdot), \mu^i(\cdot), R^i(\cdot)\}, i = 1, 2, ..., n$. Under this mechanism, if the suppliers announce their costs to be $\mathbf{x} = (x^1, \cdots, x^n)$, then supplier i is the winner with probability $q^i(\mathbf{x}) \geq 0$ and $\sum q^i(\mathbf{x}) = 1$, supplier i receives a unit price $R^i(\mathbf{x})$ from the buyer, and finally, the winner builds capacity $\mu^i(\mathbf{x})$.

To derive the buyer's optimal mechanism, we start with the suppliers' bidding behavior. Each supplier aims to maximize his own profit. Thus for supplier i we have

$$\max_{x^i} \pi_2^i = E_{\mathbf{x}^{-i}}[R^i(\mathbf{x})\lambda - q^i(\mathbf{x})c^i\mu^i(\mathbf{x})],$$

where $\mathbf{x}^{-i} = (x^1, \cdots, x^{i-1}, x^{i+1}, \cdots, x^n)$ and π_2^i is supplier i's expected profit. Again, by the revelation principle, we have the following incentive compatibility and individual rationality constraints:

$$c^i = \arg\max_{x^i} \pi_2^i(x^i) \tag{7.16}$$

and

$$\pi_2^i(c^i) \geq 0. \tag{7.17}$$

Hence the buyer's problem can be written as

$$\begin{aligned} \min_{\{q^i(\cdot),\mu^i(\cdot),R^i(\cdot)\}} \quad & E_{\mathbf{c}}\left\{\sum_i R^i(\mathbf{c})\lambda + \sum_i [q^i(\mathbf{c})C(\mu^i(\mathbf{c}))]\right\} \\ \text{s.t. (7.16) and (7.17)} & \end{aligned} \tag{7.18}$$

Theorem 3 *He buyer's optimal mechanism with $n \geq 2$ competing suppliers is as follows: The buyer offers the suppliers identical menu of contracts $\{q^o(\cdot), \mu^o(\cdot), R^o(\cdot)\}$ characterized by*

$$C'(\mu^o(x)) = -x - F(x)/f(x),$$

$$R^o(x)\lambda = (1 - F(x))^{n-1} x\mu^o(x) + \int_x^{\bar{c}} (1 - F(y))^{n-1} \mu^o(y)dy$$

$$q^o(x) = \begin{cases} 1 \text{ if } x = \min(x^1, \cdots, x^n) \\ 0 \text{ otherwise} \end{cases}.$$

The suppliers announce their true costs and the most efficient supplier will be chosen.

A couple of observations can be made about the above optimal mechanism with $n \geq 2$ suppliers. First, the capacity function $\mu^o(x)$ is independent of the number of suppliers, n. This suggests that as in the single-supplier case, the optimal mechanism results in less capacity than optimal for the supply chain when there are multiple competing suppliers. Second, this optimal mechanism does not have an intuitive format. Note that every losing bidder will receive a payment even though they do not build any capacity for the buyer. However, it can be shown that such a strange mechanism can be implemented with an equivalent auction where only the winner receives a payment (see [65] for details).

A natural extension of the lead-time mechanism with a single supplier to $n \geq 2$ potential suppliers is as follows: The buyer announces the lead time that the winning supplier must deliver and then ask the suppliers to bid on price. In particular, we assume the buyer uses a second-bid auction (i.e., the supplier with the lowest bid wins but he only needs to fulfil the second-lowest bid). Zhang [65] considers first-bid auctions and shows that they yield the same expected payoff for the buyer (i.e., revenue equivalence holds in this case). However, the revenue equivalence may fail to hold if the buyer specifies a price and asks the suppliers to bid on lead time; this is because the lead-time bids have different variances in the first-bid and second-bid auctions. Since there is a one-to-one relationship between the average lead time and the supplier's capacity, it is equivalent for the buyer to announce a required capacity, μ^t, rather than a lead time.

Theorem 4 *In the lead-time mechanism with $n \geq 2$ competing suppliers, the weakly dominant strategy for a supplier with capacity cost x is to bid $R^t(x) = \mu^t x/\lambda$. The buyer's expected total cost is convex in μ^t.*

Alternatively, the buyer may specify a late fee η^f for each outstanding order and ask the suppliers to bid on price. Again we focus on the second-bid auctions. Because the suppliers' capacity choice depends on the late fee but not the price bid, the winner with a cost x will choose capacity $\hat{\mu}(x)$. As with one potential supplier, we assume the buyer sets $\eta^f = \alpha/\lambda$. Although this is not the buyer's optimal late fee, later we show that it performs very well.

Theorem 5 *In the late-fee mechanism with $n \geq 2$ competing suppliers, the suppliers' dominant strategy is to bid $R^f(x) = \hat{C}^*_{sc}(x)/\lambda$ and the winner chooses capacity $\mu^f(x) = \hat{\mu}(x)$.*

Table 7.3 Buyer's percentage cost increase relative to the optimal mechanism (with $n = 2$ suppliers)

		Average (%)	90th percentile (%)	Maximum (%)
Uniform cost distribution	LT	0.18	0.31	0.59
	LF	0.32	0.31	3.27
Normal cost distribution	LT	0.02	0.06	0.12
	LF	0.23	0.27	3.40

Finally, a numerical study is used to compare the mechanisms with $n \geq 2$ suppliers. Since the qualitative results are similar for different n values, here we focus on $n = 2$. The parameter values are the same as in Sect. 7.3.3.2. Table 7.3 summarizes the results from this numerical study. We see that the simple mechanisms again are nearly optimal when there are multiple competing suppliers.

7.3.4.2 Generalized Holding Cost

We have assumed in the basic model that the buyer's inventory holding cost is a constant and independent of the procurement price. Now we consider a generalized holding cost. Let h be a function of the unit cost, $h = h_0 + rv$, where h_0 is a constant representing the physical holding cost, r is the interest rate and v is the buyer's effective unit cost. Notice that the effective unit cost may not equal the unit price R. For example, the buyer's effective unit cost with a late fee is the unit price minus the late fee. The buyer's operating cost is

$$C(\mu, v) = (h_0 + rv)\left[\frac{1 - \ln\left(\left(\frac{h_0+rv}{h_0+rv+b}\right)\left(\frac{\mu/\lambda-1}{\ln \mu/\lambda}\right)\right)}{\ln \mu/\lambda} - \frac{1}{\mu/\lambda - 1}\right]. \tag{7.19}$$

The evaluation of the optimal mechanism with this new holding cost structure is quite difficult. There are two major complications. First, $C(\mu, v)$ is not always jointly convex. Second, the payment and the capacity functions are interdependent in the optimality conditions (i.e., they cannot be solved separately). As a result, we need to evaluate the optimal mechanism via a full enumeration over the contract space. Specifically, we can only determine the optimal mechanism when the suppliers' costs are drawn from a discrete distribution and the suppliers are only allowed to choose capacities from a discrete set.

The process for evaluating the lead-time mechanism does not require an adjustment due to the generalized holding cost. On the other hand, the late-fee mechanism requires an adjustment because the supply chain optimal capacity with the exponential approximation, $\hat{\mu}(c)$, no longer takes a simple form proportional to $\sqrt{1/c}$. We first find the capacity that minimizes the supply chain's cost when $c = \theta$ (recall θ is the mean of the cost distribution):

$$\mu_\theta = \arg\min_\mu (C(\mu, \theta) + \theta\mu).$$

Table 7.4 Buyer's percentage cost increase relative to the optimal mechanism (with generalized holding cost $h = h_0 + rv$)

		Average (%)	90th percentile (%)	Maximum (%)
Single suppliers	LT	0.15	0.66	2.22
	LF	0.18	0.64	3.70
n=2 suppliers	LT	2.61	4.53	12.19
	LF	2.61	4.60	11.87

Because there is no closed-form solution for μ_θ, a one-dimensional search is needed to find μ_θ. Given the late fee η^f!, the supplier's optimal capacity is $\mu^f = \lambda + \sqrt{\eta^f \lambda/\theta}$. Equating μ_θ with μ^f yields

$$\eta^f = \theta(\mu_\theta - \lambda)^2/\lambda.$$

Hence, we set the late fee to coordinate the supply chain with the average cost supplier. With a single potential supplier, R is chosen so that the high-cost supplier earns zero profit:

$$R^f = \bar{c} + 2\sqrt{\eta^f \bar{c}/\lambda}.$$

With two or more suppliers, an auction sets the price R^f. To test this version of the late-fee mechanism, we take the original set of 144 scenarios and add three interest rate levels, $r = \{0.05, 0.1, 0.2\}$, to arrive at 432 scenarios. For each scenario, we divide the cost support $[\theta(1-\delta), \theta(1+\delta)]$ into $m-1$ equal intervals and assume each are equally likely (i.e., a discrete uniform distribution). Similarly, we divide the range $[1.1\lambda, 10\lambda]$ into $l-1$ equal intervals and use the l interval boundaries as the feasible capacities. As m and l are increased, our discrete problem approaches the continuous problem we studied with a fixed holding cost. However, as we already mentioned, the computational burden increases rapidly with m and l. In our numerical study, we set m = 5 and l = 20.

Table 7.4 shows the performance of the lead-time and late-fee mechanisms relative to the optimal mechanisms. Even with this holding cost, both mechanisms are nearly optimal with a single supplier. With multiple suppliers the mechanisms perform well, but now the average cost increase is a noticeable 2.61% with either mechanism. We suspect that this gap with the optimal mechanism is in large part due to the coarse discretization because the gap decreases quickly as the number of supplier cost realizations (m) increases (a sample of the scenarios $m = 7$ and $m = 9$ have been tested). Overall, we conclude that the lead-time and late-fee mechanisms perform reasonably well when the holding cost is a linear function of the buyer's procurement cost.

Table 7.5 Buyer's percentage cost increase relative to the optimal mechanism (with uniform cost distribution and $s = 0$)

		Average (%)	90th percentile (%)	Maximum (%)
Single supplier	LT	0.13	0.40	0.55
	LF	0.03	0.08	0.12
$n = 2$ suppliers	LT	0.21	0.43	0.51
	LF	0.14	0.20	0.24

7.3.4.3 Make-to-Order Buyer

In the basic model the buyer can hold inventory as a buffer to mitigate the consequence of slow delivery from the supplier. One may wonder whether the main results continue to hold when the buyer cannot hold inventory (e.g., the buyer is a make-to-order manufacturer). This also covers an important situation where the buyer is a service provider. Hence, in this section we investigate the performance of the lead-time and late-fee mechanisms when s = 0.

Consider the single-supplier case. Now the buyer's operating cost only consists of the waiting cost: $C(\mu) = \lambda b/(\mu - \lambda)$. We can derive the optimal mechanism as in Theorem 2. The simple mechanisms for the buyer are as follows. The buyer's optimal lead-time mechanism specifies capacity $\mu^t = \lambda + \sqrt{b\lambda/\bar{c}}$ and charges unit price $R^t(\mu^t) = \bar{c} + \sqrt{b\bar{c}/\lambda}$. In the optimal late-fee mechanism, the late fee η^f and R^f are given by

$$\eta^f = \left(\frac{E\left(\sqrt{c}\right)}{2\sqrt{\bar{c}} - E\left(\sqrt{c}\right)}\right)b, \tag{7.20}$$

$$R^f = \bar{c} + 2\sqrt{\eta^f \bar{c}/\lambda}. \tag{7.21}$$

The mechanisms with multiple suppliers can be derived similarly. The details can be found in Cachon and Zhang [16].

Table 7.5 reports the results from a numerical study using the same scenarios defined in Sect. 7.3.3.2 and uniform cost distribution. We see that both the lead-time and the late-fee mechanisms perform well relative to the optimal mechanism. Thus the simple mechanisms continue to perform well even if the buyer is unable to use inventory to buffer the supplier's lead-time performance.

7.4 Contracting with Make-to-Stock Supplier

We proceed in this section to study another common situation that may arise in practice: The supplier adopts a make-to-stock strategy and thus may hold inventory to improve customer service. Specifically, we model the supply chain as a two-echelon inventory system, which requires quite different analysis from that in Sect. 7.3

7.4.1 Basic Model

A buyer needs an input (say, a product) from a supplier to satisfy market demand. The supplier's marginal production cost c is private information and modeled as a random draw from a log-concave distribution F (density f) with support $[\underline{c}, \bar{c}]$. The buyer knows the distribution function F, but not the exact cost at the supplier. Thus F represents the cost uncertainty faced by the buyer. This information structure is the same as in Sect. 7.3 except that now c is the production cost rather than the cost rate for maintaining a certain capacity.

We consider a two-echelon inventory system under periodic review. Again we use subscripts 1 and 2 for the buyer and supplier, respectively. The supplier either manufactures the product or obtains it from an ample external source. Assume there is a constant production or replenishment lead time L_2 for the supplier. The transportation time from the suppliers to the buyer is L_1, which may also represent the assembly time at the buyer. To maintain tractability, assume both the supplier and the buyer adopt stationary base-stock policies. Let s_1 and s_2 denote the local base-stock levels adopted by the buyer and the supplier. We define the service level as the fill rate provided by the supplier (i.e., the expected percentage of an order that can be fulfilled immediately). It is clear that there is a one-to-one relationship between the base-stock level s_2 and the fill rate that is, all else being equal, a more responsive delivery performance is equivalent to a higher base-stock level s_2.

The firms incur linear inventory holding cost. The buyer's holding cost rate is h_1, and the supplier's holding cost rate is $h_2 = h_0 + rc$, where h_0 is a constant and r is the interest rate. Here we follow the basic model in Sect. 7.3.1 to assume that h_1 is independent of the procurement price. This applies to consignment arrangements under which the buyer pays the supplier is sold. Section 7.3.4.2 demonstrates that generalizing this holding cost assumption does not change the qualitative insight (with a make-to-order supplier).

The buyer faces a random demand in each period and the demand distribution is i.i.d. across periods. Use D^τ to denote the demand over τ periods, so D^{L_j} is the lead-time demand for stage j $(j = 1, 2)$. Define $\omega^\tau = E(D^\tau)$ as the mean demand over τ periods. Let ω denote the single-period mean demand. Let Φ^τ and ϕ^τ be the cumulative distribution function and density function, respectively, of demand over τ periods. Assume $\Phi^\tau(x)$ is differentiable for all positive integers τ. In other words, the demand has continuous density. Furthermore, assume $\Phi^1(0) = 0$, so only positive demand occurs in each period. Unmet demand in each period is backlogged and the buyer incurs a backorder cost b for each backlogged unit.

The timing of the model is similar to that with a make-to-order supplier (see Sect. 7.3.1). The only difference is that now the supplier chooses the base-stock level s_2 instead of the capacity level μ. Both firms are risk-neutral in this model. The buyer wants to minimize the expected total cost in each period, i.e., the procurement price plus the operating cost. The supplier's objective is to maximize the expected profit per period, which equals the procurement price paid by the buyer minus the production and inventory holding costs. For concision, define $x \wedge y = \min(x, y)$ and $x \vee y = \max(x, y)$.

This model is based on the two-echelon inventory system studied by Zhang [67]. He considers a more general setting where the buyer can set price to influence market demand. In this section we focus on a special case with exogenous demand. This simplifies the analysis but sharpens the insight we would like to emphasize. We will follow the cost accounting scheme proposed by Chen and Zheng [23] for the analysis.

7.4.2 Buyer's Procurement Mechanisms

7.4.2.1 Optimal Mechanism

As preparation, we first derive the buyer's optimal base-stock level s_1 given the supplier's base-stock level s_2. Define

$$\tilde{G}_1(x) = h_1(x)^+ + b(x)^-.$$

Let y be the buyer's inventory position in period t. Then the cost incurred at the buyer in period $t + L_1$ is:

$$\begin{aligned} G_1(y) &= E[\tilde{G}_1(y - D^{L_1+1})] \\ &= h_1(y - \omega^{L_1+1}) + (h_1 + b) \int_y^{\infty} (x - y)\phi^{L_1+1}(x)dx. \end{aligned}$$

It can be readily shown that $G_1(y)$ is strictly convex. Let $H_1(s_1, s_2)$ denote the buyer's operating cost function (inventory holding cost plus backorder cost). Using function G_1, $H_1(s_1, s_2)$ can be written as

$$H_1(s_1, s_2) = E[G_1(s_1 \wedge (s_1 + s_2 - D^{L_2}))], \tag{7.22}$$

where $s_1 + s_2$ is the echelon base-stock level for the supplier. It is straightforward to show that $H_1(s_1, s_2)$ is convex in s_1 and, hence, there is a unique optimal base-stock level $s_1(s_2)$ for given s_2. We have the following Lemma.

Lemma 1

(i) *$s_1(s_2)$ is decreasing in s_2;*
(ii) *$s_1(s_2) \to \hat{s}_1$ as $s_2 \to \infty$, where $\hat{s}_1 = (\Phi^{L_1+1})^{-1}(b/(h_1 + b))$.*

It can be shown that $H_1(s_1(s_2), s_2)$ may not be convex in s_2. However, we can prove the following useful properties about $H_1(s_1(s_2), s_2)$.

Lemma 2

(i) $\frac{dH_1(s_1(s_2),s_2)}{ds_2} = 0$ *for* $s_2 = 0$ *and* $\frac{dH_1(s_1(s_2),s_2)}{ds_2} < 0$ *for* $s_2 > 0$;
(ii) $\left(\frac{dH_1(s_1(s_2),s_2)}{ds_2}\right) / \Phi^{L_2}(s_2)$ *is increasing in* s_2.

Now we derive the buyer's optimal mechanism. Suppose the buyer offers a menu of contracts $\{s_2(\cdot), T(\cdot)\}$ to the supplier. If the supplier announces the cost to be x (x is not necessarily the true cost c), then he is supposed to receive a unit price $T(x)$ and adopt a base-stock level $s_2(x)$. Define

$$H_2(c, s_2) = (h_0 + rc)\{E[s_2 - D^{L_2}]^+ + \omega^{L_1}\}$$

as the supplier's operating cost (i.e., inventory holding cost). Given the above contract, the supplier's profit is as follows:

$$\pi_2(x) = \omega(T(x) - c) - H_2(c, s_2(x)).$$

According to the revelation principle, we only need to consider the truth-inducing contracts (i.e., the IC constraint):

$$c = \arg\max_x \pi_2(x). \tag{7.23}$$

In addition, the buyer needs to make sure that the supplier will accept the contract even with the highest cost $\bar{c}$ (i.e., the IR constraint):

$$\pi_2(\bar{c}) \geq 0. \tag{7.24}$$

The buyer's problem is then given by

$$\min_{\{s_2(\cdot),T(\cdot)\}} \int_{\underline{c}}^{\bar{c}} [H_1(s_1, s_2(x)) + \omega T(x)] f(x)dx \tag{7.25}$$

s.t. (7.23) and (7.24).

Theorem 6 *The buyer's optimal menu of contracts* $\{s_2^o(\cdot), T^o(\cdot)\}$ *(i.e., the solution to* (7.25)*) is characterized by*

$$\frac{dH_1(s_1(s_2^o), s_2^o)/ds_2^o}{\Phi^{L_2}(s_2^o)} = -[h_0 + rx + rF(x)/f(x)] \tag{7.26}$$

$$\omega T^o(x) = \omega\bar{c} + H_2(\bar{c}, s_2^o(\bar{c})) - \int_x^{\bar{c}} (h_0 + ry)s_2^{o\prime}(y)\Phi^{L_2}(s_2^o(y))dy. \tag{7.27}$$

Because the left-hand side of Eq. (7.26) is increasing in s_2 (see Lemma 2) and the right-hand side is decreasing in x, we know that $s_2^o(x)$ is a decreasing function in the optimal mechanism. This implies that in the optimal mechanism, a less efficient supplier will use a lower stocking level. Although a closed-form solution to the stocking level function $s_2^o(x)$ is not available, we are able to show in the next theorem that in the optimal mechanism, the buyer tends to induce a fill rate that is lower than the supply chain optimal. This distortion is because the buyer wants to reduce the information rent paid to the supplier to minimize cost. Let (s_1^*, s_2^*) denote the supply chain's optimal inventory policy, where s_1^* and s_2^* are the local base-stock levels at the buyer and the supplier, respectively.

Theorem 7 *The stocking level specified by (7.26) in the optimal mechanism is lower than the supply chain's optimal stocking level, i.e., $s_2^o(x) \leq s_2^*(x)$ for all x.*

7.4.2.2 Fixed Service-level Contract

A widely observed supply agreement in practice is that the buyer specifies a service-level requirement and compensates the supplier with a price. Call this a fixed service-level contract (FS). Practical examples of contracts involving inventory service-level agreements can be found in Katok et al. [42], Thomas [58], Thonemann et al. [59]. In such a contract, the buyer needs to propose a price so that the supplier is willing to participate. Note that specifying a fill rate is equivalent to specifying the base-stock level s_2. Given a base-stock level s_2, the supplier's total cost per period is

$$C_2(s_2) = c\omega + H_2(c, s_2).$$

To ensure supplier participation, the buyer needs to pay the supplier

$$\omega T = \bar{c}\omega + H_2(\bar{c}, s_2).$$

The buyer's total cost in fixed service-level contract is then given by

$$C_1(s_2) = H_1(s_1(s_2), s_2) + \bar{c}\omega + H_2(\bar{c}, s_2).$$

It can be readily shown that $C_1(s_2)$ is quasiconvex and has a unique global minimizer.

To minimize total cost, the buyer will choose an s_2 which is optimal for the supply chain with production cost $\bar{c}$. Hence in the fixed service-level contract, the buyer essentially sells the supply chain to the supplier and charges a price that is equal to the supply chain's optimal profit with cost $\bar{c}$.

7.4.3 Comparison of Mechanisms

In this subsection, we compare the performances of the two procurement mechanisms in Sect. 7.4.2. Because obtaining closed-form expressions for the buyer's profit is difficult in such an inventory system, we concentrate on a numerical study for the

Table 7.6 Buyer's percentage cost increase relative to the optimal mechanism (with a single supplier)

		Average (%)	90th percentile (%)	Maximum (%)
Uniform cost distribution	FS	0.01	0.03	0.07
Normal cost distribution	FS	0.08	0.18	0.42

comparison. The following parameter values are used. Fix $h_0 = 1$ (the result of the numerical study depends only on the relative magnitude of the parameter values). We assume the supplier's cost c follows either a uniform distribution or normal distribution on the support $[\underline{c}, \bar{c}]$, where $\underline{c} = \theta(1-\delta)$ and $\bar{c} = \theta(1+\delta)$, $\theta \in \{10, 100, 1000\}$ and $\delta \in \{0.1, 0.2, 0.3\}$. With normal distribution, we set the standard deviation to be $\delta\theta/4$. Note that with $\delta = 0.3\theta$, the ratio $\bar{c}/\underline{c} \approx 1.86$, which represents an unusually high uncertainty in the supplier's production cost. As to the supplier's holding cost, we set $r \in \{1\%, 5\%, 10\%\}$. In view of the 15% annual opportunity cost rate commonly found in textbooks, $r = 10\%$ is quite large for a biannually reviewed inventory model, not to mention even shorter review periods. In the numerical study, we set $h_1 = \alpha_h(h_0 + r\bar{c})$ and choose $\alpha_h \in \{2, 4\}$ (this implies $h_1 > h_2$, which is commonly assumed in the literature). Similarly, we let $b = \alpha_b h_1$ and choose $\alpha_b \in \{1, 10, 40\}$. Finally, the single-period demand follows a normal distribution with a mean of 20 and a standard deviation of 5 (there is a negligible probability for negative demand), and the lead times are $L_1 = 4$ and $L_2 \in \{2, 4, 8\}$. There are 486 combinations in total in this numerical study. Among the 486 parameter combinations, the expected fill rate in the optimal mechanisms ranges from 64% to 98%.

Table 7.6 presents the percentage cost increase of the FS mechanism relative to the optimal mechanism (OM). We see that the simple mechanism is nearly optimal: The maximum cost increase among all tested scenarios is only 0.42%. This observation has been confirmed by additional numerical experiments conducted in [67] (e.g., the demand follows a gamma distribution and the supplier's cost has a highly skewed two-point distribution).

Table 7.7 provides data on the supply chain efficiency in different procurement mechanisms (OM and FS). (The analysis for $n=2$ suppliers is given in the next section.) The data represents the percentiles of the percentage cost increase in the supply chain for different mechanisms, as compared to the supply chain optimal solution. Interestingly, both mechanisms create negligible inefficiency to the supply chain: The maximum efficiency loss for the supply chain is 0.09% in this numerical study.

In sum, we have extended the results from the previous (Sect. 7.3 with a make-to-order supplier) to the situation with a make-to-stock supplier that is, the simple mechanism is quite attractive from both the buyer's and the system's perspectives.

Table 7.7 The expected supply chain inefficiency in different procurement mechanisms (with uniform cost distribution)

		Average (%)	90th percentile (%)	Maximum (%)
Single supplier	OM	0.01	0.03	0.08
	FS	0.01	0.04	0.09
n = 2 suppliers	OM	0.01	0.02	0.06
	FS	0.01	0.03	0.07

7.4.4 Multiple Competing Suppliers

As in Sect 7.3.4.1, we may extend the basic model to consider $n \geq 2$ potential suppliers. In this case, the buyer can design a competition mechanism for the suppliers and select the most efficient one to source from. The same notations in Sect 7.3.4.1 will be used here.

First we derive the optimal mechanism with $n \geq 2$ suppliers. Consider the following menu of contracts: $\{q^i(\cdot), s_2^i(\cdot), T^i(\cdot)\}, i = 1, 2, ..., n$. That is, if the suppliers announce their costs to be $\mathbf{x} = (x^1, ..., x^n)$, then supplier i is the winner with probability $q^i(\mathbf{x}) \geq 0$; supplier i receives a payment $\omega T^i(x^i)$; the winner serves the buyer by using a base-stock level $s_2^i(x^i)$; and the losers do nothing but enjoy the payment.

The analysis starts with the supplier's bidding behavior. Supplier i maximizes the expected profit:

$$\max_{x^i} \pi_2^i = E_{\mathbf{x}^{-i}}[\omega T^i(x^i) - q^i(\mathbf{x})(c^i\omega + H_2(c^i, s_2^i(x^i))].$$

We focus on truth inducing mechanisms, i.e.,

$$c^i = \arg\max_{x^i} \pi_2^i(x^i). \tag{7.28}$$

The participation constraint or individual rationality constraint is

$$\pi_2^i(c^i) \geq 0. \tag{7.29}$$

Then the buyer's problem is

$$\min_{\{q^i(\cdot), s_2^i(\cdot), T^i(\cdot)\}} E_{\mathbf{c}} \left\{ \sum_i^n \omega T^i(c^i) + \sum_i^n [q^i(\mathbf{c}) H_1(s_1, s_2^i(c^i))] \right\}. \tag{7.30}$$

s.t. (7.28) and (7.29)

The following theorem gives the solution to (7.30).

Theorem 8 *The buyer's optimal mechanism with $n \geq 2$ suppliers is as follows: The buyer offers the suppliers identical menu of contracts $\{q^o(\cdot), s_2^o(\cdot), T^o(\cdot)\}$ characterized by*

Table 7.8 Buyer's percentage cost increase relative to the optimal mechanism (with n = 2 suppliers)

		Average (%)	90th percentile (%)	Maximum (%)
Uniform cost distribution	FS	0.58	0.97	1.06
Normal cost distribution	FS	0.15	0.26	0.28

$$\frac{dH_1(s_1(s_2^o), s_2^o)/ds_2^o}{\Phi^{L_2}(s_2^o)} = -[h_0 + rx + rF(x)/f(x)]$$

$$\omega T^o(x) = (1 - F(x))^{n-1}[\omega\bar{c} + H_2(\bar{c}, s_2^o(\bar{c}))] - \int_x^{\bar{c}} (1 - F(x))^{n-1}[(h_0 + ry)s_2^{o\prime}(y)\Phi^{L_2}(s_2^o(y))]dy$$

$$q^o(x) = \begin{cases} 1 \text{ if } x = \min(x^1, \cdots, x^n) \\ 0 \text{ otherwise} \end{cases}.$$

The suppliers announce their true costs and the most efficient supplier will be chosen.

Next we derive the fixed service-level mechanism with $n \geq 2$ suppliers. In this mechanism, the buyer specifies the fill rate (or s_2 equivalently) and asks the suppliers to bid on price. Without losing generality, we focus on the second-bid auction (it is straightforward to verify that the first-bid auction yields the same total cost for the buyer). Under this auction, we can show that each supplier will bid the break-even price (i.e., a supplier with cost x will bid price $x + H_2(x, s_2)/\omega$) and the buyer's total cost is quasiconvex in s_2.

Finally, we compare the performances of the above mechanisms with $n = 2$ suppliers. The same scenarios in Sect. 7.4.3 are considered and the results are summarized in Table 7.8. Again, the simple mechanism is nearly optimal in this numerical study. The results indicate that when both procurement price and logistics performance are taken into consideration, buyers can simply post a target service level and then ask suppliers to bid only on their costs. This seems to be consistent with practical observations: In some B2B industrial exchanges, buyers post-product specifications and logistics requirements and then ask suppliers to bid on contracts [53]. It helps us understand the challenging online procurement auction design problems, especially when multiple factors that may affect the buyer-supplier relationship are taken into account ([31] for detailed discussion).

7.5 Discussion

This section discusses several issues related to our procurement problem. First, we have assumed that the buyer is willing to contract with all supplier types including the least efficient one. This assumption is reasonable when the supplier has gone through a rigorous screening process and his cost will be within an acceptable range. However, in many other situations, the buyer may choose not to transact with the supplier if his cost is prohibitively high. For instance, the buyer may utilize a cut-off policy to exclude highly inefficient supplier types. Zhang [67] considers the impact of such a cut-off policy in the setting described in Sect. 7.4. It has been found that a fixed service-level (FS) contract continues to perform very well. Similar findings can be obtained for the setting in Sect. 7.3 Thus the results from Sects. 7.3 and 7.4 are robust when the cut-off policy is allowed.

The observation that using a fixed service level is nearly optimal gives rise to an interesting question: Can we generalize such a result to other procurement attributes? That is, can we always use a fixed attribute rather than a complex menu in supply contract design under asymmetric information? The answer is not positive. Zhang [67] includes price-sensitive demand in the above two-echelon inventory system. In his model, the buyer can set price to influence market demand; or equivalently, the buyer needs to determine the purchase quantity in each period. He shows that using a fixed purchase quantity may cause significant profit loss for the buyer, which implies that a complex menu for the quantity attribute could be highly valuable. He then proceeds to investigate why the service-level and quantity attributes have distinct implications for supply contract design. The following intuition has been offered. For both model settings in this chapter, the market demand is exogenously given; and, because using a cost-contingent service level only reduces the supply chain's operating cost for satisfying the fixed demand, the supplier has little incentive to share the true cost information unless he will keep the majority from a low-cost realization. In contrast, when the market demand is endogenously determined, the buyer will be able to set a low price to induce more demand when the supplier's cost is low. Thus a cost-contingent quantity attribute will increase the demand and revenue for the supply chain, which benefits both the buyer and the supplier. As a result, the supplier is more willing to share true cost information and its associated benefits with the buyer. This explains why the optimal menu for the quantity attribute is more valuable for the buyer.

An implicit assumption underlying the analysis in this chapter is that both supply chain firms' actions are enforceable. In Sect. 7.3, the two parties can enforce a contract based on the supplier's capacity (or lead time), and in 4 the supplier must commit to the specified base-stock level. This is not a restrictive assumption when the firms care about long-term relationship or when there is a third party who is able to verify and enforce the firms' actions. When enforcement becomes an issue, the buyer has to take it into consideration and modify the supply contract accordingly. An incentive scheme has been proposed in Zhang [67] for the buyer to induce desired supplier actions. More discussion of the impact of the enforcement issue can be found in Bolandifar et al. [8], Cachon and Lariviere [15].

7.6 Conclusion and Future Research

Outsourcing/offshoring has been increasingly used by the industry during the past few decades. This trend brings supply management under the spotlight because many companies today depend critically on their suppliers for the delivery of components, products, and services. The purpose of this chapter is to study a buyer's procurement problem while taking a key operational element into consideration: supplier's delivery performance. It is quite clear that superior delivery performance is beneficial to the buyer because more responsive deliveries reduce the buyer's operating cost (e.g., inventory and backorder costs). Responsive deliveries require capacity or inventory investments at the supplier, both of which are costly. As a result, the buyer needs to carefully design incentive schemes to induce the right action from the supplier. Most studies in the literature on supply contracting assume there is complete information in the supply chain. However, in practice, firms in a decentralized supply chain are independent organizations and may not have perfect information about each other. It is not uncommon that the buyer faces uncertainties about the supplier's cost structure when negotiating the supply contract. Therefore, in this chapter we focus on the buyer's supply contract design problem under both asymmetric cost information and delivery performance consideration. In particular, we aim to derive some useful managerial guidelines for practitioners when making procurement decisions.

Two problem scenarios have been considered in this chapter. In the first scenario, the supplier is a make-to-order manufacturer or a service provider; so we model the supplier as a queueing server (the delivery performance is determined by the average delivery lead time). In the second, the supplier is a make-to-stock manufacturer and may hold inventory to improve customer service (the delivery performance is the fill rate at the supplier). We present these two scenarios in separate sections because their model settings and analyses are quite different. For each scenario, we first derive the buyer's optimal (profit-maximizing) mechanism, which consists of a nonlinear menu of contracts. Then we propose some simple, but sub-optimal mechanisms for the buyer. The simple mechanisms only specify a target delivery performance and do not require the supplier's cost information as an input. By comparing the simple mechanisms to the optimal benchmark, we find that fixing the delivery performance attribute yields nearly optimal outcome for the buyer. Such a finding is quite robust: It applies to both scenarios and remains unchanged in various extensions of the basic model. The explanation of this result is as follows. Using a complex menu on the delivery performance attribute (i.e., a cost-contingent lead time or fill rate) may reduce the supply chain's operating cost for satisfying market demand. However, the buyer is ultimately responsible for such a cost, so the supplier has little incentive to share his true cost information with the buyer unless he can keep the majority of the benefit from the cost reduction. This means that for the delivery performance attribute, even using the optimal menu cannot extract much information rent from a low-cost supplier. Therefore, the simple mechanisms perform very well (relative to the optimal mechanism), although they do not extract any information rent from the supplier.

On one hand, the above finding suggests that in the presence of asymmetric information, the buyer only needs to use a fixed number (rather than a complex menu) to ensure satisfactory delivery performance from the supplier. It bodes well for procurement managers because a relatively simple contract term can be used to take care of the delivery performance attribute in the procurement process. This is consistent with the practical observations in various industries where buyer–supplier contracts often involve simple service-level agreements [42, 47, 59]. On the other hand, our results indicate that instead of resorting to the complex, optimal procurement contract, the buyer should try to reduce the information disadvantage by learning more about the supplier's cost structure. A better understanding of the supplier's operating process, technology, and other cost-related activities will bring significant savings in the supply contracting process.

This research can be extended in several directions. First, it has been assumed in both scenarios that unsatisfied demand at the buyer can be backlogged. An alternative assumption is that there are lost sales. It would be worthwhile to investigate whether the lost-sales assumption will change the results. Second, in many situations, the supplier's delivery performance may depend on the random yield at the supplier. For instance, due to quality-related problems, only a fraction of the production at the supplier will be useful and delivered to the buyer. Supply disruption may occur as an extreme case if the yield is sufficiently low. How the buyer should design supply contract to induce yield-improving effort is an interesting and open research topic. Finally, this chapter focuses on sole-sourcing where the buyer selects only one supplier with whom to transact. A potential direction for future research is to study a similar contracting problem with multi-sourcing.

References

1. Anupindi R, Akella R (1993) Diversification under supply uncertainty. Manag Sci 39(8): 944–963
2. Bagnoli M, Bergstrom T (2005) Log-concave probability and its applications. Econ Theory 26(2):445–469
3. Beil D, Wein L (2003) An inverse-optimization-based auction mechanism to support a multiattribute RFQ process. Manag Sci. 49(11):1529–1545
4. Benjaafar S, Elahi E, Donohue K (2007) Outsourcing via service quality competition. Manag Sci 53(2):241–259
5. Bernstein F, DeCroix GA (2006) Inventory policies in a decentralized assembly system. Oper Res 54(2):324–336
6. Beth S, Burt D, Copacino W et al (2003) Supply chain challenges: building relationships. Harv Bus Rev July
7. Bhattacharyya S, Lafontaine F (1995) Double-sided moral hazard and the nature of shared contracts. Rand J Econ 26(4):761–781
8. Bolandifar E, Feng T, Zhang F (2010) Asymmetric information and enforcement in supply contract design. Working paper, Washington University in St. Louis.
9. Bower AG (1993) Procurement policy and contracting efficiency. Int Econ Rev 34(4):873–901
10. Branco F (1997) The design of multidimensional auctions. Rand J Econ 28(1):63–81
11. Bulkeley W (2003) Plexus strategy: small runs of more things. Wal Str J 8 Oct
12. Burt D (1989) Managing suppliers up to speed. Harv Bus Rev July–Aug
13. Burnetas A, Gilbert SM, Smith C (2007) Quantity discounts in single period supply contracts with asymmetric demand information. IIE Trans 39(5):465–480
14. Cachon GP (2003) Supply chain coordination with contracts. In: Graves S, Kok T (eds) The handbook of operations research and management science: supply chain management.. Kluwer, USA

15. Cachon GP, Lariviere M (2001) Contracting to assure supply: how to share demand forecasts in a supply chain. Manag Sci 47(5):629–646
16. Cachon GP, Zhang F (2006) Procuring fast delivery: sole-sourcing with information asymmetry. Manag Sci 52(6):881–896
17. Cachon GP, Zhang F (2007) Obtaining fast delivery in a queueing system via performance-based allocation of demand. Manag Sci 53(3):408–420
18. Cachon GP, Zipkin PH (1999) Competitive and cooperative inventory policies in a two-stage supply chain. Manag Sci 45(7):936–953
19. Caldentey R, Wein LM (2003) Analysis of a decentralized production-inventory system. MSOM 5(1):1–17
20. CAPS Research (2006) Services spend management: outsourcing/offshoring your services spend. Center for advanced purchasing studies, Tempe, AZ
21. Che YK (1993) Design competition through multi-dimensional auctions. Rand J Econ 24(4):668–680
22. Chen F, Federgruen A, Zheng YS (2001) Coordination mechanisms for a distribution system with one supplier and multiple retailers. Manag Sci 47(5):693–708
23. Chen F, Zheng YS (1994) Lower bounds for multiechelon stochastic inventory systems. Manag Sci 40(11):1426–1443
24. Chen R, Roundy R, Zhang R et al (2005) Efficient auction mechanisms for supply chain procurement. Manag Sci 51(3):467–482
25. Chu LY, Sappington DEM (2007) Simple cost-sharing contracts. Am Econ Rev 97(2): 419–428
26. Corbett CJ (2001) Stochastic inventory systems in a supply chain with asymmetric information: cycle stocks, safety stocks, and consignment stock. Oper Res 49(4):487–500
27. Corbett CJ, de Groote X (2000) A supplier's optimal quantity discount policy under asymmetric information. Manag Sci 46(3):444–450
28. Corbett CJ, Zhou D, Tang CS (2004) Designing supply contracts: contract type and information asymmetry. Manag Sci 50(4):550–559
29. Department of Commerce (2006) Statistics for industry groups and industries: Annual survey of manufactures. http://www.census.gov/prod/2006pubs/am0531gs1.pdf
30. Elmaghraby WJ (2000) Supply contract competition and sourcing strategies. MSOM 2(4): 350–371
31. Elmaghraby WJ (2004) Auctions and pricing in E-markeplaces. In: Simchi-Levi D, Wu SD, Shen ZJ (eds) Handbook of quantitative supply chain analysis: modeling in the e-business era. Kluwer, USA
32. Farlow D, Schmidt G, Tsay A (1996) Supplier management at Sun Microsystems (A). Stanford business school case
33. Federgruen A, Yang N (2008) Selecting a portfolio of suppliers under demand and supply risks. Oper Res 56(4):916–936
34. Federgruen A, Yang N (2009) Optimal supply risk diversification under general supply risks. Oper Res 57(6):1451–1468
35. Friedman TL (2005) The world is flat: a brief history of the twenty-first century. Farrar Straus and Giroux, New York
36. Gilbert SM, Weng ZK (1998) Incentive effects favor nonconsolidating queues in a service system: the principle-agent perspective. Manag Sci 44(12):1662–1669
37. Gümüş M, Ray S, Gurnani H (2010) Supply side story: risks, guarantees, competition and information asymmetry. Working paper, McGill University.
38. Ha AY (2001) Supplier-buyer contracting: asymmetric cost information and cutoff level policy for buyer participation. Nav Res Logist 48:41–64
39. Ha AY, Li L, Ng SM (2003) Price and delivery logistics competition in a supply chain. Manag Sci 49(9):1139–1153
40. Hausman WH, Lee HL, Subramaniam U (2005) Global logistics indicators, supply chain metrics and bilateral trade patterns. World bank policy research working paper, November

41. Holmstrom B, Milgrom P (1987) Aggregation and linearity in the provision of intertemporal incentives. Econometrica 55(2):303–328
42. Katok E, Thomas D, Davis A (2008) Inventory service-level agreements as coordination mechanisms: the effect of review periods. MSOM 10(4):609–624
43. Kay E (2005) Ways to measure supplier performance. http://www.purchasing.com/
44. Kayiş E, Erhun F, Plambeck EL (2007) Delegation vs. control of component procurement. Working paper, Stanford University
45. Klemperer P (1999) Auction theory: a guide to the literature. J Econ Surv 13(3):227–286
46. Kostamis D, Beil D, Duenyas I (2009) Total-cost procurement auctions: impact of suppliers' cost adjustments on auction format choice. Manag Sci 55:1985–1999
47. Lacity MC, Willcocks LP (1998) An empirical investigation of information technology sourcing practices: Lessons from experience. MIS Q 22(3):363–408
48. Laffont JJ, Martimort D (2002) The theory of incentives: the principal-agent model. Princeton University Press, Princeton
49. Li CL, Kouvelis P (1999) Flexible and risk-sharing supply contracts under price uncertainty. Manag Sci 45(10):1378–1398
50. Özer Ö, Wei W (2006) Strategic commitments for an optimal capacity decision under asymmetric forecast information. Manag Sci 52(8):1238–1257
51. Pinker E, Seidmann A, Vakrat Y (2003) Managing online auctions: current business and research issues. Manag Sci 49(11):1457–1484
52. Pyke D, Johnson E (2003) Sourcing strategies and supplier relationships: alliances vs. eprocurement. In: Billington C, Lee H, Neale J (eds) The practice of supply chain management. Kluwer Publishers, USA
53. Rangan VK (1998) FreeMarkets Online. Harv Bus Sch case #9-598-109.
54. Ramasesh RV, Ord JK, Hayya JC et al (1991) Sole versus dual sourcing in stochastic lead-time (s, Q) inventory models. Manag Sci 37(4):428–443
55. Ray S, Li S, Song Y (2005) Tailored supply chain decision making under price-sensitive stochastic demand and delivery uncertainty. Manag Sci 51(12):1873–1891
56. Tang CS (2006) Perspectives in supply chain risk management. Int Prod Econ 103(2):451–488
57. Taylor TA, Plambeck EL (2007) Simple relational contracts to motivate capacity investment: price-only versus price-and-quantity. MSOM 9(1):94–113
58. Thomas DJ (2005) Measuring item fill-rate performance in a finite horizon. MSOM 7(1):74–80
59. Thonemann U, Behrenbeck K, Kupper J et al (2005) Supply chain excellence im handel (in German). Financial Times Deutschland/Gabler, Wiesbaden, Germany
60. Thurm S (1998) Some US manufacturers prosper by easing rise of 'virtual' firm. Wall Str J 18 August
61. Tomlin B (2006) On the value of mitigation and contingency strategies for managing supply chain disruption risks. Manag Sci 52(5):639–657
62. Tunca T, Wu Q (2009) Multiple sourcing and procurement process selection with bidding events. Manag Sci 55(5):763–780
63. Wise R, Morrison D (2000) Beyond the exchange: the future of B2B. Harv Bus Rev Nov–Dec 86–96
64. Yang Z, Aydin G, Babich V et al (2009) Supply disruptions, asymmetric information and a backup production option. Manag Sci 55(2):192–209
65. Zhang F (2004) Coordination of lead times in supply chains. Ph.D. Dissertation. University of Pennsylvania
66. Zhang F (2006) Competition, cooperation and information sharing in a two-echelon assembly system. MSOM 8(3):273–291
67. Zhang F (2010) Procurement mechanism design in a two-echelon inventory system with price-sensitive demand. MSOM 12(4): 608–626

Chapter 8
Risk, Financing and the Optimal Number of Suppliers

Volodymyr Babich, Göker Aydın, Pierre-Yves Brunet, Jussi Keppo and Romesh Saigal

8.1 Introduction

Should firms in developed economies work with more or fewer suppliers than firms in developing economies? More generally, how does the number of suppliers for a firm depend on the firm's economic environment? To answer these questions we identify several economic and business factors that might affect the number of suppliers (and that separate developed and developing economies): supply risk, fixed costs of working with suppliers, and access to financing (particularly trade-credit financing).

V. Babich
McDonough School of Business, Georgetown University,
37th and O Streets NW,
Washington, DC 20057, USA
e-mail: vob2@georgetown.edu

G. Aydın (✉)
Kelley School of Business, Indiana University,
Bloomington, IN 47405, USA
e-mail: ayding@indiana.edu

P.-Y. Brunet · J. Keppo · R. Saigal
Industrial and Operations Engineering, University of Michigan,
1205 Beal Ave, Ann Arbor, MI 48109, USA
e-mail: brunetp@umich.edu

P.-Y. Brunet
Credit Risk Office, Barclays Bank PLC,
27-11 Queens Plaza North, Long Island City, NY 10016, USA
e-mail: pierre.brunet@barclaycard.co.uk

J. Keppo
e-mail: keppo@umich.edu

R. Saigal
e-mail: rsaigal@umich.edu

H. Gurnani et al. (eds.), *Supply Chain Disruptions: Theory and Practice of Managing Risk*, DOI: 10.1007/978-0-85729-778-5_8,

Supply risk has been recognized as one of the main reasons for firms to diversify their supply base both in empirical research (see [41]) and in theory (see [1, 6, 15, 27, 38, 43, 44]). The severity of exposure to supply risks depends on where suppliers are located and a variety of other economic and business factors. For example, firms operating in developing economies may have to cope with greater uncertainty about supplier reliability due to underdeveloped production, transportation, information, and financial infrastructures, as well as insufficient legal protection and political risk.[1] The ramifications of supply risk and, hence, the need for diversification also depend on the extent to which the owners of a firm are liable for the disruption consequences. If the owners could walk away from their financial and other obligations to their financial partners and customers, the owners would be less concerned about supply risk and rely less on diversification. Although the connection between supply risk and diversification has been studied extensively, the effect of liability on this connection, which we study in this chapter, has received scant attention in the literature.

Fixed costs of working with suppliers create incentives for firms to lean out their supply base. These costs can take various forms. For example, some firms (e.g. in automotive and aerospace industries) must certify potential suppliers using rigorous and costly process before any parts can be procured. Furthermore, just the initial certification may not be sufficient. To guarantee that suppliers perform according to the manufacturer's expectations, suppliers must be continuously monitored and the quality of their parts must be checked. Costs of monitoring and contract enforcement is significantly higher in countries with less developed legal systems and information infrastructure.

With respect to access to financing, in addition to raising capital in well-functioning and mature financial markets, the majority of companies in the USA. (and other developed countries) rely on alternative sources, such as trade-credit[2] and private investments, to finance their strategic, tactical, and operational decisions. In fact, trade-credit financing is the single largest source of external short-term corporate financing in the United States (see [35, 37]). According to [37], accounts payable and creditors constituted 15%, while debt in current liabilities was only 7.4% of the total book value for an average non-financial firm in 1991.[3] Several studies

---

[1] For example, Ukrainian government seized assets of a number of firms in a reprivatization campaign following the "Orange Revolution" (see [33]).

[2] Trade-credit is the delay in payments from the buyer to the supplier of goods. One can think of trade-credit as a loan extended by the supplier to the buyer. The buyer effectively obtains the loan from the supplier by not paying for the purchase initially, but it has to repay the loan later. The typical trade-credit contracts in the United States are "net 30" and "2/10 net 30" (see [30]). According to the former, the buyer does not have to pay for the purchase for 30 days, thus obtaining, effectively, a 30-day interest-free loan. According to the latter contract, the buyer receives a 2% discount if it pays for the purchase within 10 days, and it has up to 30 days to pay for the goods. As [30] points out, the 2% discount is equivalent to the buyer obtaining a 20-day loan at the implicit interest of 43.9%. Trade-credit contracts vary by industry and country. It is not uncommon to see trade-credit terms that have maturity longer and shorter than 30 days, and higher and lower implied interest rates.

[3] Given the prevalence of trade-credit, examples of companies that rely heavily on this form of financing are easy to find. Consider, for instance, TenderCare International, Inc., which sells dis-

(see [3, 31]) find that the reliance on trade-credit financing increases when other sources of financing are restricted (for example, during times of monetary crunch in the economy). This is why, in developing economies, whose financial markets are still in their infancy, access to trade-credit has profound consequences for the firm's growth potential, competitive abilities, and survival. Several recent empirical studies emphasize the importance of trade-credit as a financing source in developing economies. Fisman and Love [19] observes that, in countries with weaker financial institutions, the industries with higher dependence on trade-credit financing exhibit higher rates of growth, relative to other industries. Fisman [18] discovers a correlation between the availability of trade-credit financing and a firm's operational performance. Using a sample of African firms, he finds that firms with access to trade-credit financing are less likely to experience stock-outs, and are more likely to have higher production-capacity utilization. To the best of our knowledge, the interactions among financing constraints, trade-credit loans, and the number of suppliers have not been studied before, a gap that this chapter aims to fill.

In this chapter we analyze how financing constraints, the dual role played by suppliers as the providers of parts and the financiers of the manufacturer, supplier risk, limited liability of the manufacturer, and fixed costs of working with suppliers affect the manufacturer's choices of order quantities and the number of suppliers. Using a one-period model with homogeneous suppliers (and later a model with non-homogeneous suppliers), we consider the joint procurement and financing decisions of a firm with access to limited internal financing and loans from suppliers, facing either an uncertain demand or an uncertain supply (random yield). In contrast to the traditional operational models, where the decision-maker is fully liable for losses, we model the decisions of the owners of the firm, who may have limited liability (e.g. they could be limited-liability partners or shareholders; we will refer to the owners of the firm as the shareholders in the sequel).

The contracts between suppliers and the manufacturer are assumed to be given exogenously and, as we perform comparative statics analysis on our solution, the contract terms, in particular, the terms of the loans extended by the suppliers to the manufacturer, do not change. There are several reasons for the contract terms to be fixed. In practice, contracts are renegotiated on a periodic basis and are not sensitive to short-term fluctuations in the manufacturer's condition. Furthermore, although we are analyzing a single product line of the firm, the firm might have other product lines and might be selling in multiple markets. Thus, the contract terms must take into account the firm's overall condition and not just the condition of one of the firm's subdivisions. We focus on firms that do not have access to stock markets for additional financing (as is the case in some developing countries) or consider equity financing costs to be too high (see [32] for a discussion about the costs of going public

---

posable baby diapers, natural formula wipes, and related products in the United States. According to this company's annual report, it had \$1.172 million in accounts payable, and \$0.007 million in short-term debt out of total \$1.218 million in total current liabilities in 2005. In the same year, the company's cost of goods sold was \$2.185 million. Thus, with $Days\ Payables\ Outstanding = Account\ Payables/COGS * 365 \approx 6\,months$, TenderCare International depends greatly on its suppliers for financing.

and trade offs between equity vs. bank financing). The lack of equity investor scrutiny exacerbates the non-transparency of the firm's operations to outside investors. This asymmetric information violates the perfect capital market assumption in [28] and prevents the lenders from reacting to changing conditions of the firm. A number of empirical studies find that loan terms (both in the form of trade-credit and bank loans) are fairly insensitive to individual firm conditions. For example, using data from a sample of small firms, [34] observes that, once a decision to extend a loan is made, the loan terms are determined based on industry practices, economy-wide factors, internal policies and conditions of the lender, and do not depend on the conditions of the borrower. Ng et al. [30] finds that, while there are significant differences between industries, the trade-credit contract terms are standardized within industries, with "net 30" and "2/10 net 30" being the most popular contracts.

Based on our analysis, we offer several testable hypotheses about the relationship between economic conditions and size of the manufacturer's supply base. Some of these hypotheses have already been confirmed in prior empirical studies; others still need to be empirically verified. Specifically, as one would expect, our analysis suggests that the alternative financing sources (internal financing and trade-credit) are substitutable that is, ceteris paribus, the firm uses more suppliers if internal financing is not available. Surprisingly, we also find that, because of limited liability of the shareholders, the optimal production quantity could be increasing in fixed costs. Furthermore, the optimal number of suppliers could be increasing in fixed costs of working with suppliers as well because working with more suppliers could relax the manufacturer's financing constraints. In addition, we observe that the limits on loans and the wholesale price affect the optimal number of suppliers in a non-monotone way. Interestingly, we find that the value of the shareholders and the optimal number of suppliers of the firm could be increasing or decreasing in the standard deviation of the supplier random yield. This is a consequence of the trade offs between the expected profit and the value of the option to default that shareholders hold. We study the effects of limited liability and find that, when suppliers are perfectly reliable, the greater the loss for which the manufacturer is responsible, the smaller the order quantity the manufacturer will place and the fewer the number of suppliers with whom the manufacturer will work. However, when suppliers are not reliable, it may happen that the greater the loss the manufacturer is responsible for, the more suppliers the manufacturer may order from in order to take advantage of supply-risk diversification benefits.

Finally, we address the question whether firms operating in developing economies must contract with more suppliers than firms operating in developed economies. The answer is "no" if the fixed cost of an extra supplier is high. However, in this case, our model predicts that financial constraints will force firms in developing economies to sub-optimal levels of production and cause higher stock-out rates. This conclusion is consistent with the results of earlier empirical studies.

The remainder of the chapter is organized as follows. In the next section we discuss the related literature. The model, which contains both financial and operational decisions, is discussed in Sect. 8.3. In Sect. 8.4 we find the optimal financial decisions, given the optimal operational decisions. We then derive a series of analytical results on the optimal operational decisions and the role of financial constraints in the supplier selection. These analytical results guide a numerical study, the results of which are presented in Sect. 8.5. Sect. 8.6 considers a model with heterogeneous suppliers. Conclusions, model limitations, and future research directions are discussed in Sect. 8.7. Finally, the Appendix contains technical lemmas and proofs.

8.2 Literature Review

If capital markets were perfect, as [28] proved in their seminal paper, managers could consider financing decisions independently from the firm's other decisions. However, imperfections of real-capital markets, such as transaction costs, taxes, information asymmetry, and bankruptcy costs imply that managers could generate value for shareholders by jointly considering financial and non-financial decisions (a survey of research on the choice of capital structure and the effects of market imperfections is offered by [23]). Several recent papers ([7, 10, 13, 25, 26, 42]) consider the value of combining financial, operational, and technology decisions. Similar to those papers, we will explicitly model the firm's ordering and financing decisions and investigate how financing terms (in our case, trade-credit terms) affect ordering decisions. For example, [13] studies how asset-based bank loans affect the ability of the manufacturer to grow (using a dynamic deterministic model) and how asset-based financing terms can be optimally set by banks (using a one-period Stackelberg game with a stochastic demand and several borrowers). Similar to the model in [13], we consider the limited liability of the borrower. Unlike their model, however, we focus on the manufacturer's choice of the number of suppliers, driven by the trade off between the fixed costs of adding a supplier and the benefits of relaxing financing constraints or diversifying supply risk. The external financing in our model is provided by the suppliers (through trade-credit), who perform a dual function by supplying components and offering financing to the manufacturer.

A number of researchers studied the effects of trade-credit on inventory policies. For example, [21] proves that, even in the presence of trade-credit, the order-up-to inventory policy remains optimal for a discrete-time, joint inventory-financing model, and suggest an algorithm for computing the optimal stock level for a continuous-time model. The effects of delayed payments on the EOQ model were investigated by [14, 22, 36]. In our chapter, in addition to determining optimal order quantities (as is done in the cited articles), we will compute the optimal number of suppliers that a firm should have.

If the supplier yields are random, then the firm may benefit by ordering from several suppliers. The benefits of diversification[4] as a remedy for supplier random yields were considered, for example, by [2, 15, 17, 39, 40]. Babich et al. [5, 6] quantify diversification benefits when suppliers are competing. However, the majority of the

[4] By diversification we mean holding a portfolio of contracts, instead of just one contract. In this chapter specifically, diversification means placing orders with several suppliers.

research in the random yield literature assumes that the supplier set is exogenously given (an extensive review of random yield research can be found in [45]). Among exceptions is [1] that studies trade offs between diversification benefits and supplier set-up costs using a one-period model with independent, multiplicative, and normally distributed supplier yields. The authors consider cases of both identical and non-identical suppliers, provide conditions that the optimal order quantities must satisfy, and suggest numerical procedures for determining the optimal number of suppliers. Agrawal and Nahmias [1] finds that the optimal order quantity is likely to be increasing and the optimal profit is decreasing in the volatility of the supplier yield. They also find that the profit is increasing and the optimal order quantity is likely to be decreasing in the yield's mean.

We extend the analysis of the identical-supplier case of Agrawal and Nahmias [1] by adding financing decisions and financing constraints, and by allowing for the limited liability of the decision-makers. These additional assumptions produce results different from those in [1]. For example, while the optimal value of the objective function is increasing in the random yield's mean both in our model and in [1], we observe that with the limited-liability assumption, the decision-makers in our model may benefit from an increase in the volatility of supplier yields. While diversification benefits provide a powerful incentive for the firm to order from several suppliers, financing constraints in our model may either hinder or encourage diversification. Unlike [1], we do not study in depth the supplier selection problem with non-identical suppliers because our focus is on financing decisions, financing constraints, limited liability, and their effects on the size of the supply base, rather than on interactions with individual suppliers. However, we do extend the analytical results of their model with heterogeneous suppliers to our setting.

A number of earlier studies observed effects of limited liability similar to our findings. Gollier et al. [20] considers the investment problem of a risk-averse firm with limited liability. They show that the optimal exposure to risk is always larger under limited liability compared to full liability. Brander [11] shows that limited liability may commit a leveraged firm to a more aggressive output stance. Because firms will have incentives to use financial structure to influence the output market, this demonstrates a new determinant of the debt-equity ratio. Faure-Grimaud [16] shows that asymmetric information between banks and firms plays a crucial role in financial decisions and output market strategies. In his model debt causes firms to compete less aggressively: the usual (positive) limited-liability effect on quantities is offset by a negative one due to (endogenous) financial costs.

8.3 Model and Assumptions

Consider a one-period model with a firm that procures a component from outside suppliers in order to meet random customer demand, denoted by D. We are modeling only one of possibly several businesses that the firm may have. There exists a number of suppliers, and the managers of the firm must decide with how many suppliers to contract (the number of suppliers contracted is denoted by N) and how much to order from each (the quantity ordered from supplier i is denoted by y_i). Similar to [1], we will assume (to simplify the analysis) that suppliers are identical. Therefore, each supplier will receive the same order quantity: $y_i = y$. The total order placed with the suppliers is $z = Ny$.

8.3.1 Demand

The customer demand D is a random variable with probability density function (p.d.f.) g and cumulative distribution function (c.d.f.) G. Define $\overline{G}(x) = 1 - G(x)$. We will also consider models with deterministic demand D.

Assumption 4 Let demand D be defined over a domain $[x_l, x_u]$ where $0 \leq x_l < x_u < \infty$. Let G be strictly increasing and twice continuously differentiable. Furthermore, assume $g(x)/\overline{G}^2(x)$ is increasing.

The requirement of a finite domain imposed by assumption (4) is needed to rule out pathological cases where the manufacturer may find it desirable to order arbitrarily large quantities. The condition that $g(x)/\overline{G}^2(x)$ is increasing is a weaker condition than requiring that the demand distribution has an increasing failure rate (IFR) and, therefore, is satisfied by all IFR distributions (and their truncated versions), including the normal, Weibull, and Gamma distributions. This assumption will be used to establish the unimodality of the objective function later in the chapter.

8.3.2 Random Yield

The yield of a supplier is random and independent across suppliers. Similar to [1], it will be convenient to assume that the yields are stochastically proportional and normally distributed. Therefore, if an order for y units is placed with a supplier at the beginning of the planning horizon, a quantity Xy will be delivered by the sales time, where $X \sim \mathcal{N}(\mu, \sigma)$. If the total order quantity $z = Ny$ is placed with N suppliers, then a quantity $Q(N, z) = \overline{X}z$ will be delivered, where $\overline{X} \sim \mathcal{N}\left(\mu, \sigma/\sqrt{N}\right)$. To guarantee that the probability of supplier yield falling outside of the range [0,1] is negligible, we will assume that $0 < \mu \pm 3\sigma < 1$.

8.3.3 Operational Costs and Revenues

Each supplier charges the firm w per unit of the component when the order is placed. The timing of payments is not of great importance when suppliers are perfectly reliable. When suppliers are unreliable, the manufacturer prefers to pay after the delivery and only for items that are actually delivered. Suppliers prefer up-front payments for the orders that have been placed. Depending on the market power of the manufacturer and the suppliers, some combination of per-ordered and per-delivered payment contracts will be adopted.[5] Even if the manufacturer enjoys full market power, there are circumstances when only contracts with per-ordered payments are possible. For example, the cost of verifying the delivery size (e.g. via quality control) could be prohibitively high, leaving the manufacturer no alternative but to accept the entire order and rely on its customers to identify defective products. Furthermore, even if the inspection costs are low, in practice the manufacturer accepts the whole order if, e.g. 98% of the products in the samples are good, since the inspection process is not error free—error rates between 20–30% are not unusual (see Chaps. 14 and 15 in [29]). We analyzed a general model with both per-ordered and per-delivered payments; however, the insights we obtained were the same as those for the model with only per-ordered payments. Therefore, for the sake of exposition, we only present the simpler model with per-ordered payments.

In addition to variable costs, the firm incurs a fixed cost of C for working with a supplier. The fixed cost in this chapter represents an amalgam of various costs that a manufacturer has to incur by working with a supplier, including supplier-selection costs, contract-monitoring costs, legal fees, quality control expenses, etc. These costs encompass both physical and financial transaction costs. Thus, the cost of operations is $NC + wz$. Unmet demand becomes lost sales, and leftover inventory has no value. Therefore, the revenue of the firm is $p \min[D, Q(N, z)]$, and the firm's operational profit is $p \min[D, Q(N, z)] - wz - NC$.

8.3.4 Financing

The firm has two financing options for its operational decisions: internal capital and trade-credit (recall that in Sect. 8.1 we argued that trade-credit is the single most important form of external short-term financing). Assume that at the beginning of the planning horizon the internal capital (which is internally generated cash available to the firm, e.g. retained earnings) is I. The firm can invest at the rate of r_I.

The firm may also use trade-credit contracts offered by the suppliers. As we discussed in the introduction, a trade-credit contract allows the buyer to delay payments for the goods received, which is equivalent to the supplier offering a loan to the buyer.[6] For example, suppose that a manufacturer places an order for y units with a supplier. The supplier offers the manufacturer the choice either to make an immediate payment of wy or to postpone paying for any part of the order until the end of the planning horizon. At the end of the planning horizon the manufacturer must pay a higher per unit amount $w(1 + r_S)$. Effectively, the manufacturer can take out a loan, S, up to the monetary value of the purchase, wy (i.e. $S \leq wy$), with the supplier. The interest rate on this loan is r_S. In general, the supplier may offer a trade-credit on only a part of the order (i.e. $S \leq \alpha y$, where $0 \leq \alpha \leq w$) and may even offer some amount, β, regardless of the order size, just for receiving an order from the manu-

[5] As [6] demonstrates, in equilibrium the suppliers and the manufacturer could be indifferent between per-ordered and per-delivered payments.

[6] See Chap. 30 (pp. 812–840) of [12] for a description of trade-credit contract terms.

facturer (i.e. $S \leq \alpha y + \beta$, where $0 \leq \alpha \leq w$). In most applications, $\beta = 0$. However, we derive our results with a more general assumption: $\beta \geq 0$. The terms offered on the supplier loans can be better or worse than those of internal financing, depending on competition among suppliers, transaction costs, information asymmetry, and other factors.[7] The absolute size of the loan from a supplier depends on the supplier's size (in financial terms), monetary supply in the economy, the ability of the supplier to access capital, and the credit-risk management activities of the supplier (lenders usually limit the size of a loan that can be offered to any single borrower). Thus, an additional constraint on the supplier loan is $S \leq \hat{S}$, where $\hat{S}$ is the upper limit on the loan, regardless of the manufacturer's order size. Putting together two constraints on the loan from a supplier, we obtain $S \leq \bar{S}(y) = \min(\hat{S}, \alpha y + \beta)$.

If the fixed cost of working with a supplier, C, is greater than the absolute limit on the supplier loan, $\hat{S}$, then the manufacturer cannot use additional suppliers to relax financing constraints. To make the problem more interesting, we will assume that $C < \hat{S}$, which, in practice, is a reasonable property. It is also natural to assume that the amount of money, β, the buyer receives from the supplier just by placing an order is less than the fixed cost of working with a supplier, C. This assumption is likely to be true in practice, because in most applications $\beta = 0$. In general, this assumption prevents the buyer from having infinite wealth at time 0 just by placing orders with an infinite number of suppliers (each supplier increasing the manufacturer's wealth by $\beta - C > 0$). To summarize, we make the following assumptions on the parameters of the supplier loans:

Assumption 5 $0 \leq \alpha < w$ and $0 \leq \beta < C < \hat{S}$.

Furthermore, we will assume (to avoid trivial solutions) that the rate of return on both financing sources, internal capital and trade-credit, is small enough that the manufacturer is able to recover the wholesale price plus the interest by selling the product. That is:

Assumption 6 $p > (1 + \max\{r_S, r_I\})w$.

In our one-period model, we assume that the loan terms are fixed. One could think, for example, that the manufacturer makes decisions after having observed the loan rates and the loan limits offered by the lenders. As we discussed in the introduction, in perfect capital markets, the loan terms would reflect the default probability of the borrower and the loans would be fairly priced. However, real markets are not perfect. For example, due to information asymmetry, suppliers may not be able to adjust their rates in response to the changing business risk of the manufacturer. Using data from a sample of small firms, [34] observes that, once a decision to extend a loan is made, the loan terms are determined based on industry practices, economy-wide factors, and internal policies and conditions of the lender, and are fairly insensitive to the conditions of the borrower. One may wonder if the loan limits are more sensitive to the changes in the borrower's state. As [8] discusses, although credit line agreements usually include clauses that allow banks to revoke credit in case of significant changes

[7] For discussions of factors affecting trade-credit terms, see [9, 35].

in the conditions of the borrower, these clauses can only be invoked based on verifiable events. Even when events are verifiable, [4] shows that banks are reluctant to invoke these clauses. Likewise, suppliers may decide to extend favorable loan terms even to their risky customers, in order to benefit in the long run by avoiding the costs of their customers' defaults.

8.3.5 Objective

The objective of the firm's managers is to maximize the value for the firm's shareholders, who have limited liability because of bankruptcy protection. We are considering a single-period model and, therefore, the value of the business is the value of its cash position at the end of the planning horizon. If the firm loses money on this business, part of the losses can be absorbed by the other businesses that the firm has. From the shareholder's perspective, they are liable for the losses from this business up to an amount $l < 0$ i.e. if the cash position of this business is x, the shareholders will receive $\max(l, x)$. Special cases of limited liability are $l = -\infty$, in which case the shareholders are liable for all losses (as is traditionally assumed in the operations literature) and $l = 0$, in which case the shareholders are not liable for losses at all (as would be the case if the firm had only one business). In the subsequent sections we will highlight the results that are driven by the limited liability assumption.

8.3.6 Timing of Events and Cash Flows

At the beginning of the planning horizon, the firm decides on the optimal number of suppliers, the total order quantity, and the financing sources and amounts. It receives loans from the suppliers, NS, and pays operational costs, $NC + wz$.

The suppliers deliver product by the end of the planning horizon. Random demand is realized and the firm collects revenue, $p\min[D, Q(N, z)]$, repays loans, $N(1 + r_S)S$, or declares bankruptcy.

8.3.7 Mathematical Formulation

With the assumptions listed above, the firm's objective function is as follows:

$$E\left[\max\left\{l,\, p\min[D,\, Q(N,z)] - (1 + r_I)(wz + NC) - (r_S - r_I)NS + (1 + r_I)I\right\}\right]. \tag{8.1}$$

The first term inside the max operator, l, is a non-positive number and it represents the amount of losses the shareholders are liable for. The second term inside the max operator is the firm's cash position at the end of the planning horizon. (Recall that if the firm incurs a loss, in which case the cash position is negative, the shareholders are liable only up to an amount l.) The cash position itself consists of four terms. The first term, $p\min[D, Q(N,z)]$, is the revenue from sales, assumed to be collected at the end of the planning horizon. The second term captures the operational costs incurred by the firm at the beginning of the planning horizon, $wz + NC$, inflated by the firm's internal rate of return, r_I. The third term, $(r_S - r_I)NS$, is the interest paid on supplier loans. The fourth term, $(1 + r_I)I$, is the firm's internal capital, I, inflated by internal rate of return, r_I.

Using $\max\{l, x\} = l + (x - l)^+$ and noting that l is a constant, we can simplify the objective (8.1), obtaining the following optimization problem for the shareholders

$$\max_{S\geq 0, z\geq 0, N\in\mathbb{N}} \quad E\{p\min[D, Q(N,z)] - (1+r_I)(wz+NC)$$

$$- (r_S - r_I)NS + (1+r_I)I - l\}^+ \tag{8.2a}$$

$$wz + NC \leq NS + I, \tag{8.2b}$$

$$S \leq \overline{S}(z/N). \tag{8.2c}$$

8.4 Model Analysis

In this section we will first investigate the optimal financing decisions when the operational decisions are already fixed. Subsequently, we will investigate the optimal operational decisions.

8.4.1 Financing Decisions

In this subsection we will describe the optimal financing decision (i.e. the amount of supplier loan, S), provided that the operational decisions (i.e. order quantity, z, and number of suppliers, N) are fixed. In order for the operational decisions to be financially feasible, we need

$$wz + NC \leq N\overline{S}(z/N) + I. \tag{8.3}$$

Using the expression for the limit on the supplier loan $\bar{S}(y) = \min(\hat{S}, \alpha y + \beta)$, one can write an expression for financial feasibility of the operational decisions given by (8.3) as follows

$$\begin{cases} (w-\alpha)z + (C-\beta)N \leq I, \\ wz - (\hat{S} - C)N \leq I. \end{cases} \tag{8.4}$$

An inspection of problem (8.2) yields the following crucial observation: depending on whether the supplier loan is more ($r_S > r_I$) or less ($r_S < r_I$) expensive than the internal rate of return, the manufacturer will either borrow the smallest feasible or the largest feasible amount from suppliers. The following proposition presents the optimal supplier loan amounts:

Proposition 1 *Suppose that the operational decisions—number of suppliers, N, and order quantity, z— are fixed and financially feasible [as in inequality (8.3)]. Then the optimal loan amounts are*

Case I $r_I < r_S$

$$NS^* = (wz + NC - I)^+. \tag{8.5}$$

Case II $r_I > r_S$

$$NS^* = N\overline{S}(z/N). \tag{8.6}$$

Now, using optimal financing decisions (8.5), (8.6) and the expression for financial feasibility of operations decisions (8.4), we can rewrite optimization problem (8.2) as follows:

Case I $r_I < r_S$

$$\max_{z\geq 0, N\in\mathbb{N}} E\{p \min[D, Q(N,z)] - (1+r_I)(wz+NC) - (r_S - r_I)(wz + NC - I)^+ + (1+r_I)I - l\}^+ \tag{8.7a}$$

s.t.

$$(w-\alpha)z + (C-\beta)N \leq I, \tag{8.7b}$$

$$wz - (\hat{S} - C)N \leq I. \tag{8.7c}$$

For *Case I*, the term $(wz + NC - I)^+$ in the objective function is zero if only internal financing is used, and it is positive if both internal and supplier financing are used. Therefore, the equation $(wz + NC) - I = 0$ defines a *threshold of dual financing* for Case I.

Case II $\boldsymbol{r_I > r_S}$

$$\max_{z\geq 0, N\in\mathbb{N}} E\{p \min[D, Q(N,z)] - (1+r_I)(wz+NC) + (r_I - r_S)\min(N\hat{S}, \alpha z + N\beta) + (1+r_I)I - l\}^+ \tag{8.8a}$$

s.t

$$(w-\alpha)z + (C-\beta)N \leq I, \tag{8.8b}$$

$$wz - (\hat{S} - C)N \leq I. \tag{8.8c}$$

For *Case II*, instead of a threshold of dual financing, we are concerned with the *threshold of exceeding supplier loan limit* i.e. whether the firm will order so much from each supplier that it will reach the limit on the supplier loan $\hat{S}$.

Instead of proceeding with the analysis of problems (8.7) (when $r_I < r_S$) and (8.8) (when $r_I > r_S$) separately, we observe that they possess the same mathematical structure. Therefore, we can formulate a generalized problem, which highlights the salient model features and streamlines the analysis. Instead of repeating the same analysis twice, we can perform it only once with the generalized model and then apply the results derived to each special case. Define a generalized objective function $f(N, z)$ as follows:

$$f(N, z) = E[\Pi_L(N, z)]^+ \cdot 1_{\{T(N,z)\leq 0\}} + E[\Pi_R(N, z)]^+ \cdot 1_{\{T(N,z)>0\}} \tag{8.9}$$

where

$$\Pi_L(N, z) = \begin{cases} p \min[D, Q(N, z)] & \text{if } r_I < r_S, \\ \quad -(1 + r_I)(wz + NC) + (1 + r_I)I - l & \\ p \min[D, Q(N, z)] - (1 + r_I)(wz + NC) & \text{if } r_I > r_S; \\ \quad +(r_I - r_S)(\alpha z + N\beta) + (1 + r_I)I - l & \end{cases} \tag{8.10}$$

$$\Pi_R(N, z) = \begin{cases} p \min[D, Q(N, z)] & \text{if } r_I < r_S, \\ \quad -(1 + r_S)(wz + NC) + (1 + r_S)I - l & \\ p \min[D, Q(N, z)] & \text{if } r_I > r_S; \\ \quad -(1 + r_I)(wz + NC) + (r_I - r_S)N\hat{S} + (1 + r_I)I - l & \end{cases} \tag{8.11}$$

$$T(N, z) = \begin{cases} wz + NC - I & \text{if } r_I < r_S, \\ \alpha z + N\beta - N\hat{S} & \text{if } r_I > r_S. \end{cases} \tag{8.12}$$

Thus, the objective function consists of two components, defined using function Π_L and Π_R, with function T defining the threshold $\widehat{z}$ seperating the domains of Π_L and Π_R. For a given N,

$$T(N, \widehat{z}) = 0. \tag{8.13}$$

Threshold $\widehat{z}$ is what we earlier called the *threshold of dual financing* and the *threshold of exceeding supplier loan limit*. Throughout the remainder of the chapter, we will denote

$$f_L(N, z) = E[\Pi_L(N, z)]^+ \text{and} f_R(N, z) = E[\Pi_R(N, z)]^+. \tag{8.14}$$

Furthermore, observe that $f_L(N, z)$ and $f_R(N, z)$ have the same structure; both can be written as $E\{p \min[D, Q(N, z)] - a_k z - b_k(N) - l\}^+$ for $k = L, R$, with a_k and $b_k(N)$ given by:

$$a_L = \begin{cases} (1+r_I)w \\ (1+r_I)w \\ \quad -(r_I - r_S)\alpha \end{cases} \qquad b_L(N) = \begin{cases} (1+r_I)NC & \text{if } r_I < r_S \\ \quad -(1+r_I)I & \\ (1+r_I)NC - (r_I - r_S)N\beta & \text{if } r_I > r_S \\ \quad -(1+r_I)I & \end{cases} \tag{8.15}$$

$$a_R = \begin{cases} (1+r_S)w \\ (1+r_I)w \end{cases} \qquad b_R(N) = \begin{cases} (1+r_S)NC & \text{if } r_I < r_S \\ \quad -(1+r_S)I & \\ (1+r_I)NC & \text{if } r_I > r_S \\ \quad -(r_I - r_S)N\hat{S} - (1+r_I)I & \end{cases} \tag{8.16}$$

There is a good reason to introduce new notation. With the original parameters, the subsequent expressions would have been difficult to parse. New parameters have simple and practical interpretations: a_k, $k = L, R$ is the manufacturer's unit variable procurement cost, and $b_k(N)$, $k = L, R$ is the manufacturer's overhead cost.

In this subsection, we explored the optimal financing decisions. The next two subsections are dedicated to analyze operational decisions.

8.4.2 Operational Decisions Under Stochastic Demand and Deterministic Yield

In this section we investigate the effect of financial constraints on the optimal operational decisions for the manufacturer when the demand, D, is stochastic and the supplier yield, X, is deterministic and perfect, i.e. $Q(N, z) = z$. That is, in this section we will make the following assumption:

Assumption 7 The yields of the suppliers are deterministic and perfect, $X = 1$.

In Sect. 8.4.3, we will turn our attention to the effect of diversification for the case with random yield and deterministic demand.

8.4.2.1 Manufacturer's Objective Function

We begin with the analysis of the objective function of our problem, while assuming that for the duration of this subsection, problem constraints are not binding. Recall definitions (8.9) and (8.14) of the manufacturer's objective function. Unfortunately, functions $f_k(N, z)$, and $k = L, R$ are not necessarily concave in z, for a fixed N. Hence, the function $f(N, z)$ is not concave in z either. Nevertheless, under assumptions (4) through (7), the function $f(N, z)$ is well-behaved, as stated in the following proposition:

Proposition 2 *Suppose N is fixed. Then, $f(N,z)$ is unimodal in z.*

The key observation that yields the result of Proposition 2 is the following: the function $f(N,z)$ is given by $f_L(N,z)$ to the left of $\widehat{z}$ (i.e. for $z < \widehat{z}$) and by $f_R(N,z)$ to the right of $\widehat{z}$ (i.e. for $z > \widehat{z}$), and the two unimodal functions $f_L(N,z)$ and $f_R(N,z)$ intersect at $z = \widehat{z}$. Given this observation, the function $f(N,z)$ will be non-unimodal only if $f_L(N,z)$ is decreasing and $f_R(N,z)$ is increasing in z at $z = \widehat{z}$. As it turns out, this cannot happen. Hence, the function $f(N,z)$ is unimodal. Furthermore, this observation leaves us with only three possibilities regarding the behavior of functions $f_L(N,z)$ and $f_R(N,z)$ at $\widehat{z}$. Either both $f_L(N,z)$ and $f_R(N,z)$ are decreasing in z at $z = \widehat{z}$ (in which case, the optimal z is given by the maximizer of $f_L(N,z)$), or both $f_L(N,z)$ and $f_R(N,z)$ are increasing in z at $z = \widehat{z}$ (in which case the optimal z is given by the maximizer of $f_R(N,z)$), or $f_L(N,z)$ is increasing and $f_R(N,z)$ is decreasing in z at $z = \widehat{z}$ (in which case, the optimal z is given by $\widehat{z}$). This observation is formalized in Lemma 4 (see Appendix), which forms the basis for an algorithm (provided in Appendix) to determine the optimal order quantity, $z^* \stackrel{\text{def}}{=} \arg\max_z\{f(N,z)\}$, for a given N.

One can show that $f_k(N,z), k = L, R$ is supermodular in order quantity, z, and overhead cost, b_k, for fixed N. Therefore,

Proposition 3 *Suppose that the number of suppliers, N, is fixed. Then, for $k = L$, R, $z_k^* \stackrel{\text{def}}{=} \arg\max_z\{f_k(N,z)\}$ is non-decreasing in the overhead cost, b_k.*

Because the overhead cost, b_k, is increasing in the fixed cost of working with a supplier, C, it follows that, if the optimal order quantity, $z^* = z_k^*$ for $k = L$ or $k = R$, then the optimal order quantity may be increasing in C. This is an effect of the limited-liability assumption. In the model with full liability (i.e. $l = -\infty$) the optimal order quantity does not depend on the fixed cost, provided that the fixed costs are small enough for the firm to be in business.

One would expect that the optimal order quantity of a limited-liability manufacturer will be higher than the optimal order quantity of a full-liability manufacturer. After all, limited liability curbs the manufacturer's overage costs, thereby inducing the manufacturer to stock larger quantities. This is, indeed, the case as stated in the following proposition:

Proposition 4 *Suppose that the number of suppliers, N, is fixed. Then, the more negative the liability level, l, the smaller the optimal order quantity, z^*.*

So far we have explored the choice of order quantity, z, assuming that the number of suppliers, N, is fixed and ignoring optimization constraints. Next, let us consider the choice of the number of suppliers, N, still ignoring optimization constraints.

Proposition 5 *Suppose that z is fixed. Then:*

1. *If $r_I < r_S$, then the objective function, $f(N,z)$ is decreasing in the number of suppliers, N, and the manufacturer will choose the optimal number of suppliers, $N^* = 1$.*

2. *If* $r_I > r_S$ *and the fixed cost of working with a supplier,* $C > (r_I - r_S)/(1 + r_I)\hat{S}$, *then the objective function,* $f(N, z)$, *is decreasing in the number of suppliers, N, and the manufacturer will choose the optimal number of suppliers,* $N^* = 1$.
3. *If* $r_I > r_S$ *and the fixed cost of working with a supplier,* $C \leq (r_I - r_S)/(1 + r_I)\beta$, *then the objective function,* $f(N, z)$ *is increasing in the number of suppliers, N, and the manufacturer will choose to work with the largest possible number of suppliers (as long as the financing constraints are ignored).*

These results are intuitive. Ignoring the financing constraints, part 1 of the proposition says that, if the internal capital is the cheaper source of financing, then the manufacturer will order from only one supplier, as there is no reason for the manufacturer to work with multiple suppliers and incur the fixed cost of C for each one (recall that, in this section, there is no supply risk). On the other hand, as shown in part 3, when the suppliers are the cheaper source of financing and the cost of working with an extra supplier is much smaller than the guaranteed supplier loan amount, then the manufacturer may choose to work with multiple suppliers because it can reinvest money borrowed from the suppliers in its other businesses. However, as shown in part 2, when the fixed cost of working with a supplier (C) is sufficiently close to the maximum loan available from a supplier ($\hat{S}$), then the additional loan from a supplier will not be worth the additional fixed cost, and the manufacturer will again choose to work with only one supplier. Finally, note that Proposition 5 does not describe the behavior of the objective function when $r_I > r_S$, and $(r_I - r_S)/(1 + r_I)\beta < C \leq (r_I - r_S)/(1 + r_I)\hat{S}$. In this case, depending on which part (f_L or f_R) of the objective function we are considering, the objective function can be either increasing or decreasing in the number of suppliers N.

To further study the optimal choice of the number of suppliers, N^*, we have to consider the effects of the financing constraints.

8.4.2.2 Manufacturer's Problem with Financial Constraints

When there is no limit on the total capital available to the manufacturer, Sect. 8.4.2.1 describes the optimal operational decisions of the manufacturer. Unfortunately, the manufacturer does not have access to unlimited capital. In this subsubsection, we consider the effect of financial constraints on the manufacturer's operational decisions.

Finding the value of optimal N for a problem with financial constraints is not a trivial task. Even when the financial constraints are not binding, as N changes, $f_L(N, z)$ and $f_R(N, z)$ may increase or decrease, and the unconstrained optimal order quantity, z^*, may switch between z_L^*, z_R^* and $\hat{z}$. This complicated relationship between N and z^*, together with the discrete nature of N, makes analytical derivation of the optimal number of suppliers, N^*, unlikely. Propositions in Sect. 8.4.2.1 provide structural properties of the objective function, $f(N, z)$, and bounds on the optimal N

and z. In addition, the following propositions describe further bounds on the optimal values of N and z due to financing constraints.

First, we will derive a bound on the optimal number of suppliers, N^*, in the presence of financing constraints.

According to Assumption (5), the loan available from each supplier is less than the cost of ordering from that supplier by at least $C - \beta$, and the difference must be made up by the internally generated capital. Therefore, the amount of internal capital imposes a limit on the number of suppliers the manufacturer can work with [formally, this is seen from the first inequality in (8.4)].

Proposition 6 *The optimal number of suppliers, N^*, is limited by $N^* \leq I/C - \beta$.*

Next, we will present bounds on the optimal order quantity. The number of suppliers, N, restricts the feasible choices for the order quantity, z, since the number of suppliers affect the amount of loans available to the manufacturer. The following proposition (which follows from system (8.4) by fixing N) formalizes this relationship.

Proposition 7 *Suppose that the number of suppliers, N, is fixed and satisfies $N \leq I/C - \beta$. Then:*

1. *If $N \leq \frac{\alpha I}{w(\hat{S}-\beta)-\alpha(\hat{S}-C)}$, then the optimal order quantity satisfies*

$$z \leq z_{\max}(N) \stackrel{\text{def}}{=} \frac{I + (\hat{S} - C)N}{w}. \tag{8.17}$$

2. *If $\frac{\alpha I}{w(\hat{S}-\beta)-\alpha(\hat{S}-C)} \leq N \leq \frac{I}{C-\beta}$, then the optimal order quantity satisfies*

$$z \leq z_{\max}(N) \stackrel{\text{def}}{=} \frac{I - (C - \beta)N}{w - \alpha}. \tag{8.18}$$

In Sect. 8.4.2 we focused on the manufacturer's operational decisions when the demand is stochastic and the yield is deterministic. Next we turn our attention to the case in which the demand is deterministic and the yield is stochastic.

8.4.3 Operational Decisions Under Deterministic Demand and Stochastic Yield

The manufacturer may decide to use several suppliers not only to acquire access to a larger capital pool (as we discussed in Sect. 8.4.2), but also to diversify risk, if suppliers are not perfectly reliable. The trade offs between diversification benefits and set-up costs, without financing constraints and with full manufacturer's liability, have been studied in [1]. This subsection extends their analysis by considering a model with limited liability and financing constraints. This is a difficult model to analyze. Therefore, in this section, we will make the following assumption:

Assumption 8 *The demand, D, is deterministic.*

Besides simplifying the analysis, this assumption may be useful for a problem where the demand uncertainty is much smaller than the supply uncertainty, for example, when the manufacturer has long-term contracts with the customers.

8.4.3.1 Manufacturer's Objective Function

Recall that the manufacturer's objective function is given by Expressions (8.9) and (8.14). We will only consider operational decisions: number of suppliers, N, and order quantity, z, which satisfy

Assumption 9 $pD > a_k z + b_k(N), k = L, R$.

Assumption (9) can be made without loss of generality because, if it is violated, the firm is guaranteed to have negative profit. The following lemma offers a convenient expression for $f_k, k = L, R$ (see Eq. 8.14 for definitions of $f_k, k = L, R$):

Proposition 8 *Suppose that the number of suppliers, N, is fixed. Let $\overline{X}$ be a random variable with c.d.f. Φ and p.d.f. ϕ and let $f_k, k = L, R$ be defined by Eq. 8.14. Define*

$$\Gamma(m) = \int_{m}^{\infty} x\phi(x)\mathrm{d}x, \tag{8.19}$$

$$\gamma(m) = \Gamma'(m) = -m\phi(m) \tag{8.20}$$

Then,

$$f_k(z) = (pD - a_k z - b_k - l)\int_{\frac{D}{z}}^{\infty} \phi(x)\,\mathrm{d}x + \int_{\frac{a_k z + b_k + l}{pz}}^{\frac{D}{z}} [pxz - a_k z - b_k - l]\phi(x)\,\mathrm{d}x \tag{8.21}$$

$$\begin{aligned} f_k'(z) &= p\int_{\frac{a_k z + b_k + l}{pz}}^{\frac{D}{z}} x\phi(x)\,\mathrm{d}x - a_k \Pr\left[\overline{X} \geq \frac{a_k z + b_k + l}{pz}\right] \\ &= p\left[\Gamma\left(\frac{a_k z + b_k + l}{pz}\right) - \Gamma\left(\frac{D}{z}\right)\right] - a_k \Pr\left[\overline{X} \geq \frac{a_k z + b_k + l}{pz}\right] \end{aligned} \tag{8.22}$$

and, if $\gamma(\cdot)/\Gamma(\cdot)$ is decreasing, $f_k, k = L, R$ are unimodal in z.

Using this lemma, one can prove that the objective function of the model with the random supplier yield is unimodal in the total order quantity.

Proposition 9 *If the mean of supplier yields is normally distributed,* $\overline{X} \sim N\left(\mu, \sigma/\sqrt{N}\right)$, *and Assumptions (5), (6), and (9) hold, then the manufacturer's objective function,* $f(z)$, defined by (8.9), is unimodal in the order quantity, z.

Thus, the optimization problem in this section has the same structure as the optimization problem with certain yield and random demand in Sect. 8.4.2. If the number of suppliers, N, is fixed, we can find the optimal order quantity, z^*, using the analog of Lemma 4 in Appendix 8.8.2. Unlike the model in Sect. 8.4.2, each component $(f_k(N, z), k = L, R)$ of the objective function is submodular in the order quantity, z, and the overhead cost, b_k. Therefore,

Proposition 10 *Suppose that the number of suppliers, N, is fixed. For* $k = L, R$, *if* $b_k + l > 0$, *then* $z_k^* \overset{\text{def}}{=} \arg\max_z\{f_k(N, z)\}$ *is non-increasing in the overhead cost,* b_k, *and as the liability level* l *becomes more negative,* z_k^* *increases (non- strictly). If* $b_k + l < 0$, *then* $z_k^* \overset{\text{def}}{=} \arg\max_z\{f_k(N, z)\}$ *is non- decreasing in the overhead cost,* b_k, *and as the liability level* l *becomes more negative,* z_k^* *decreases (non-strictly).*

Finding the optimal number of suppliers, N^*, for the model in this section is as difficult as finding the optimal number of suppliers for the model in Sect. 8.4.2. To obtain further managerial insights into the manufacturer's optimal decisions, in particular, the optimal number of suppliers, we conduct a numerical study, which is discussed in the next section.

8.5 Numerical Study

Propositions in Sect. 8.4 provide structural properties of the objective function, $f(N, z)$, and bounds on the optimal N and z. Such analytical result allows us to devise an efficient search algorithm to find the optimal solution, thus facilitating a numerical study. As we discuss in this section, the numerical study provides several valuable insights into the choice of optimal operational decisions under financing constraints.

8.5.1 Stochastic Demand, Deterministic Yield

We first focus on the case in which the demand is random, but the yield is perfect and deterministic. The numerical study uses the following default values of model parameters: the rate of internal financing and the rate of supplier loans are $(r_I, r_S) = (0.2, 0.1)$,[8] the per unit revenue is $p = 3$, the wholesale price is $w = 0.5$, the fixed cost of working with a supplier is $C = 20$, the parameters of the supplier loans are $\alpha = 0.48$ and $\beta = 4$, the internal capital is $I = 75$, the limit on the supplier loan is

[8] The graphs look either identical or similar when $(r_I, r_S) = (0.1, 0.2)$

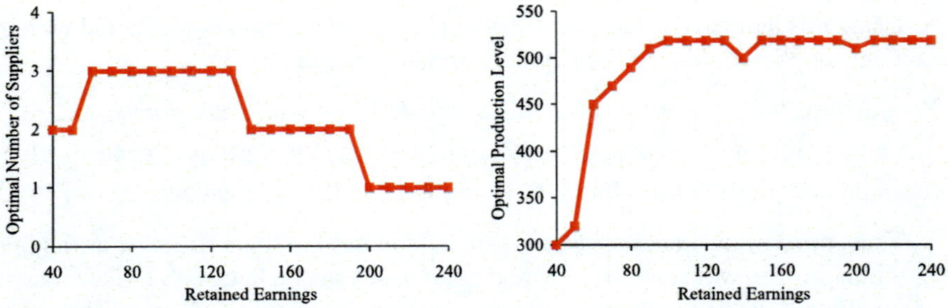

Fig. 8.1 The effect of the internal capital

$\hat{S} = 75$, the demand is normal with mean $\mu_D = 500$ and variance $\sigma_D^2 = 500$, and the liability level is $l = 0$.

8.5.1.1 Effects of Internal Capital

Consider the effects of the internal capital first, depicted in Fig. 8.1.

As the internal capital, *I*, decreases, financing constraint forces the firm to order smaller quantities (as shown in the right panel of Fig. 8.1), which, in turn, causes a decrease in the firm's revenues. Rather than suffer from a further decline in revenues, the firm may prefer to incur the fixed cost of working with an extra supplier who will provide the firm with additional financing. Therefore, we observe in the left panel of Fig. 8.1 that, for high *I*, the optimal number of suppliers, *N*, and the optimal order quantity, *z*, may increase as *I* decreases. However, once the internal capital (*I*) becomes too small, the firm will have to reduce the number of suppliers again. To see why, recall that the limit on the trade-credit available from a supplier, $\bar{S}(z/N) = \min\{\hat{S}, \alpha z/N + \beta\}$, is less than the cost of ordering from that supplier, $wz/N + C$. Therefore, for each supplier the firm works with, the difference between the trade-credit and the cost of ordering must be covered through the use of internal financing. If *I* is too small, the firm cannot afford the fixed cost of working with an additional supplier, which forces the firm to reduce the number of suppliers. Similar behavior is observed in the supplier loan-limit study.

8.5.1.2 Effects of Fixed Cost

An increase in the supplier fixed cost, *C*, may cause an increase in the optimal number of suppliers, *N*. One example of such behavior is depicted in Fig. 8.4 by curves marked 'Developing,' which correspond to the default parameter set for these numerical examples.

This surprising result is observed for moderate *C*, when the firm is borrowing the absolute maximum amount, $\hat{S}$, from each supplier. The following is the intuitive explanation for this phenomenon. As the fixed cost of working with a supplier, *C*, increases, because of the financing constraints [inequalities (8.4)], the firm has to

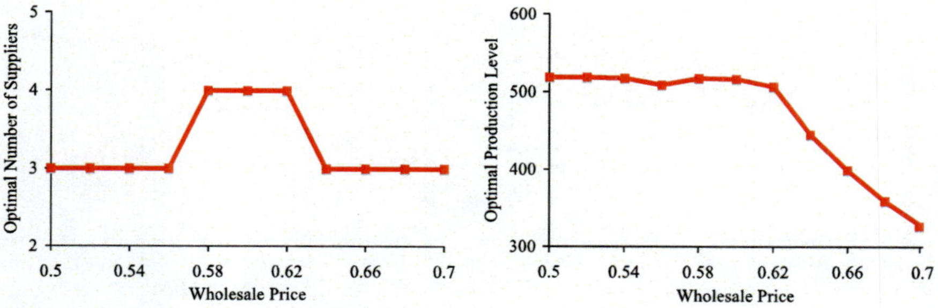

Fig. 8.2 The effect of the wholesale price

reduce the order quantity (and, hence, future revenues) in order to pay for the increased cost, C. If the firm is borrowing the absolute maximum amount, $\hat{S}$, (that is the second inequality in (8.4) is binding), the firm can relax its financial constraints (and increase revenues) by adding suppliers.

How many suppliers the firm will add depends on the extra supplier cost, C, which appears in the objective function, and also on the value of N for when the manufacturer runs out of internal capital to finance additional fixed costs of working with suppliers (i.e. the first inequality in constraint (8.4) becomes binding).

8.5.1.3 Effects of Wholesale Price

The number of suppliers may also be non-monotone in the wholesale price, w, as depicted in Fig. 8.2 (in this numerical example, $C = 5$).

When the wholesale price, w, is large, the business is barely profitable, and the firm works with few suppliers as expected. As w becomes smaller, the firm would like to order a larger quantity (since the profit margin on each unit sold to the customer is larger), and in an effort to order a larger quantity, the firm may find it preferable to work with additional suppliers so that it can raise the necessary cash. As we have already explained in our discussion on the effects of the fixed cost of working with a supplier, the firm can relax its financial constraints by working with more suppliers, if the second inequality in (8.4) is binding. Once the wholesale price, w, is sufficiently small, the firm reduces the number of suppliers again to save on the fixed cost of working with a supplier. Although this reduces the cash available for purchases, which, in turn, drives the order quantity down, the firm still prefers saving the fixed cost because the extent of the reduction in order quantity is dampened by the small wholesale price.

8.5.1.4 Effects of the Standard Deviation of the Demand

The left panel of Fig. 8.3 demonstrates that, depending on the value of unit revenue, p, the optimal number of suppliers could be either increasing or decreasing in the standard deviation of the demand. The key to understand this behavior is the right

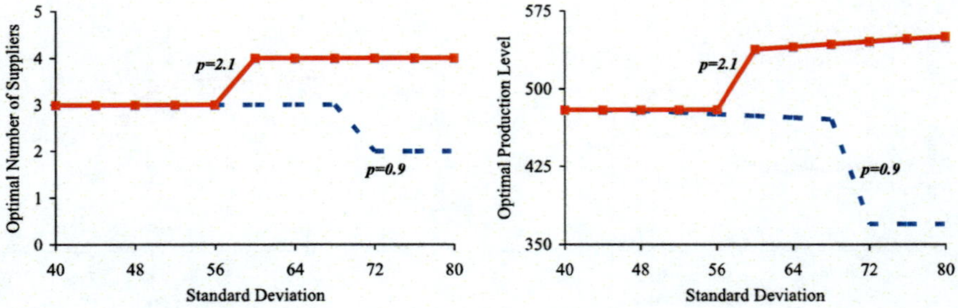

Fig. 8.3 The effect of the demand standard deviation

panel of Fig. 8.3 and what we know about the relationship between the standard deviation of the demand and the optimal order quantity for the newsvendor problem (although, because of the limited liability, we do not have a newsvendor problem, the behavior of the optimal order quantity for our problem is similar).

For the normally distributed demand, the optimal order quantity for the newsvendor problem is $z^{\text{news}} = \mu_D + \sigma_D \mathcal{N}^{-1}(1 - a/p)$, where $\mathcal{N}$ is the c.d.f. of a standard normal random variable, a is the variable cost, μ_D is the demand's mean, and σ_D is the demand's standard deviation. From this expression, if $a/p < 1/2$, then z^{news} increases in σ_D, and if $a/p > 1/2$, then z^{news} decreases in σ_D. The unconstrained order quantity for our problem [derived from the first order condition (8.34)] also depends on the value of ratio a/p. If a/p is small (e.g. when $p = 2.1$, for the case $r_I > r_S$, $a_L/p = [(1 + r_I)w - (r_I - r_S)\alpha]/p = 0.263$ and $a_R/p = (1 + r_I)w/p = 0.286$), the optimal order quantity increases and, to finance this increase (to relax financing constraints), the manufacturer may increase the number of suppliers. Conversely, if a/p is large (e.g. when $p = 0.9$, for the case $r_I > r_S$, $a_L/p = [(1 + r_I)w - (r_I - r_S)\alpha]/p = 0.613$ and $a_R/p = (1 + r_I)w/p = 0.667$), the unconstrained optimal order quantity decreases and the manufacturer can reduce the number of suppliers because financing constraints are no longer binding.

8.5.1.5 Developing versus Developed Economies

Finally, let us contrast the effect of financial constraints on firms operating in developing and developed economies. Firms in developed economies enjoy access to much more capital compared to firms in developing economies.[9] In this numerical example the loan limits for the developing economy are set to $I = 75$, $\hat{S} = 75$. Let loan limits for the developed country be $I = 75$, $\hat{S} = 300$.

[9] In addition to the capital availability, firms operating in developed and developing economies may also face different costs of capital. Our numerical experiments showed that that the effects of capital costs are predictable: for instance, higher internal financing rate encourages the manufacturer to work with more suppliers.

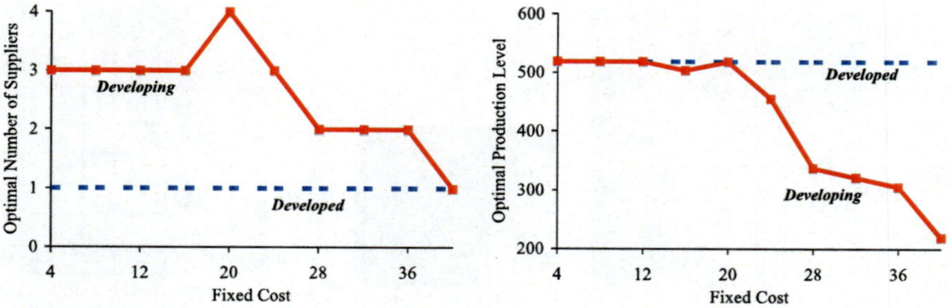

Fig. 8.4 Developing versus developed economies

Therefore, as Fig. 8.4 illustrates, we would expect that for any given level of the fixed cost, C, firms in developing economies will tend to have a greater number of suppliers (ceteris paribus).

Does this mean that we should expect to observe this disparity empirically? Not necessarily. Because firms in developing economies will also tend to have greater fixed costs, C (due to the operational inefficiencies), they may optimally keep the number of suppliers low. However, as Fig. 8.4 illustrates, if C is very large, our model predicts that these firms will place smaller orders, have lower inventory (and hence, experience a higher frequency of stock-outs), an effect that has been empirically observed in [18].

Here we focused on the availability of external financing and fixed costs as the main differentiators between developed and developing countries. One could make further comparisons by focusing on other differentiators, e.g. supplier yield characteristics (suppliers in developed countries may be more reliable).

8.5.1.6 Interactions Between Parameter Values

Figure 8.5 shows the optimal number of suppliers as functions of the limit on supplier loans and the fixed cost of working with a supplier, the limit on supplier loan and the wholesale price, and the internal capital of the manufacturer and the demand mean. Lighter shades correspond to a higher number of suppliers. The black shade corresponds to $N^* = 1$; the white shade corresponds to $N^* = 3$.

From Fig. 8.5 we observe that the higher number of suppliers corresponds to low-fixed cost and low-supplier loan limit (left panel) or low-wholesale price and low-supplier loan limit (center panel). The picture in the right panel of Fig. 8.5 shows that limited internal capital may prevent the manufacturer from increasing the number of suppliers even as the demand for products (and hence demand for financing) increases.

8.5.2 Deterministic Demand, Stochastic Yield

We next discuss the numerical results when the yield is random. In this numerical study, we observed that the effects of the financial constraints, the fixed cost, and the

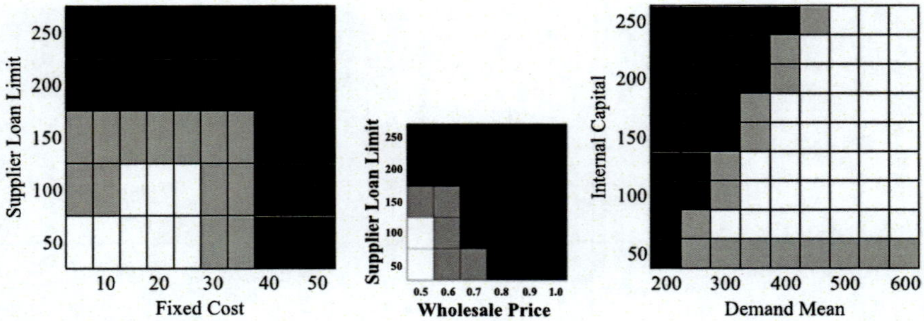

Fig. 8.5 The optimal number of suppliers. Interactions between parameter values

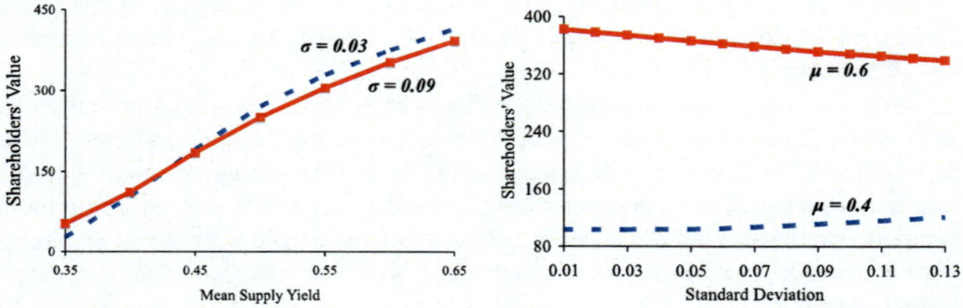

Fig. 8.6 Shareholder value as a function of yield parameters

wholesale price are the same under this model as in the model with deterministic yield. Therefore, in what follows, we only focus on the effects of the random yield, which is characterized by its mean, μ, and its standard deviation, σ, on the shareholders' value, the optimal number of suppliers, and the optimal order level. The presentation will focus on the case where the internal rate is lower than suppliers' interest rate ($r_I < r_S$). The results for the other case ($r_I > r_S$) are similar.

The following parameter values were used in the numerical examples in this subsection: unit revenue, $p = 1.75$, wholesale price, $w = 0.5$, interest rate on supplier loans, $r_S = 0.2$, rate on internal capital, $r_I = 0.1$, fixed cost of working with a supplier, $C = 5$, $\beta = 4.9$, $\alpha = 0.48$, absolute limit for supplier loans, $\hat{S} = 225$, internal capital, $I = 20$, and demand, $D = 500$.

According to the left panel of Fig. 8.6, the shareholders' value is increasing as the expected yield, μ, increases. This behavior is to be anticipated because the firm benefits if the average reliability of its suppliers increases.

Surprisingly, as shown in the right panel of Fig. 8.6, the increase in the standard deviation, σ, of the supplier yield may result in either an increase or a decrease in the shareholders' value. If the expected yield is high (low) the shareholders' value decreases (increases) in the standard deviation of the yield. To understand this phenomenon, consider how functions f_L and f_R depend on mean yield $\overline{X} = \sum_{k=1}^{N} X_k/N$. Recall that $f_k(\overline{X}) = E\left[\Pi_k(\overline{X})\right]^+$, $k = L, R$ [see Eq. 8.14]. We can rewrite this expression as

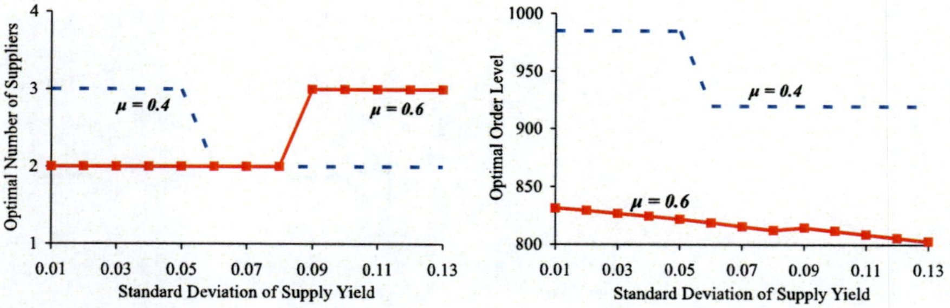

Fig. 8.7 Effect of a supplier's yield uncertainty

$$f_k(\overline{X}) = E\left[\Pi_k(\overline{X})\right] + E\left[-\Pi_k(\overline{X})\right]^+ \quad k = L, R \tag{8.23}$$

The first term in (8.23) represents the firm's expected profit. Because Π_k is a concave function, as the standard deviation of yield increases, this term decreases. The second term in (8.23) represents the value of the option to default that shareholders hold (because they have limited liability). Function $[-\Pi_k(\cdot)]^+$ is convex and, therefore, this term increases as the standard deviation of the yield increases. Thus, the change in the shareholders' value, as the volatility of the yield, σ, increases, comes from the decrease in the expected profit and the increase in the value of the option to default. When the expected yield is small (e.g. when $\mu = 0.4$), the firm is close to bankruptcy and, therefore, the value of the option provides the largest contribution to the shareholders' value. This means that the convex part of the objective function dominates, and the decision-maker behaves as a risk-seeking agent and responds to the increasing volatility of the supplier yield by decreasing the number of suppliers, as shown in the left panel of Fig. 8.7. When the expected yield is high (e.g. when $\mu = 0.6$), the expected profit provides the largest contribution to the shareholders' value that is, the concave part of the objective function dominates, and the decision-maker behaves as a risk-averse agent and responds to the increasing volatility in supplier yield by increasing the number of suppliers.

The left panels of Figs. 8.7 and 8.8 confirm our intuition about diversification and the risk-averse behavior of the decision-maker. We use the standard deviation of the order delivered as a proxy of risk and observe that, for the curves corresponding to $\mu = 0.6$, an increase in the optimal number of suppliers coincides with a decrease in risk. Thus, the decision-maker is acting as a risk-averse agent. Similarly, for the curves corresponding to $\mu = 0.4$, a decrease in the number of suppliers corresponds to an increase in risk, indicating that the decision-maker is acting as a risk-seeking agent. We defer the discussion of the right panel of Fig. 8.8 until our discussion of Fig. 8.9.

Next, consider the right panel of Fig. 8.7, which illustrates the effect of the standard deviation, σ, of supplier's yield on the optimal order quantity, z^*. We will focus on the curve corresponding to the expected yield, $\mu = 0.4$.

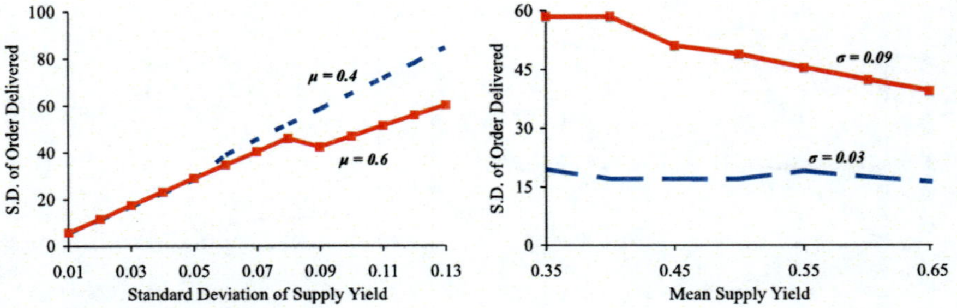

Fig. 8.8 The standard deviation of the order delivered

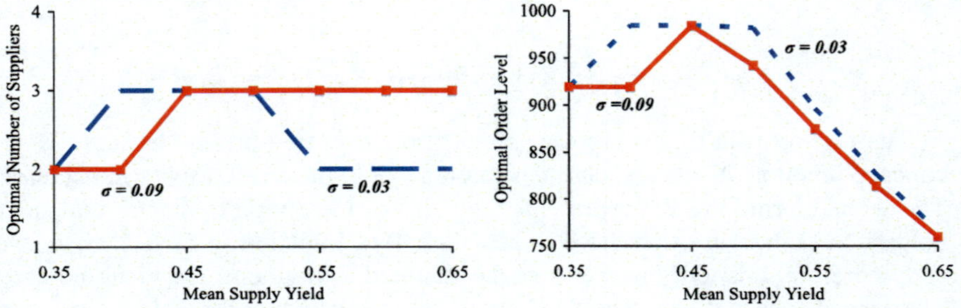

Fig. 8.9 Effect of a supplier's expected yield

Observe that the optimal order quantity decreases in σ. To understand this behavior, note that the profit of the firm can be written as

$$\Pi = B - A\overline{X}z - p\max(D - \overline{X}z, 0), \tag{8.24}$$

for some constants, A and B. This is a payoff on a portfolio consisting of a safe bank account (B), a short position in the underlying asset ($\overline{X}z$), and a short position in a put option ($\max(D - \overline{X}z, 0)$) on the underlying asset.[10] The distribution of the underlying asset is $\mathcal{N}\left(\mu z, \sigma/\sqrt{N}z\right)$. From the option theory, the value of the put option increases in the underlying asset's variance (in this case σ^2/Nz^2). Therefore, assuming that the optimal number of suppliers, N^*, is constant, an increase in the standard deviation of the yield, σ, leads to an increase in the put option value ($E[\max(D - \overline{X}z, 0)]$) and a decrease in the expected profit ($E[\Pi]$). Shareholders can hedge against the effects of increasing σ by reducing the optimal order quantity, z^*. Finally, observe that for $\mu = 0.6$ the optimal order level increases when the number of suppliers changes from 2 to 3. The explanation for this behavior is akin to the results in Proposition 10. As the number of suppliers increases, the overhead

---

[10] See [24] for definitions and discussion of option contracts.

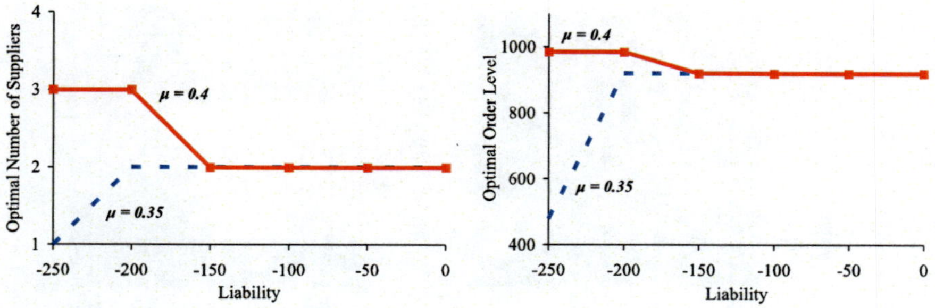

Fig. 8.10 Effect of limited liability

cost increases and the maximizers of each of the two parts of the objective function increases.

Curves in the left panel of Fig. 8.9 are formed due to the (now familiar) trade off between the risk-seeking and risk-averse behavior of the shareholders, and also due to a trade off between the benefits of diversification and the costs of working with suppliers. Comparing the left panel of Fig. 8.9 and the right panel of Fig. 8.8, observe that an increase in the number of suppliers corresponds to a decrease in risk, and a decrease in the number of suppliers corresponds to an increase in risk. When the expected yield, μ, is small, the convex part of the shareholders' objective dominates and the shareholders behave as risk-seeking agents. As the expected supplier yield increases, the risk-seeking behavior is replaced by the risk-averse one and the number of suppliers increases. As the expected yield continues to grow, that is, as the suppliers become more reliable, the need for diversification becomes less pressing and the firm can start saving on fixed costs by reducing the number of suppliers. Curves in the right panel of Fig. 8.9 follow from the observation that, as the suppliers become more reliable, the manufacturer does not have to order as much to compensate for possible losses.

Figure 8.10 illustrates the effect of limited liability. In this numerical study the standard deviation of the supplier yields is $\sigma = 0.1$. First, while conducting numerical experiments, we observed that the limited liability manifests itself only when the manufacturer is fairly close to bankruptcy and, hence, the option to default on part of its obligations is valuable. Therefore, Fig. 8.10 contains curves for two cases which bring the manufacturer close to bankruptcy: mean supplier yields $\mu = 0.35$ and $\mu = 0.4$. If supply were certain, the more negative liability level, l, becomes, the less valuable the business becomes, the smaller the order that will be placed, the less financing will be needed, and the fewer suppliers the manufacturer will work with. However, when supply is uncertain, the manufacturer benefits from diversification by working with more suppliers. Whether the number of suppliers increases or decreases as l decreases depends on which of the two forces (financing required vs. diversification) dominates. For the graph corresponding to $\mu = 0.4$ in Fig. 8.10 the diversification force prevailed.

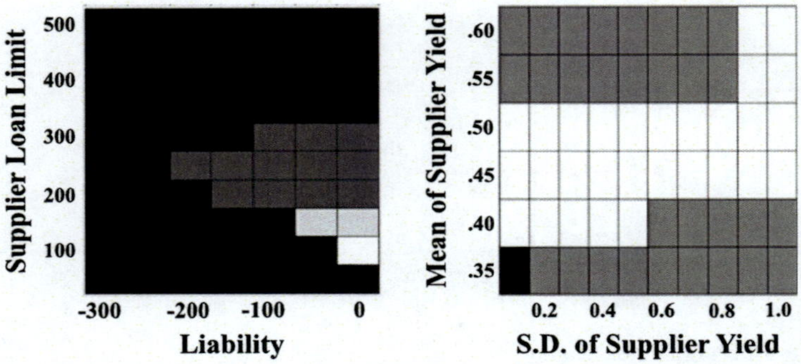

Fig. 8.11 The optimal number of suppliers as a function of limited liability and supplier loan limit (*left panel*), and mean and standard deviation of the supplier yield (*right panel*)

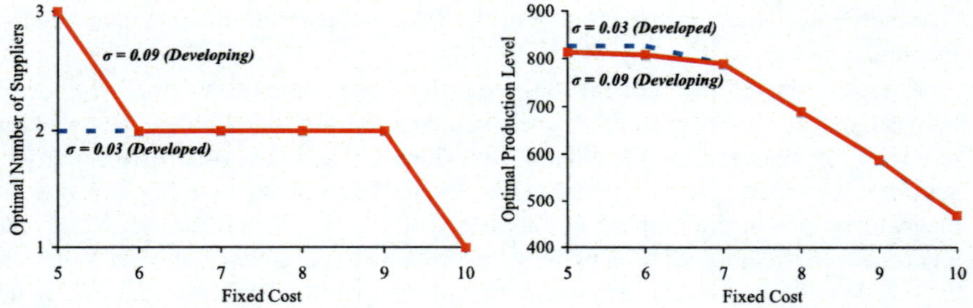

Fig. 8.12 Developing versus developed economies. Random yield. Mean value of the supplier yield is $\mu = 0.6$

Figure 8.11 highlights the interaction between limited liability and financing constraints due to supplier loan limits (left panel), and diversification choices for different values of yield parameters μ and σ. In this figure we plot the optimal number of suppliers. The lighter shades correspond to the greater number of suppliers. The black color corresponds to $N^* = 1$.

Finally, Fig. 8.12[11] illustrates that a greater volatility of supplier yields will encourage firms in developing economies to have a greater number of suppliers if the fixed cost of working with the suppliers is not too high.

8.6 Heterogeneous Suppliers

In this section we will discuss the implications of relaxing the assumption of a homogeneous supplier base. In general, suppliers could differ in a number of attributes: (μ_i, σ_i), distribution parameters of supplier i yield; w_i, wholesale price charged by suppler i; C_i, fixed cost of working with supplier i; $\widehat{S_i}$, absolute limit on supplier i loan amount; (α_i, β_i), parameters of the supplier i trade-credit loan; r_i, interest rate

[11] In this example, $\mu = 0.6$.

on supplier i loan. The multi-attribute problem of selecting a subset of suppliers for the manufacturer to work with can only be solved numerically, except for special cases.

To begin, assume that the manufacturer can only work with a single supplier and, hence, the supplier selection problem becomes: which supplier should win the manufacturer's business. For each of the suppliers one needs to solve problem (8.2) with an additional constraint, $N = 1$, and then select the solution which offers the highest value of the objective function. It is not difficult to see that, everything else being equal, the manufacturer favors the supplier with the lowest r_i, C_i, or w_i, and the highest $\widehat{S}_i$, β_i, α_i, or μ_i. The effect of σ_i is not immediately obvious. As we discussed in Sect. 8.4.3, the presence of the option to default may encourage the manufacturer to take more risk, by working with suppliers whose yield distribution has a higher standard deviation.

A general model, where the manufacturer can work with any number of suppliers, has the following mathematical form:

$$\max_{\{\mathbf{S},\mathbf{y}\}} E\left\{ p\min[D, Q(\mathbf{y})] - (1+r_I)\sum_i \left(w_i y_i + C_i \mathbf{1}_{\{y_i>0\}}\right) - \sum_i (r_i - r_I)S_i + (1+r_I)I \right\}^+ \tag{8.25a}$$

$$\sum_i \left(w_i y_i + C_i \mathbf{1}_{\{y_i>0\}}\right) \leq \sum_i S_i + I, \tag{8.25b}$$

$$0 \leq S_i \leq \overline{S}_i(y_i), \tag{8.25c}$$

$$\overline{S}_i(y) = \min\left(\widehat{S}_i, \beta_i + \alpha_i y\right). \tag{8.25d}$$

Where $Q(\mathbf{y})$ is the quantity received by the manufacturer from the suppliers (possibly a random variable). Similar to the analysis in Sect. 8.4.1, we can determine the optimal financing decisions, given a particular choice of operational decisions. Specifically, suppose that $\mathbf{y}$ is given. Consider only the suppliers that received positive orders ($y_i > 0$) from the manufacturer and sort them according to the value of the rates on their loans, r_i, in increasing order. Let $k = \min\{j : r_j \geq r_I\}$ be the index of the first supplier whose rate exceeds the rate on internal capital for the manufacturer. Then, for $1 \leq i < k$, the optimal loan amount is

$$S_i^* = \overline{S}_i(y_i). \tag{8.26}$$

For $i \geq k$, the optimal loan amount is

$$S_i^* = \min\left\{ \overline{S}_i(y_i), \left[\sum_j \left(w_j y_j + C_j\right) - I - \sum_{j=1}^{i-1} S_j^*\right]^+ \right\}. \tag{8.27}$$

Substituting the optimal loan amounts, $\mathbf{S}^*$, into Problem (8.25) we derive an optimization problem with operational decisions, $\mathbf{y}$, only. Similar to the analysis in Sect. 8.4.1, the domain of the objective function of this optimization problem can be divided into regions such that, within each region, k, the objective function is written as $f_k(\mathbf{y}) = E\left\{p \min\left[D, Q(\mathbf{y})\right] - \sum_i (a_i y_i + b_i)\right\}^+$, where b_i may depend on $\mathbf{1}_{y_i>0}$. The constraints for this optimization problem are

$$\sum_i w_i y_i - \sum_i \left(\widehat{S}_i - C_i\right) \mathbf{1}_{y_i>0} \le I, \tag{8.28a}$$

$$\sum_i (w_i - \alpha_i)\, y_i + \sum_i (C_i - \beta_i)\, \mathbf{1}_{y_i>0} \le I. \tag{8.28b}$$

While in the model with homogeneous suppliers we had only two parts (one per financing source) of the objective function, here there are as many parts as there are suppliers plus one (corresponding to financing from internal capital). This makes the problem with external financing too complex to analyze.

Let us focus on the effect of supplier random yield. To simplify, let us assume that the interest rate on internal financing is the lowest (i.e. $r_I \le r_i$, for all i), the manufacturer has more than sufficient internally generated capital to run the firm (i.e. $I > \sum_i (w_i y_i + C_i)$ for all reasonable values of $\mathbf{y}$, implying, in particular, that no loans from the suppliers are needed and limited liability is never used), and demand, D, is deterministic. With these assumptions, we obtain a model similar to the non-identical suppliers model in [1]. The essential difference between their model and ours lies in the assumption about payment from the manufacturer to the suppliers. We assumed that the manufacturer pays for the items ordered, while they assumed (effectively) that the manufacturer pays only for the items delivered. Still, the similarity between models allows us to replicate the results in [1]. Specifically, we can show that the objective function can be represented as follows:

$$\begin{aligned} &E\left\{p \min[D, Q(\mathbf{y})] - (1+r_I)\sum_i \left(w_i y_i + C_i \mathbf{1}_{y_i>0}\right) + (1+r_I) I\right\} \\ &\quad = p\left[D + (\overline{\mu} - D)\Phi(D) - \overline{\sigma}^2 \phi(D)\right] \\ &\quad - (1+r_I)\sum_i \left(w_i y_i + C_i \mathbf{1}_{y_i>0}\right) + (1+r_I) I, \end{aligned} \tag{8.29}$$

where ϕ and Φ are p.d.f. and c.d.f. of the normal random variable $Q(\mathbf{y})$, whose mean is $\overline{\mu} = \sum_i \mu_i y_i$, and variance $\overline{\sigma}^2 = \sum_i \sigma_i^2 y_i^2$. Using Theorem 3 in [1], the objective function is concave. The first order condition for the order quantity with supplier i is:

$$p\mu_i \Phi(D) - p y_i \sigma_i^2 \phi(D) = (1+r_I) w_i. \tag{8.30}$$

Suppose that $\mu_i = \mu$ and $w_i = w$ for all i. Then, conditions (8.30) imply that the optimal order quantities to suppliers i and j are inversely proportional to the variances of supplier yields:

$$\frac{y_i^*}{y_j^*} = \frac{\sigma_j^2}{\sigma_i^2}. \tag{8.31}$$

Alternatively, one may argue that the supplier market is in equilibrium where the prices w_i are proportional to the expected fraction the order suppliers deliver, that is, $w_i = A\mu_i$. In this case, the orders to suppliers will be:

$$\frac{y_i^*}{y_j^*} = \frac{\sigma_j^2/\mu_j}{\sigma_i^2/\mu_i}. \tag{8.32}$$

Agrawal and Nahmias [1] solves a general non-identical supplier model only numerically. Because of the financing constraints, piecewise-defined objective function, and limited liability, our model is even more complex than that in Agrawal and Nahmias [1]. Therefore, we also cannot solve a general model analytically.

8.7 Conclusions, Limitations, and Extensions

Numerous empirical studies report that trade-credit (supplier financing) is the number one source of short-term financing in developed countries, accounting for as much as twice the amount of short-term bank loans. The role of trade-credit is even more prominent in developing countries, where access to traditional sources of financing is severely limited. Other differences in business environments between developed and developing countries are the costs of working with suppliers, and supplier reliability.

In this chapter, we study how trade-credit financing, internal financing, cost of working with a supplier, and the supplier yield affect the optimal number of suppliers and the optimal order size. For our study we use a stylized model where the salient problem features are joint operational and financial decisions and financing constraints.

Our extensive theoretical and numerical analysis generated several testable hypotheses. Some of these hypotheses have already been confirmed in prior empirical studies. Others can be verified in future empirical work. We derived both anticipated and surprising results. These results can be explained by considering trade offs between the main elements of the model: financing constraints and their effect on feasible order quantity, cost of working with the suppliers and its effect on the objective function, order quantities and their effects on revenues, decomposition of the objective function into a concave profit term (which encourages diversification) and a convex option to default term (which encourages reducing the number of suppliers). For example, as the availability of either internal financing or supplier loans diminishes, the optimal number of suppliers may increase. To understand this, consider that the manufacturer, by paying the extra cost of working with additional suppliers, benefits by relaxing financing constraints and increasing order quantity, to earn higher expected revenues.

More surprising are the observations that an increase in the cost of working with a supplier or the wholesale price may result in an increase in the optimal number of suppliers.

Also surprisingly, we observe that, as the standard deviation of the supplier yields increases, the optimal number of suppliers could either increase or decrease. The intuitive explanation for this behavior is the trade off between the concave part of the objective function (which induces risk-averse preferences on the manufacturer) and the convex option to default (which encourages risk-seeking actions by the manufacturer).

The initial motivation for this research was the question: “Should one expect to observe empirically that firms in developing countries work with more suppliers?” The answer to this question is “it depends.” For example, everything being equal, our model predicts that firms in developing countries will have more suppliers than comparable firms in developed countries. But if, in a developing country, the cost of working with a supplier is very high or the manufacturer is close to bankruptcy, then that manufacturer may actually have fewer suppliers than its counterpart in a developed country. In this case, our model predicts that firms in developing economies will place lower order quantities and will have higher stock-out probabilities, which matches perfectly the observations of the earlier empirical studies.

Thus, to answer the question: “Should one expect to observe empirically that firms in developing countries work with more suppliers?" one needs a sophisticated empirical analysis, which carefully accounts for the factors that we considered in this chapter. We believe that one of the main contributions of this chapter is in providing a set of testable hypotheses for future empirical studies.

To focus on the essential problem features, we have assumed away many other practical concerns. For example, in comparative statics analysis, financing terms (interest rates and loan limits) do not change as the manufacturer’s financial conditions change. This presents an opportunity for the manufacturer to take advantage of its lenders. In practice, as long as the markets are not perfect (as defined by [28]), this type of “mispricing” is possible.[12] We selected the simplest functional form (i.e. no changes in financing terms) to represent this phenomenon. Other assumptions are possible. However, as long as the functional form of mapping between manufacturer’s financial state and financing terms is exogenously given, some form of mispricing will be present and the essential predictions of our model will not be altered, but the analysis will be more complex. For instance, we considered an extension to our model, where financing terms depend on the number of suppliers. For this more general model, we derived numerically the same insights as the ones presented in this chapter.

[12] This mispricing need not constitute an arbitrage because market imperfections preclude market participants from creating an arbitrage portfolio.

We assumed that the firm uses only trade-credit as the source of external financing. The analysis and the results are easily extended to the model where, in addition, the firm can borrow from a bank. However, this adds unnecessary details to the presentation. For instance, instead of two parts in the objective function we would have to consider three parts.

In this chapter we presented a model where the manufacturer pays up-front for the entire order placed with the supplier. As we discussed in Sect. 8.3, there are many real-life systems where this is a good assumption. However, payments for items actually delivered are also common. We considered a more general model, where the manufacturer pays both for the order placed and for the parts received. The analysis of this more general model did not yield additional insights, and therefore, for the sake of exposition, we chose to use the simpler model.

To focus on the question about the number of suppliers, we assumed that suppliers were homogeneous. To address the question of supplier selection, a different model, emphasizing the differences among suppliers, is needed. While we have analyzed a model with heterogeneous suppliers, an in-depth research of the supplier selection problem, for example, extending analysis in [15] and [17], is important and should be addressed in future studies.

Other generalizations, such as the dynamic relationship between lenders and borrowers and the role of asymmetric information, are also subjects for future research.

Appendix

Proofs of Propositions 2 and 3:

These two proofs utilize Lemmas 1, 2, and 3, which are stated and proven following the propositions' proofs.

Proof of Proposition 2:

Recall that $f(N, z)$ is given by $f_L(N, z)$ for $z \leq \widehat{z}$ and by $f_R(N, z)$ for $z > \widehat{z}$, where $f_L(N, z) = f_R(N, z)$ for $z = \widehat{z}$. Furthermore, observe from Lemma 2(i), (ii) that the function $f_k(N, z)$ is zero for z values outside the range $[b_k + l/p - a_k, px_u - b_k - l/a_k]$, $k = L, R$. Therefore, we can divide the proof into two cases:

Case 1 $f(N, \widehat{z}) > 0$. In this case, $\widehat{z}$ must be in the ranges $\left[\frac{b_k+l}{p-a_k}, \frac{px_u-b_k-l}{a_k}\right]$, $k = L, R$.

Case 2 $f(N,\widehat{z}) = 0$. In this case, $\widehat{z}$ must be outside the ranges $\left[\frac{b_k+l}{p-a_k}, \frac{px_u-b_k-l}{a_k}\right]$, $k = L, R$.

In this proof, we deal with Case 1, which is more interesting. The second case is a degenerate case where either $f(N,z) = 0$ for all $z \leq \widehat{z}$ or $f(N,z) = 0$ for all $z \geq \widehat{z}$, and the result could be proven similarly for that case.

Lemma 2 shows that $f_L(N,z)$ and $f_R(N,z)$ are unimodal. Observe that, by definition, $f(N,z) = f_L(N,z)$ for $z \leq \widehat{z}$ and $f(N,z) = f_R(N,z)$ for $z > \widehat{z}$. Now, notice that the function $f(N,z)$ will not be unimodal only if it is decreasing as z approaches $\widehat{z}$ from below and starts increasing once z exceeds $\widehat{z}$. Equivalently, the function $f(N,z)$ will not be unimodal only if $\partial f_L(N,\widehat{z})/\partial z < 0$ and $\partial f_R(N,\widehat{z})/\partial z > 0$. Now, we will prove by contradiction that this cannot happen. Suppose $\partial f_L(N,\widehat{z})/\partial z < 0$ and $\partial f_R(N,\widehat{z})/\partial z > 0$. This, coupled with the fact that $f_L(N,\widehat{z}) = f_R(N,\widehat{z})$, implies that there must exist $z > \widehat{z}$ such that $f_L(N,z) < f_R(N,z)$. However, by Lemma 3, we must have $f_L(N,z) \geq f_R(N,z)$ for $z \geq \widehat{z}$, which yields a contradiction. Therefore, we can never have $\partial f_L(N,\widehat{z})/\partial z < 0$ and $\partial f_R(N,\widehat{z})/\partial z > 0$, and $f(N,z)$ is unimodal in z.

Proof of Proposition 3:

The result follows from the supermodularity of f_k in (z, b_k). To show supermodularity, consider z such that $pz - a_k z - b_k - l \geq 0$. The derivative of f_k with respect to z is derived in Lemma 2 (see Eq. 8.36). Taking a derivative of expression in (8.36) with respect to b_k, we find that

$$\frac{\partial^2 f_k}{\partial z_k \partial b_k} = \frac{a_k}{p} g\left(\frac{a_k z_k + b_k + l}{p}\right) \geq 0$$

Lemma 1 *Consider the model with stochastic demand and deterministic supplier yield from Sect. 8.4.2. We can write the function $f_k(N,z)$ as follows:*

$$f_k(N,z) = 1_{\{pz-a_k z-b_k-l\geq 0\}}\left[-(a_k z + b_k + l)\overline{G}\left(\frac{a_k z + b_k + l}{p}\right) + pz\overline{G}(z) + p\int_{\frac{a_k z+b_k+l}{p}}^{z} xg(x)\,\mathrm{d}x\right], \quad (8.33)$$

where $\overline{G}(x) = 1 - G(x)$.

Proof For notational convenience, define $\lambda_k = a_k z + b_k + l$ and observe that $f_k(N,z) = E[p\min(D, Q(N,z)) - \lambda_k]^+$, $k = L, R$ (see Eqs. 8.14, 8.15, and 8.16). Recall that demand D is stochastic with c.d.f. G, and the quantity delivered, $Q(N,z)$, is deterministic and equal to z by Assumption 7. Then:

$$
\begin{aligned}
f_k(N,z) &= -\lambda_k E\left[1_{\{p\min(D,z)-\lambda_k\ge 0\}}\right] + pE\left[\min(D,z)1_{\{p\min(D,z)-\lambda_k\ge 0\}}\right] \\
&= -\lambda_k 1_{\{pz-\lambda_k\ge 0\}}\Pr[pD-\lambda_k\ge 0] + p1_{\{pz-\lambda_k\ge 0\}}E\left[\min(D,z)1_{\{pD-\lambda_k\ge 0\}}\right] \\
&= -\lambda_k 1_{\{pz-\lambda_k\ge 0\}}\overline{G}\left(\frac{\lambda_k}{p}\right) + p1_{\{pz-\lambda_k\ge 0\}}zE\left[1_{\{pD-\lambda_k\ge 0, D\ge z\}}\right] \\
&\quad + p1_{\{pz-\lambda_k\ge 0\}}E\left[D1_{\{pD-\lambda_k\ge 0, D<z\}}\right] \\
&= -\lambda_k 1_{\{pz-\lambda_k\ge 0\}}\overline{G}\left(\frac{\lambda_k}{p}\right) + p1_{\{pz-\lambda_k\ge 0\}}z\overline{G}(z) + p1_{\{pz-\lambda_k\ge 0\}}\int_{\frac{\lambda_k}{p}}^{z} xg(x)\,\mathrm{d}x \\
&= 1_{\{pz-\lambda_k\ge 0\}}\left[-\lambda_k\overline{G}\left(\frac{\lambda_k}{p}\right) + pz\overline{G}(z) + p\int_{\frac{\lambda_k}{p}}^{z} xg(x)\,\mathrm{d}x\right].
\end{aligned}
$$

Lemma 2 *Consider the model with stochastic demand and deterministic supplier yield from Sect. 8.4.2. If $px_u - a_k x_u - b_k - l < 0$, $k = L, R$, then the function $f_k(N,z) = 0$ for all order quantities z in the domain of demand, $[x_l, x_u]$. Otherwise:*

(i) $f_k(N,z) = 0$ *for* $z < \frac{b_k+l}{p-a_k}$.
(ii) $f_k(N,z) = 0$ *for* $z > \frac{px_u-b_k-l}{a_k}$.
(iii) $f_k(N,z)$ *is unimodal in* z.
(iv) *There exists a unique* $z_k \in \left[\frac{b_k+l}{p-a_k}, \frac{px_u-b_k-l}{a_k}\right]$ *that maximizes* $f_k(N,z)$ *and satisfies*

$$
p\overline{G}(z_k) = a_k\overline{G}\left(\frac{a_k z_k + b_k + l}{p}\right) = a_k \Pr\left[pD > a_k z_k + b_k + l\right]. \tag{8.34}
$$

Proof Recall that the random variable D has a density function defined over the domain $[x_l, x_u]$. First, suppose that $px_u - a_k x_u - b_k - l < 0$. Then, for any order quantity $z \in [x_l, x_u]$, we have $pz - a_k z - b_k - l < 0$ (because, by Assumption 6, $p > a_k$ for $k = L, R$). Therefore, for any order quantity $z \in [x_l, x_u]$, we have $1_{\{pz-a_k z-b_k-l\ge 0\}} = 0$. It now follows that $f_k(z) = 0$ for any $z \in [x_l, x_u]$. (See (8.33) in Lemma 1.)

Now, we turn to the more interesting case where $px_u - a_k x_u - b_k - l \ge 0$, $k = L, R$.

Proof of (i) If $z < \frac{b_k+l}{p-a_k}$, then $1_{\{pz-a_k z-b_k-l\ge 0\}} = 0$, and the result follows from Lemma 1.

Proof of (ii) The inequality $z > \frac{px_u-b_k-l}{a_k}$ is equivalent to $\frac{a_k z+b_k+l}{p} > x_u$. Hence, the first term in brackets in (8.33) is zero. Furthermore, if $z > \frac{px_u-b_k-l}{a_k}$, then one can verify that $z > x_u$ (using also the current assumption that $px_u - a_k x_u - b_k - l \ge 0$). Hence, the second and third terms in (8.33) are also zero. It now follows that $f_k(z) = 0$.

Proof of (iii) By parts (i) and (ii), $f_k(z) = 0$ for $z < \frac{b_k+l}{p-a_k}$ and $z > \frac{px_u-b_k-l}{a_k}$. Therefore, we will conclude the proof if we can show that $f_k(z)$ is unimodal for $z \in \left[\frac{b_k+l}{p-a_k}, \frac{px_u-b_k-l}{a_k}\right]$. In this range, by Lemma 1, we have:

$$f_k(z) = -(a_k z+b_k+l)\overline{G}\left(\frac{a_k z + b_k + l}{p}\right)+pz\overline{G}(z)+p\int\limits_{\frac{a_k z+b_k+l}{p}}^{z} xg(x)\,\mathrm{d}x \quad (8.35)$$

It is not difficult to check the following claim is true:

Claim **(a)** $f_k'(z) > 0$ at $z = b_k + l/p - a_k$. In addition, we will now prove the following claim:

Claim **(b)** $f_k''(z) < 0$ whenever $f_k'(z) = 0$.

The first derivative of f_k is

$$f_k'(z) = -a_k\overline{G}\left(\frac{a_k z + b_k + l}{p}\right) + p\overline{G}(z) \quad (8.36)$$

The first order condition is

$$a_k\overline{G}\left(\frac{a_k z + b_k + l}{p}\right) = p\overline{G}(z) \quad (8.37)$$

The second derivative of f_k is

$$f_k''(z) = \frac{a_k^2}{p}\, g\left(\frac{a_k z + b_k + l}{p}\right) - pg(z) = p\left[g\left(\frac{a_k z + b_k + l}{p}\right)\frac{a_k^2}{p^2} - g(z)\right] \quad (8.38)$$

Let z_0 satisfy the first order condition (8.37). Then,

$$f_k''(z_0) = p\overline{G}^2(z_0)\left[\frac{g\left(\frac{a_k z_0+b_k+l}{p}\right)}{\overline{G}^2\left(\frac{a_k z_0+b_k+l}{p}\right)} - \frac{g(z_0)}{\overline{G}^2(z_0)}\right] \le 0, \quad (8.39)$$

where the inequality follows from the facts that $z_0 > a_k z_0 + b_k + l/p$ and $g(x)/\overline{G}^2(x)$ is increasing. Hence, we have shown that *Claim* (*b*) holds. Now, *claim* (*a*) implies that the function $f_k(z)$ starts increasing from zero at $z = b_k + l/p - a_k$. Furthermore, the function goes back to zero at $z = px_u - b_k - l/a_k$ and any stationary point of the function $f_k(\cdot)$ in the range $[b_k + l/p - a_k, px_u - b_k - l/a_k]$ is a local maximum by *claim* (*b*). Therefore, we conclude that there exists only one stationary point of the function $f_k(\cdot)$ in the range $[b_k + l/p - a_k, px_u - b_k - l/a_k]$, and this stationary point is a maximizer. Hence, the function $f_k(\cdot)$ is unimodal. (If there were two stationary points, both of them would have to be local maxima by *claim* (*b*), which would require the existence of a local minimum in between these two local maxima, which contradicts *claim* (*b*).)

Proof of (iv) This follows from parts (i) through (iii) of the lemma.

Lemma 3 *If* $z \geq \widehat{z}$, *then* $f_L(N, z) \geq f_R(N, z)$.

Proof We first prove the lemma for the case where $r_I < r_S$. From Eqs. 8.10, 8.11, and 8.12, we observe that, when $r_I < r_S$, we have $\Pi_L(N, z) - \Pi_R(N, z) = (r_S - r_I)T(N, z)$. Furthermore, $\widehat{z}$ is defined in (8.13) such that $T(N, z) \geq 0$ for any $z \geq \widehat{z}$. Therefore, we conclude that $\Pi_L(N, z) - \Pi_R(N, z) \geq 0$ for any $z \geq \widehat{z}$. Hence, $f_L(N, z) = E[\Pi_L(N, z)]^+ \geq E[\Pi_R(N, z)]^+ = f_R(N, z)$ for any $z \geq \widehat{z}$. In the other case where $r_I > r_S$, we have $\Pi_L(N, z) - \Pi_R(N, z) = (r_I - r_S)T(N, z)$ from Eqs. 8.10, 8.11, and 8.12. The lemma follows similarly for this case.

Proof of Proposition 4

This proof utilizes Lemmas 4 and 5, which are stated and proven after the proposition's proof.

Proof of Proposition 4

We again focus on the more interesting case where $f(N, \widehat{z}) > 0$ (as opposed to the degenerate case where $f(N, \widehat{z}) = 0$.) In this proof, we write $z^*(l)$ to make explicit the dependence of the optimal order quantity, z^*, on the maximum liability, l. Similarly, we add l to the list of arguments for functions $f(N, z)$ and $f_k(N, z), k = L, R$. We define $z_k^*(l) = \arg\max\{f_k(N, z, l)\}$. Suppose $l_1 > l_2$. Our goal is to prove $z^*(l_1) \geq z^*(l_2)$. We will prove the result by considering four different cases, each one corresponding to one of the cases in the statement of Lemma 4.

Case 1 $\widehat{z} \leq 0$. In this case, from Lemma 4(i), it follows that $z^*(l_1) = z_R^*(l_1)$ and $z^*(l_2) = z_R^*(l_2)$. Now, the result follows since $z_R^*(l_1) \geq z_R^*(l_2)$ by Lemma 5.

Case 2 $\widehat{z} > 0$ and $\partial f_L(N, \widehat{z}, l_1)/\partial z < 0$. In this case, from Lemma 4(ii), we know that $z^*(l_1) = z_L^*(l_1)$. Furthermore, since $f_k(N, z, l)$ is supermodular in (z, l) (as shown in Lemma 5), it must be that $\partial f_L(N, \widehat{z}, l_2)/\partial z < 0$. Therefore, from Lemma 4(ii), we know that $z^*(l_2) = z_L^*(l_2)$. The result now follows since $z_L^*(l_1) \geq z_L^*(l_2)$ by Lemma 5.

Case 3 $\widehat{z} > 0$, $\partial f_L(N, \widehat{z}, l_1)/\partial z > 0$ and $\partial f_R(N, \widehat{z}, l_1)/\partial z > 0$. In this case, we have $z^*(l_1) = z_R^*(l_1)$ by Lemma 4(iii). Furthermore, note that $z^*(l_1) \geq \widehat{z}$ (since $\partial f_R(N, \widehat{z}, l_1)/\partial z > 0$ and $f_R(N, z)$ is unimodal), which will be used in the rest of the proof. We will consider a number of subcases depending on the signs of $\partial f_L(N, \widehat{z}, l_2)/\partial z$ and $\partial f_R(N, \widehat{z}, l_2)/\partial z$.

Case 3(a) $\partial f_L(N, \widehat{z}, l_2)/\partial z < 0$. In this case, from Lemma 4(ii), we know that $z^*(l_2) = z^*_L(l_2)$. Furthermore, note that $z^*(l_2) \leq \widehat{z}$ (since $\partial f_L(N, \widehat{z}, l_2)/\partial z < 0$ and $f_L(N, z, l)$ is unimodal). The result now follows since $z^*(l_2) \leq \widehat{z} \leq z^*(l_1)$.

Case 3(b) $\partial f_L(N, \widehat{z}, l_2)/\partial z > 0$ and $\partial f_R(N, \widehat{z}, l_2)/\partial z > 0$. In this case, from Lemma 4(iii), we know that $z^*(l_2) = z^*_R(l_2)$. Since $z^*(l_1) = z^*_R(l_1)$, the result follows from Lemma 5.

Case 3(c) $\partial f_L(N, \widehat{z}, l_2)/\partial z > 0$ and $\partial f_R(N, \widehat{z}, l_2)/\partial z < 0$. In this case, from Lemma 4(iv), we know that $z^*(l_2) = \widehat{z}$. The result follows since $z^*(l_2) = \widehat{z} \leq z^*(l_1)$.

Case 4 $\widehat{z} > 0$, $\partial f_L(N, \widehat{z}, l_1)/\partial z > 0$, and $\partial f_R(N, \widehat{z}, l_1)/\partial z < 0$. In this case, from Lemma 4(iv), we know that $z^*(l_1) = \widehat{z}$. Furthermore, since $f_k(N, z, l)$ is supermodular in (z, l), it must be that $\partial f_R(N, \widehat{z}, l_2)/\partial z < 0$. Again, we consider a number of subcases depending on the signs of $\partial f_L(N, \widehat{z}, l_2)/\partial z$ and $\partial f_R(N, \widehat{z}, l_2)/\partial z$.

Case 4(a) $\partial f_L(N, \widehat{z}, l_2)/\partial z < 0$. By Lemma 4(ii), we have $z^*(l_2) \leq \widehat{z}$. The desired result follows since $z^*(l_1) = \widehat{z}$.

Case 4(b) $\partial f_L(N, \widehat{z}, l_2)/\partial z > 0$. Given that we also have $\partial f_R(N, \widehat{z}, l_2)/\partial z < 0$, it follows from Lemma 4(iv) that $z^*(l_2) = \widehat{z}$. The result now follows since $z^*(l_1) = \widehat{z}$ as well.

Lemma 4 *Consider the model with stochastic demand and deterministic supplier yield from Sect. 8.4.2. Suppose N is fixed. Then, the optimal order quantity*

$$z^* \in \left\{ \arg\max_z \{f_L(N, z)\}, \arg\max_z \{f_R(N, z)\}, \widehat{z} \right\}.$$

Furthermore, to find the optimal order quantity, z^*, one could use the following properties:

(i) *If* $\widehat{z} \leq 0$, *then* $z^* = \arg\max_z\{f_R(N, z)\}$.
(ii) *If* $\widehat{z} > 0$ *and* $\partial f_L(N, \widehat{z})/\partial z < 0$, *then* $z^* = \arg\max_z\{f_L(N, z)\}$.
(iii) *If* $\widehat{z} > 0$, $\partial f_L(N, \widehat{z})/\partial z > 0$ *and* $\partial f_R(N, \widehat{z})/\partial z > 0$, *then* $z^* = \arg\max_z\{f_R(N, z)\}$.
(iv) *If* $\widehat{z} > 0$, $\partial f_L(N, \widehat{z})/\partial z > 0$ *and* $\partial f_R(N, \widehat{z})/\partial z < 0$, *then* $z^* = \widehat{z}$.

Proof Once again, we focus on the more interesting case where $f(N, \widehat{z}) > 0$ (as opposed to the degenerate case where $f(N, \widehat{z}) = 0$.) For the purposes of this proof, define $z^*_k = \arg\max_z f_k(N, z)$ for $k = L, R$. We first prove Properties (i) through (iv):

Proof of (i) If $\widehat{z} < 0$, then $f(N, z) = f_R(N, z)$ for all $z \geq 0$ and $z^* = z^*_R$, which concludes the proof.

Proof of (ii) If $\widehat{z} > 0$ and $\partial f_L(N, \widehat{z})/\partial z < 0$, then it must be that $\partial f_R(N, \widehat{z})/\partial z < 0$ as well. (Otherwise, we would obtain a contradiction to the unimodality of $f(N, z)$, which was proven in Proposition 2.) Now, since $f(N, z) = f_R(N, z)$ for $z \geq \widehat{z}$, it follows that $f(N, z)$ must be decreasing in z for $z \geq \widehat{z}$ (since $\partial f_R(N, \widehat{z})/\partial z < 0$ and

$f_R(N, z)$ is unimodal.) Therefore, it must be that $z_R^* \leq \widehat{z}$. In addition, we know that $f(N, z) = f_L(N, z)$ for $z \leq \widehat{z}$, and, furthermore, $z_L^* < \widehat{z}$ (since $\partial f_L(N, \widehat{z})/\partial z < 0$ and $f_L(N, z)$ is unimodal.) Therefore, we have $z^* = z_L^*$, which concludes the proof.

Proof of (iii) Since $f(N, z) = f_L(N, z)$ for $z \leq \widehat{z}$, it follows that $f(N, z)$ must be increasing in z for $z \leq \widehat{z}$ (since $\partial f_L(N, \widehat{z})/\partial z > 0$ and $f_L(N, z)$ is unimodal.) Therefore, it must be that $z_L^* \geq \widehat{z}$. In addition, we know that $f(N, z) = f_R(N, z)$ for $z \geq \widehat{z}$, and, furthermore, $z_R^* > \widehat{z}$ (since $\partial f_R(N, \widehat{z})/\partial z > 0$ and $f_R(N, z)$ is unimodal.) Therefore, we have $z^* = z_R^*$, which concludes the proof.

Proof of (iv) Since $f(N, z) = f_L(N, z)$ for $z \leq \widehat{z}$, it follows that $f(N, z)$ must be increasing in z for $z \leq \widehat{z}$ (since $\partial f_L(N, \widehat{z})/\partial z > 0$ and $f_L(N, z)$ is unimodal.) Furthermore, since $f(N, z) = f_R(N, z)$ for $z \geq \widehat{z}$, it follows that $f(N, z)$ must be decreasing in z for $z \geq \widehat{z}$ (since $\partial f_R(N, \widehat{z})/\partial z < 0$ and $f_R(N, z)$ is unimodal.) Hence, we have $z^* = \widehat{z}$, which concludes the proof.

The statement that $z^* \in \{\arg\max_z\{f_L(N, z)\}, \arg\max_z\{f_R(N, z)\}, \widehat{z}\}$ now follows as a corollary to Properties (i) through (iv).

Algorithm for computing the optimal operational decisions — N^* and z^* :

1. For each $N \leq I/C - \beta$
   a. Compute $\widehat{z}: \ T(N, \widehat{z}) = 0$
   b. Compute $z_{\max}(N)$
   c. If $\widehat{z} < 0$, then find $z^*(N)$ that maximizes $f_R(N, z)$ by searching over all $z \in [0, z_{\max}(N)]$
   d. If $\widehat{z} \geq 0$, then
      i. If $\partial f_L(N, \widehat{z})/\partial z \leq 0$, then find $z^*(N)$ that maximizes $f_L(N, z)$ by searching over all $z \in [0, \widehat{z}]$
      ii. If $\partial f_L(N, \widehat{z})/\partial z > 0$ and $\partial f_R(N, \widehat{z})/\partial z > 0$, then find $z^*(N)$ that maximizes $f_R(N, z)$ by searching over all $z \in [\widehat{z}, z_{\max}(N)]$
      iii. If $\partial f_L(N, \widehat{z})/\partial z > 0$ and $\partial f_R(N, \widehat{z})/\partial z \leq 0$, then $z^*(N) = \widehat{z}$
2. Pick N for which $f(N, z^*(N))$ is the largest.

Lemma 5 *Suppose that Assumptions (4), (5), (6), (7) hold. For $k = L, R$, let $z_k^* = \arg\max_z\{f_k(N, z)\}$. Then, z_k^* is increasing in l.*

Proof The result follows from the supermodularity of f_k in (z, l). Consider z such that $pz - a_k z - b_k - l \geq 0$. The derivative of f_k with respect to z is given by Eq. 8.36. Taking the derivative of the expression in (8.36) with respect to l, we find that

$$\frac{\partial^2 f_k}{\partial z \partial l} = \frac{a_k}{p} g\left(\frac{a_k z + b_k + l}{p}\right) \geq 0$$

Proofs of Propositions 5, 6, 7

In this subsection, we provide the proofs of Propositions 5 through 7 followed by a lemma that is useful for these proofs.

Proof of Proposition 5

Proof of (i) In order to prove the result, we will show that for any two integers N_1 and N_2 such that $N_1 < N_2$, we have $f(N_1, z) - f(N_2, z) \geq 0$ when $r_I < r_S$. The desired result will then follow. Consider the following four cases:

Case 1 $T(N_1, z) \leq 0$ and $T(N_2, z) \leq 0$. In this case, by (8.9), $f(N_1, z) - f(N_2, z) = f_L(N_1, z) - f_L(N_2, z)$. The result now follows from Lemma 6(i).

Case 2 $T(N_1, z) > 0$ and $T(N_2, z) > 0$. In this case, $f(N_1, z) - f(N_2, z) = f_R(N_1, z) - f_R(N_2, z)$. The result now follows from Lemma 6(i).

Case 3 $T(N_1, z) \leq 0$ and $T(N_2, z) > 0$. In this case, $f(N_1, z) - f(N_2, z) = f_L(N_1, z) - f_R(N_2, z)$. Since $T(N_2, z) > 0$, we have $z \geq \widehat{z}$ at N_2, and, therefore, $f_L(N_2, z) \geq f_R(N_2, z)$ by Lemma 3. Furthermore, note that $f_L(N_1, z) \geq f_L(N_2, z)$ by Lemma 6(i). Hence, $f_L(N_1, z) \geq f_R(N_2, z)$, which yields the desired result.

Case 4 $T(N_1, z) > 0$ and $T(N_2, z) \leq 0$. This case cannot occur, since $T(N, z)$ is increasing in N.

Proof of (ii) The proof is similar to that of (i) and uses Lemma 6(ii) where the proof of (i) uses Lemma 6(i).

Proof of (iii) In order to prove the result, we will show that for any two integers N_1 and N_2 such that $N_1 < N_2$, we have $f(N_1, z) - f(N_2, z) \leq 0$ when $r_I > r_S$. It will then follow that $N^* > 1$. Consider the following four cases:

Case 1 $T(N_1, z) \leq 0$ and $T(N_2, z) \leq 0$. In this case, $f(N_1, z) - f(N_2, z) = f_L(N_1, z) - f_L(N_2, z)$. The result now follows from Lemma 6(iii), since $f_L(N, z)$ is increasing N.

Case 2 $T(N_1, z) > 0$ and $T(N_2, z) > 0$. In this case, $f(N_1, z) - f(N_2, z) = f_R(N_1, z) - f_R(N_2, z)$. The result now follows from Lemma 6(iii), since $f_R(N, z)$ is increasing N.

Case 3 $T(N_1, z) \leq 0$ and $T(N_2, z) > 0$. Notice from (8.12) that when $r_I > r_S$, $T(N, z)$ is decreasing in N (since $\beta < \widehat{S}$ by Assumption 5). Therefore, this case cannot occur, since $T(N, z)$ is decreasing in N.

Case 4 $T(N_1, z) > 0$ and $T(N_2, z) \leq 0$. In this case, $f(N_1, z) - f(N_2, z) = f_R(N_1, z) - f_L(N_2, z)$. Since $T(N_1, z) > 0$, we have $z > \widehat{z}$ at N_1 and, therefore, $f_L(N_1, z) \geq f_R(N_1, z)$ by Lemma 3. Furthermore, note that $f_L(N_2, z) \geq f_L(N_1, z)$

by Lemma 6(iii). Combining these last two observations, we obtain $f_L(N_2, z) \geq f_R(N_1, z)$, which yields the desired result.

Proof of Proposition 6

From the first of the two constraints stated in (8.4), it follows that we must have $(C - \beta)N \leq I$, which yields the desired result.

Proof of Proposition 7

Proof of (i) When $N \leq \alpha I/w(\hat{S} - \beta) - \alpha(\hat{S} - C)$, it is not difficult to check any $z \geq 0$ that satisfies the second of the two constraints stated in (8.4) will satisfy the first one as well. Therefore, when $N \leq \alpha I/w(\hat{S} - \beta) - \alpha(\hat{S} - C)$, z is bounded by the second constraint in (8.4), which yields the desired result.

Proof of (ii) When $\alpha I/w(\hat{S} - \beta) - \alpha(\hat{S} - C) \leq N \leq \frac{I}{C-\beta}$, it is not difficult to check that any $z \geq 0$ that satisfies the first of the two constraints stated in (8.4) will satisfy the second one as well. Therefore, in this case, z is bounded by the first constraint in (8.4), which yields the desired result.

Lemma 6 *Consider the model with stochastic demand and deterministic supplier yield from Sect. 8.4.2. At a fixed z:*

(i) *If $r_I < r_S$, then both $f_L(N, z)$ and $f_R(N, z)$ are decreasing in N.*
(ii) *If $r_I > r_S$ and $C > (r_I - r_S)/(1 + r_I)\hat{S}$, then both $f_L(N, z)$ and $f_R(N, z)$ are decreasing in N.*
(iii) *If $r_I > r_S$ and $C < r_I - r_S/1 + r_I\beta$, then both $f_L(N, z)$ and $f_R(N, z)$ are increasing in N.*

Proof of (i) Notice from (8.10) that $\Pi_L(N, z)$ is decreasing in N when $r_I < r_S$. (Note that $Q(N, z) = z$ in the model of Sect. 8.4.2.) Since $f_L(N, z)$ is defined in (8.14) to be a monotonic transformation of $\Pi_L(N, z)$, we observe that $f_L(N, z)$ is decreasing in N when $r_I < r_S$. Similarly for $f_R(N, z)$.

Proof of (ii) Note that $C > (r_I - r_S)/(1 + r_I)\hat{S}$ is equivalent to $(1 + r_I)C > (r_I - r_S)\hat{S}$. Now, when $r_I > r_S$ and $(1 + r_I)C > (r_I - r_S)\hat{S}$, we notice from (8.10) and (8.11) that both $\Pi_L(N, z)$ and $\Pi_R(N, z)$ are decreasing in N. The result follows, since $f_L(N, z)$ and $f_R(N, z)$ are monotonic transformations of $\Pi_L(N, z)$ and $\Pi_R(N, z)$, respectively.

Proof of (iii) Note that $C < r_I - r_S/1 + r_I\beta$ is equivalent to $(r_I - r_S)\beta > (1+r_I)C$. Now, when $r_I > r_S$ and $(r_I - r_S)\beta > (1+r_I)C$, observe from (8.10) and (8.11) that both $\Pi_L(N, z)$ and $\Pi_R(N, z)$ are increasing N. The result follows because $f_L(N, z)$ and $f_R(N, z)$ are monotonic transformations of $\Pi_L(n, z)$ and $\Pi_R(n, z)$, respectively.

Proofs of Propositions 8, 9, and 10

Proof of Proposition 8

Let us first prove that $f_k(z)$, defined by (8.14), can be written as in (8.21). For notational convenience, define $\lambda_k = a_k z + b_k + l$. Observe that $f_k(N, z) = E[p \min(D, Q(N, z)) - \lambda_k]^+,\ k = L, R$. Recall that demand D is deterministic and the quantity delivered, $Q(N, z)$, is given by $\overline{X}z$. Then:

$$\begin{aligned} f_k(N,z) &= -\lambda_k E\left[1_{\{p\min(D,\overline{X}z)-\lambda_k\geq 0\}}\right] + pE\left[\min(D,\overline{X}z)1_{\{p\min(D,\overline{X}z)-\lambda_k\geq 0\}}\right] \\ &= -\lambda_k 1_{\{pD-\lambda_k\geq 0\}} \Pr[p\overline{X}z - \lambda_k \geq 0] \\ &\quad + p1_{\{pD-\lambda_k\geq 0\}} E\left[\min(D,\overline{X}z)1_{\{p\overline{X}z-\lambda_k\geq 0\}}\right] \end{aligned}$$

Notice that $pD - \lambda_k \geq 0$ by Assumption 9. Hence, $1_{\{pD-\lambda_k\geq 0\}} = 1$ and we have:

$$\begin{aligned} f_k(N,z) &= -\lambda_k \Pr[p\overline{X}z - \lambda_k \geq 0] + pE\left[\min(D,\overline{X}z)1_{\{p\overline{X}z-\lambda_k\geq 0\}}\right] \\ &= -\lambda_k \Pr[p\overline{X}z - \lambda_k \geq 0] + pE\left[\overline{X}z1_{\{p\overline{X}z-\lambda_k\geq 0,\overline{X}z\leq D\}}\right] \\ &\quad + pE\left[D1_{\{p\overline{X}z-\lambda_k\geq 0,\overline{X}z\geq D\}}\right] \end{aligned}$$

Furthermore, by Assumption 9, we have $\frac{D}{z} \geq \frac{\lambda_k}{pz}$. Using this observation, we can write:

$$\begin{aligned} f_k(N,z) &= -\lambda_k \Pr[p\overline{X}z - \lambda_k \geq 0] + pE\left[\overline{X}z1_{\{p\overline{X}z-\lambda_k\geq 0,\overline{X}z\leq D\}}\right] + pD\Pr[p\overline{X}z \geq D] \\ &= -\lambda_k \Pr\left[\overline{X} \geq \frac{\lambda_k}{pz}\right] + pE\left[\overline{X}z1_{\{\frac{\lambda_k}{pz}\leq\overline{X}\leq\frac{D}{z}\}}\right] + pD\Pr\left[\overline{X} \geq \frac{D}{pz}\right] \end{aligned}$$

The expression above can now be re-written as the expression presented in (8.21). The derivative presented in (8.22) can be verified by taking the derivative of (8.21).

Next, we prove the unimodality of $f_k(z)$ when $\frac{\gamma(\cdot)}{\Gamma(\cdot)}$ is decreasing. We start by writing (8.22) as follows:

$$f_k'(z) = -a_k\overline{\Phi}\left(\frac{a_k z + b_k + l}{pz}\right) - p\Gamma\left(\frac{D}{z}\right) + p\Gamma\left(\frac{a_k z + b_k + l}{pz}\right) \tag{8.41}$$

By taking the derivative of (8.41), we obtain:

$$f_k''(z) = -\phi\left(\frac{a_k z + b_k + l}{pz}\right)\frac{a_k(b_k+l)}{pz^2} + p\frac{D}{z^2}\gamma\left(\frac{D}{z}\right) - \frac{b_k+l}{z^2}\gamma\left(\frac{a_k z + b_k + l}{pz}\right) \tag{8.42}$$

In what follows, we let $K = a_k z + b_k + l/pz$ for convenience. In order to prove unimodality, it will be sufficient to prove that $f_k''(z) < 0$ whenever $f_k'(z) = 0$. To that end, we start by restating (8.42) as follows:

$$f_k''(z) = -\phi(K)\frac{a_k(b_k+l)}{pz^2} + p\frac{D}{z^2}\gamma\left(\frac{D}{z}\right) - \frac{b_k+l}{z^2}\frac{\gamma(K)}{\Gamma(K)}\Gamma(K)$$

We then substitute for $\Gamma(K)$ from (8.41) in the above equation to obtain:

$$f_k''(z)\big|_{f_k'(z)=0} = -\phi(K)\frac{a_k(b_k+l)}{pz^2} + p\frac{D}{z^2}\gamma\left(\frac{D}{z}\right) - \frac{b_k+l}{z^2}\frac{\gamma(K)}{\Gamma(K)}\frac{a_k\overline{\Phi}(K) + p\Gamma\left(\frac{D}{z}\right)}{p}$$

Rearranging the terms, we can write:

$$f_k''(z)\big|_{f_k'(z)=0} = \frac{p}{z^2}\Gamma\left(\frac{D}{z}\right)\left[D\frac{\gamma\left(\frac{D}{z}\right)}{\Gamma\left(\frac{D}{z}\right)} - \frac{b_k+l}{p}\frac{\gamma(K)}{\Gamma(K)}\right]$$
$$-\phi(K)\frac{a_k(b_k+l)}{pz^2} - \frac{\gamma(K)\overline{\Phi}(K)}{\Gamma(K)}\frac{a_k(b_k+l)}{pz^2}$$

Substituting $\gamma(K) = -K\phi(K)$ in the above equation yields:

$$f_k''(z)\big|_{f_k'(z)=0} = \frac{p}{z^2}\Gamma\left(\frac{D}{z}\right)\left[D\frac{\gamma\left(\frac{D}{z}\right)}{\Gamma\left(\frac{D}{z}\right)} - \frac{b_k+l}{p}\frac{\gamma(K)}{\Gamma(K)}\right]$$
$$-\phi(K)\frac{a_k(b_k+l)}{pz^2}\left[1 - \frac{K\overline{\Phi}(K)}{\Gamma(K)}\right] \tag{8.43}$$

Observe that, by Assumption 9, we have $pD > b_k + l$ and $\frac{D}{z} \geq K$. Also, note that $\gamma(\cdot) < 0$ (by definition). Now, invoke our assumption that $\gamma(\cdot)/\Gamma(\cdot)$ is decreasing to observe that the first term in (8.43) is negative. In addition, observe that $\Gamma(K) = K\overline{\Phi}(K) + \int_K^\infty \overline{\Phi}(x)\,\mathrm{d}x$ (by integration by parts), so $1 - K\overline{\Phi}(K)/\Gamma(K) > 0$; hence, the second term in (8.43) is negative. Thus, we conclude that $f_k''(z)\big|_{f_k'(z)=0} < 0$, which concludes the proof of the lemma.

Proof of Proposition 9

First, let us prove that $f_k, k = L, R$ are unimodal in z. To that end, we need to prove that, for $\overline{X} \sim N\left(\mu, \sigma/\sqrt{N}\right)$, $\gamma(\cdot)/\Gamma(\cdot)$ is decreasing, where γ and Γ are defined by (8.20) and (8.19). We can then apply Proposition 8. ϕ is the p.d.f. of a normal random variable with mean μ and standard deviation $\sigma/\sqrt{N}$. Define $t(m) = (m-\mu)\sqrt{N}/\sigma$ and let ϕ_S and Φ_S denote the p.d.f. and c.d.f. of the standard normal distribution, respectively. The following observations are standard:

$$\gamma(m) = -m\phi_S[t(m)]; \qquad \Gamma(m) = \mu\overline{\Phi}_S[t(m)] + \frac{\sigma}{\sqrt{N}}\phi_S[t(m)] \tag{8.44}$$

From (8.44), it follows that

$$\frac{\Gamma(m)}{\gamma(m)} = -\frac{\mu\overline{\Phi}_S[t(m)]}{m\phi_S[t(m)]} - \frac{\sigma}{\sqrt{N}m} \tag{8.45}$$

Note that $\overline{\Phi}_S(\cdot)/m\phi_S(\cdot)$ is decreasing, since standard normal distribution has IFR. Using this observation, we note from (8.45) that $\Gamma(m)/\gamma(m)$ is increasing in m. Hence, $\gamma(m)/\Gamma(m)$ is decreasing in m.

Now the proof that the objective function is unimodal is similar to the proof of Proposition 2.

Proof of Proposition 10

Take the derivative of the Expression (8.22) with respect to b

$$\frac{\partial}{\partial b}h'(z) = \frac{\partial}{\partial b}\left[p\int_{\frac{az+b+l}{pz}}^{\infty} x\phi(x)\,\mathrm{d}x - a\int_{\frac{az+b+l}{pz}}^{\infty}\phi(x)\,\mathrm{d}x\right]$$

$$= -\frac{1}{pz}\phi\left(\frac{az+b+l}{pz}\right)\left[p\,\frac{az+b+l}{pz} - a\right] \leq 0 \qquad (8.46)$$

References

1. Agrawal N, Nahmias S (1997) Rationalization of the supplier base in the presence of yield uncertainty. Prod Oper Manag 6(3):291–308
2. Anupindi R, Akella R (1993) Diversication under supply uncertainty. Manag Sci 39(8):944–963
3. Atanasova CV, Wilson N (2003) Bank borrowing constraints and the demand for trade credit: evidence from panel data. Manag Decis Econ 24(6):503–514
4. Avery RB, Berger AN (1991) Loan commitments and bank risk exposure. J Bank Financ 15:173–192
5. Babich V (2006) Vulnerable options in supply chains: effects of supplier competition. Nav Res Logist 53(7):656–673
6. Babich V, Burnetas AN, Ritchken PH (2007) Competition and diversication effects in supply chains with supplier default risk. Manuf Serv Oper Manag 9(2):123–146
7. Babich V, Sobel MJ (2004) Pre-IPO operational and nancial decisions. Manag Sci 50(7): 935–948
8. Berger AN, Udell GF (1995) Relationship lending and lines of credit in small rm nance. J Bus 68(3):351–381
9. Biais B, Gollier C (1997) Trade credit and credit rationing. Rev Financial Studies 10(4): 903–937
10. Boyabatli O, Toktay B (2007) Stochastic capacity investment and technology choice in imperfect capital markets. INSEAD Working Paper, Fontainebleau, France
11. Brander JA, Lewis TR (1986) Oligopoly and financial structure: the limited liability effect. Am Econ Rev 76(5):956–970
12. Brealey RA, Myers SC, Allen F (2006) Principles of corporate finance. McGraw-Hill/Irwin, New York
13. Buzacott JA, Zhang RQ (2004) Inventory management with asset- based nancing. Manag Sci 50(9):1274–1292
14. Chapman CB, Ward SC, Cooper DF et al (1984) Credit policy and inventory control. J Oper Res Soc 35(12):1055–1065
15. Dada M, Petruzzi NC, Schwarz LB (2007) A newsvendors procurement problem when suppliers are unreliable. Manuf Serv Oper Manag 9(1):9–32

16. Faure-Grimaud A (2000) Product market competition and optimal debt contracts: the limited liability effect revisited. Eur Econ Rev 44(10):1823–1840
17. Federgruen A, Yang N (2007) Selecting a portfolio of suppliers under demand and supply risks. Oper Res, (Forthcoming)
18. Fisman R (2001) Trade credit and productive effciency in developing countries. World Dev 29(2):311–321
19. Fisman R, Love I (2003) Trade credit, nancial intermidary development, and industry growth. J Finance 58(1):353–374
20. Gollier C, Koehl PF, Rochet JC (1997) Risk-taking behavior with limited liability and risk aversion. J Risk Insur 64(2):347–370
21. Gupta D, Wang L (2008) A stochastic inventory model with trade credit. Manag Serv Oper Manag. Forthcoming
22. Haley CW, Higgins RC (1973) Inventory policies and trade credit financing. Manag Sci 20(4):464–471
23. Harris M, Raviv A (1991) The theory of capital structure. J Finance 46(1):297–355
24. Hull J (2000) Options, futures, and other derivatives. 4th edn. Prentice Hall, Upper Saddle River
25. Lederer PJ, Singhal VR (1994) The effect of financing decisions of the choice of manufacturing technologies. Int J Flexible Manuf Syst 6:333–360
26. Li L, Shubik M, Sobel M (2005) Control of dividends, capital subscriptions, and physical inventories. Working paper
27. Minner S (2003) Multiple-supplier inventory models in supply chain management: a review. Int J Prod Econ 81(-82):265–279
28. Modigliani F, Miller M (1958) The cost of capital, corporation nance, and the theory of investment. Am Econ Rev 48:261–297
29. Montgomery DC (2005) Introduction to statistical quality control. Wiley, New York
30. Ng CK, Smith JK, Smith RL (1999) Evidence on the determinants of credit terms used in interrm trade. J Finance 54(3):1109–1129
31. Nilsen JH (2002) Trade credit and the bank lending channel. J Money Credit Bank 34(1): 226–253
32. Pagano M, Panetta F, Zingales L (1998) Why do companies go public? an empirical analysis. J Finance 53(1):27–63
33. Paskhaver A, Verkhovodova L (2007) Privatization before and after the orange revolution. Probl Econ Transit 50(3):5–40
34. Petersen MA, Rajan RG (1994) The benets of lending relationships: evidence from small business data. J Finance 49(1):3–37
35. Petersen MA, Rajan RG (1997) Trade credit: theories and evidence. Rev Financial Stud 10: 661–692
36. Rachamandugu R (1989) Effect of delayed payments (trade credit) on order quantities. J Oper Res Soc 40(9):805–813
37. Rajan RG, Zingales L (1995) What do we know about capital structure? some evidence from international data. J Finance 50(5):1421–1460
38. Tang CS (2006) Perspectives in supply chain risk management. Int J Prod Econ 103(2):451–488
39. Tomlin BT (2006) On the value of mitigation and contingency strategies for managing supply-chain disruption risks. Manag Sci 52(5):639–657
40. Tomlin BT (2007) The impact of supply learning when suppliers are unreliable. Manag Serv Oper Manag. (Forthcoming)
41. Wu Z, Choi T (2005) Supplier-supplier relationships in the buyer-supplier triad: building theories from eight case studies. J Oper Manag 24:27–52
42. Xu X, Birge J (2004) Joint production and financing decisions: modeling and analysis. Working paper. Northwestern University, Evanston, IL

43. Yang Z, Aydin, Babich V et al (2008) Supply disruptions, asymmetric information, and a dual sourcing option. Working paper. Industrial and Operations Engineering, University of Michigan.
44. Yang Z, Aydin G, Babich V et al (2009) Supply disruptions, asymmetric information, and a backup production option. Manag Sci 55(2):192–209
45. Yano CA, Lee HL (1995) Lot sizing with random yields: a review. Oper Res 43(2):311–334

Chapter 9
A Reengineering Methodology for Supply Chain Networks Operating Under Disruptions

Alain Martel and Walid Klibi

9.1 Introduction

Goods are procured, produced and distributed to customers using supply chain networks (SCN) involving several facilities owned by a company, or a set of collaborating companies. These networks are engineered or reengineered through strategic decisions on the number, location, capacity, and mission of their production–distribution facilities. Decisions on the selection of suppliers, subcontractors, and 3PLs, and on the offers to make to product-markets, may also be involved. These strategic decisions shape the structure of the network but, once implemented, the SCN is used on a daily basis to respond to customers demands, and possibly to unforeseen disruptions. Day-to-day procurement, production, warehousing, transportation, and demand-management decisions trigger product flows in the network, with associated costs, revenues, and service levels. The adequate design of a SCN requires the anticipation of these future demands, flows, costs, revenues, and service levels. An important issue is the performance measures used to evaluate the quality of the network designed. Return-on-investment measures, such as the *economic value added* (EVA), are often used by strategic decision-makers in this context, but the SCN design robustness is also an important dimension to consider.

A major preoccupation of contemporary businesses is the consideration of risk when designing SCNs. Since SCNs must be designed to last for several years, it is

A. Martel (✉)
Département Opérations et systémes de décision, Faculté des sciences de l'administration, Interuniversity Research Centre on Enterprise Networks, Logistics and Transportation (CIRRELT), Université Laval, Quebec, G1V 0A6, Canada
e-mail: alain.martel@cirrelt.ca

W. Klibi
Operations Management and Information Systems Department, BEM Bordeaux Management School, 680 Cours de la Libération, 33405 Talence Cedex, France
e-mail: walid.klibi@bem.edu

H. Gurnani et al. (eds.), *Supply Chain Disruptions: Theory and Practice of Managing Risk*, DOI: 10.1007/978-0-85729-778-5_9,

clear that they should be robust enough to cope with all the random environmental factors (supply, demand, prices, exchange rates) affecting the normal operations of a company. In addition, SCNs should perform well under major disruptions. In view of recent events, such as the 9/11 terrorist attacks on WTC, hurricane Katrina, and the 2010 Haiti and Chile earthquakes, companies are aware that they should prepare for the next disaster, but in reality only a few do [13, 37]. At a point in time when management strives to make supply chains as lean as possible such disruptions may have serious impacts on company performances [15]. [6, 31] investigate SCNs vulnerability to extreme unforeseen events such as natural disasters and strikes, and [37] examines the case of several companies who suffered from fires, earthquakes, floods, intentional attacks, etc. SCNs can be geographically dispersed across large regions which increase their exposure to extreme events and, in order to design robust networks, the impact of such plausible events must be considered. More specifically, SCNs should be designed to avoid risks as much as possible and to be resilient. Looking at the current SCN design literature from this point of view, [20] highlights its major drawbacks.

SCNs are very complex organizational systems, and their reengineering, in real life, gives rise to major projects which must be carefully planned and managed. These projects must follow a comprehensive analysis and design methodology taking into account all the problem facets previously discussed, and they must be supported with appropriate computer-aided analysis and design tools. This text proposes such a comprehensive methodology, and it is organized as follows. Section 9.2 presents the key concepts needed to design SCNs under uncertainty. It also introduces a specific SCN design problem, known as the location-transportation problem, which is used to illustrate the methodology. Section 9.3 explains the steps of the SCN reengineering methodology proposed, and it illustrates them for the location-transportation problem. Finally, conclusions are drowned. Note that although much of the following discussion is cast in a business context, the reengineering process proposed applies as well to other situations.

9.2 Reengineering SCNs Under Uncertainty

9.2.1 Designing Value-Creating SCN

SCN design problems deal with strategic decisions such as facility location, technology selection, and capacity acquisition that require large investments. These investments must be weighted against projected resulting improvements to the future value of the firm. SCN reengineering decisions impose resource availability and utilization constraints on the users of the network which, through their daily supply, production, and distribution actions, in response to customer demands, determine the return that will be obtained from these investments. It can be argued that the paramount goal of a business should be the sustainable creation of shareholder value [48]. This can be partially measured using static financial performance indicators such as the EVA or the return on capital employed (ROCE) but, under uncertainty, coherent risk measures must also be used to evaluate SCN robustness. All these facets of the problem are considered in the SCN reengineering methodology proposed by [17]. Their methodology involves the explicit modeling of design and user response decisions over a multi-period planning horizon, and it takes the temporal hierarchy between the decisions made at the design and user response levels into account. The reengineering approach presented in this text is based directly on this methodology.

Consider a planning horizon covering several *planning periods* (years, seasons) $t \in T$, as illustrated in Fig. 9.1. At the beginning of the planning horizon, SCN reengineering decisions are made and after an *implementation period* the network reengineered becomes available for use during a *usage period* which may last several years. At the user level, managers make daily or weekly supply, production, distribution, and demand management decisions to serve customers, and they react to disruptions on an ongoing basis. In order to model user response decisions, time is usually divided into short *working periods* (days or weeks) $\tau \in T^u$. The concatenation of the implementation and usage periods following a reengineering decision defines a SCN *reengineering cycle*. Several such cycles unfold in time as SCN reengineering decisions are made. Reengineering decisions are usually made on a rolling horizon basis that is, the only decision implemented when the problem is solved at the beginning of the horizon is the first reengineering decisions. Subsequent design decisions are considered as future opportunities to adapt the network to its evolving environment. Also, it is important to understand that it is through the users short-term response decisions that revenues and expenditures are generated during the SCN usage period, and therefore that value is created. Unfortunately, at the beginning of the planning horizon, the future is not known. The best that can be done is to anticipate, with the information available, what the users and the designer would subsequently do to respond and adapt to plausible future business environments, and thus to model the SCN value-creation process. The approach used to anticipate the value of future operational and reengineering decisions has a major impact on the quality of the SCN designed. Also, an adequate characterization of the future SCN environment is required to obtain good value anticipations.

9.2.2 Taking Uncertainty into Account

Uncertainty is defined here as the inability to determine the true state of the future business environment which may be partially known or completely unknown. When some information is available, three types of uncertainties can be distinguished: randomness, hazard, and deep uncertainty. Accordingly, three types of events should be considered to characterize SCN environments: random, hazardous and deeply uncertain events. The SCN reengineering methodology presented in this paper takes all these types of events into account and is based on recent work on catastrophe modeling [11], scenarios planning [46], and risk analysis [14]. It builds on the fact that

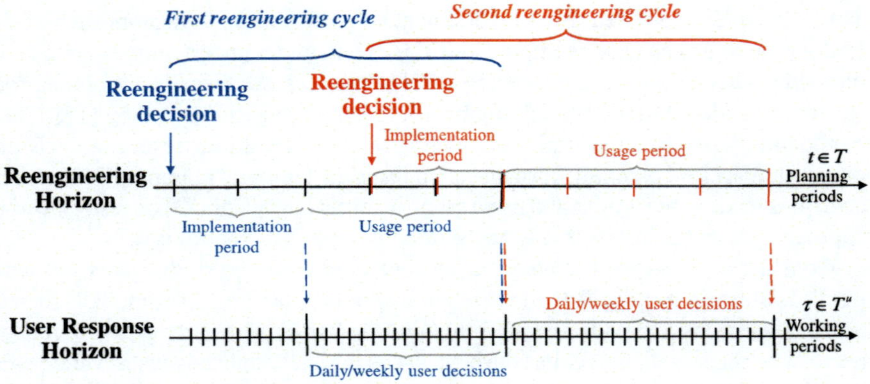

Fig. 9.1 Decision-time hierarchy for two reengineering cycles

the information available on the upcoming business environment can be presented in the form of a set of scenarios about how the future may unfold.

Random events describe factors with a probability of occurrence that can be estimated. Historic information on supply, demand, costs, lead times, exchange rates, etc., can be used to estimate the probability distribution of the random variables related to the business-as-usual operations of a SCN. These events include certain events, a particular case obtained when perfect information is available. *Hazardous events* describe factors or incidents affecting a number of adjacent working periods and leading to SCN disruptions. Hazards are rare but repetitive events which may be characterized by formal location, severity, and occurrence processes. Hazardous events involve natural, accidental, or wilful incidents affecting SCN resources. They include accidental disruptions in operations such as major equipment breakdowns, strikes, and discontinuities in supply due to supplier bankruptcy. They also include disruptions arising from natural disasters affecting a geographical region, such as earthquakes, floods, windstorms, volcanic eruptions, droughts, forest fires, heat waves, freezes, and cold waves. In the recent past, events such as the Kobe earthquake, hurricane Katrina, Haiti, and Chile earthquakes have provoked major disruptions to companies SCNs and to countries critical infrastructures. For such events, catastrophe models have been used to provide likelihood of occurrence and/or likelihood of associated monetary losses, based on historical data and/or professional expert opinions [11]. *Deeply uncertain events* are incidents affecting a number of adjacent working periods for which no directly relevant information exists. These events include isolated, non-repetitive, and extreme events for which a likelihood of occurrence cannot be evaluated [1]. Events related to terrorism (sabotage, bombing) and political instability (sudden currency devaluation, coup), with unpredictable time of occurrence, severity, and location, are usually considered as deeply uncertain. In the recent past, some of these disruptions, such as the 9/11 WTC attack, the SARS epidemic, and the US financial system crisis have lead to major business failures.

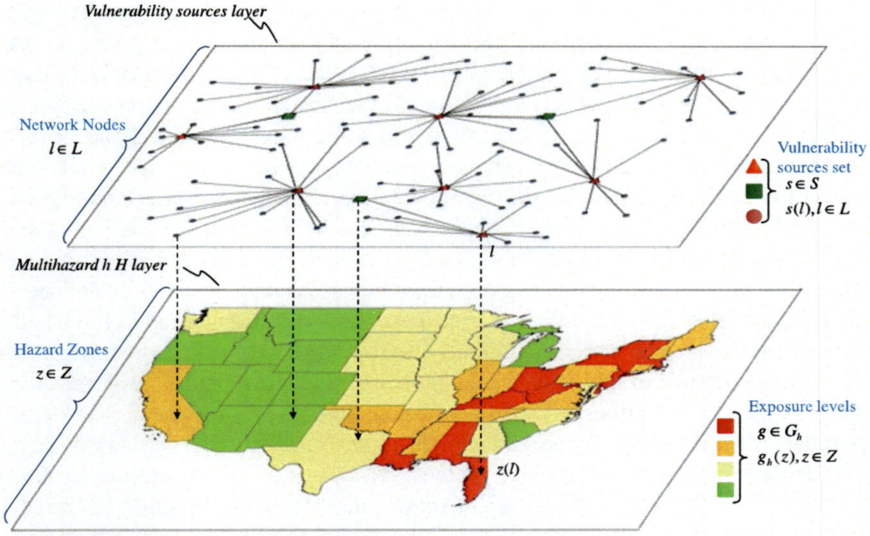

Fig. 9.2 SCN exposure modeling

[25] suggests the elaboration of narrative scenarios for deep-uncertainty situations, and show how to use these scenarios to enhance solution robustness.

Under randomness, some SCN variables (demands, prices, exchange rates) are considered as random and their probability distributions can be estimated. The joint-events associated to the possible values of the random variables define a set of plausible future scenarios Ω^R, and a scenario $\omega \in \Omega^R$ has a probability of occurrence $p(\omega)$. High-impact hazardous events must also be taken into account but they cannot be treated the same way as low-impact business-as-usual events. Moreover, identifying potential threats and assessing their risk are very challenging undertakings. This is the domain of risk analysis which addresses three fundamental questions: (1) What can go wrong? (2) What is the likelihood of that happening? (3) What are the consequences? These questions are implicitly embedded in the SCN reengineering methodology proposed in the next section.

The first question leads to the identification of a set S of SCN *vulnerability sources*. These vulnerability sources are used to partition the set L of SCN locations into location subsets $L_s \subset L,\ s \in S$, affected in the same way by extreme events. For example, locations could thus be partitioned into supply sources, production-distribution sites and demand zones. This question also leads to the identification of the type of hazards threatening the SCN. Considering every possible type of hazard separately is too cumbersome, however, which brings forth the definition of a set H of *multihazards*, i.e. aggregate extreme events incorporating classes of relevant hazards [9, 34]. Typical multihazards would be: natural disasters, geopolitical failures, market failures, and industrial accidents. In order to answer the second question, the geographical territory in which the SCN performs must be partitioned into a set of *hazard zones* Z. Using geographical coordinates, the hazard zone $z(l) \in Z$ of a location $l \in L$ can be identified. Hazard zones delineate areas with similar geological, meteorological, political, economical, and critical infrastructure characteristics. For each multihazard $h \in H$, a set G_h of zone aggregates called *exposure level* are then introduced. Exposure levels can be defined top-down but they are usually constructed by evaluating an *exposure index* for each zone, and associating levels to adjacent index value intervals. Each zone $z \in Z$ is then assigned to a level $g \in G_h$ based on its index value, thus defining a membership relationship $g_h(z)$ between zones and exposure levels. As illustrated in Fig. 9.2, at this point, each SCN location $l \in L$ is associated to a vulnerability source $s(l) \in S$, a hazard zone $z(l) \in Z$, and exposure levels $g_h(z(l)) \in G_h$, $h \in H$. A compound stochastic process then needs to be defined for each exposure level to describe how multihazards occur in space and in time, and to specify incidents intensity and duration. The third question arises when the SCN is hit. The occurrence of an incident in a hazard zone does not necessarily result in a hit of all the SCN locations in that zone. Attenuation probabilities need to be defined to reflect hits likelihood. When a location is hit, the impact on the network resources and demand can be modeled using recovery functions based on intensity and time to recover variables. The application of these concepts to the modeling of Canadian Forces humanitarian, peace keeping, and peace enforcement missions, in response to natural catastrophes and armed conflicts, is found in [27].

9.2.3 Defining Plausible Futures

The superposition, during the planning horizon, of a specific instance of such an hazard process, over specific instances of the business-as-usual random variables described previously, yields a probabilistic scenario $\omega \in \Omega^P$ with probability $p(\omega)$. Some of these plausible future scenarios may involve only a few multihazard over the planning horizon but others may be much more chaotic. An intuitive measure to assess the risk associated to a scenario $\omega \in \Omega^P$ is the number of hits it undergoes during the planning horizon. An alternative measure would be the cumulative damage level during the planning horizon. Figure 9.3 illustrates the distribution of the number of hits for a large sample of scenarios with exponential multihazard inter-arrival times. In order to distinguish between the scenarios a decision-maker would consider as acceptable, in term of the risks involved, and those that would raise a serious concern, we define a hazard tolerance level κ. This level is the maximum number of hits (or the maximum cumulative damage level) the decision-maker can tolerate without serious concern. This tolerance level is used to partition the set of probabilistic scenarios Ω^P in two subsets: Ω^A a set of acceptable-risk scenarios, with associated acceptable-risk probability $\pi^A = \sum_{\omega \in \Omega^A} p(\omega)$ and conditional scenario probabilities $\pi^A(\omega)$, $\omega \in \Omega^A$, and Ω^S a set of serious-risk scenarios, with associated serious-risk probability $\pi^S = \sum_{\omega \in \Omega^S} p(\omega)$ and conditional scenario probabilities $\pi^S(\omega)$, $\omega \in \Omega^S$. This distinction is required to take into account the decision-maker attitude to risk when formulating a design model.

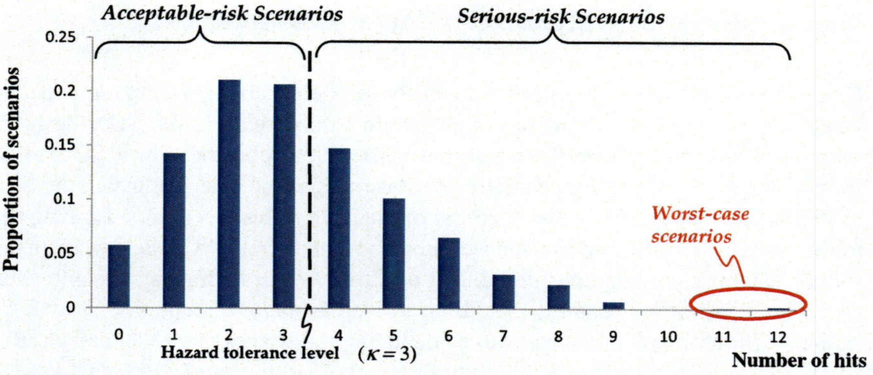

Fig. 9.3 Distribution of the number of hits for a large scenario sample

As indicated previously, probabilistic scenarios may completely overlook some potential extreme events for which no information and experience exist. It is to cope with these potential threats that imaginative deeply uncertain scenarios can be considered. However, for our purposes, these scenarios must be expressed quantitatively in terms of parameters which can be incorporated in a SCN design model. Moreover, these scenarios necessarily include random events and they may also include hazards so they are most easily elaborated by perturbating probabilistic scenarios. In what follows, our interest in the set Ω^U of deep-uncertainty scenarios will be mainly related to our need to consider worst-case scenarios. These would typically be probabilistic scenarios in the tail of the distribution of the number of hits (see Fig. 9.3) perturbed by deep-uncertainty events imagined by experts.

Businesses and organizations operate in a complex world and, when considering a long planning horizon, it cannot be assumed that the future will unfold in the tracks of the past. When developing plausible scenarios, companies like Shell study significant events, they analyze political, social, and economic actors and their motivations, they explore what the world might look like over the next twenty years, and the impact of alternative views of the future on their business environment [39]. In other words, they define possible *evolutionary paths*. The scenarios in $\Omega = \Omega^P \cup \Omega^U$ are possible realizations of a set of underlying stochastic processes with known (for Ω^P) or unknown (for Ω^U) parameters. In what follows, it is assumed that a set K of evolutionary paths with probability p_k, $k \in K$, can be defined and that the parameters of the scenario generating stochastic processes depend on evolutionary paths. It is thus seen that the set of scenarios Ω is the union of the scenario sets Ω^{Pk}, Ω^{Uk} associated to the evolutionary paths $k \in K$, and that the probabilistic scenario probabilities $p_k(\omega)$ also depend on k. Our challenge now is to take all this into account in our SCN reengineering methodology.

9.2.4 Fostering SCN Robustness Under Disruptions

Given the uncertainties discussed previously, it is clear that maximizing expected value is not sufficient. To achieve sustainable value creation, the SCN designed must also be robust. Several authors have discussed robustness in a SC context [8, 13, 28, 32, 37, 43]. They characterize robustness as the extent to which the SCN is able to carry its functions for a variety of plausible future scenarios. Linking this to the evaluation of SCN performance, it can be stated that a SCN design is robust, for the planning horizon considered, if it is capable of providing sustainable value creation under normal business conditions as well as major disruptions.

This definition provides means to evaluate the robustness of a SCN design. But, what kind of SCN structure is likely to be robust? More specifically, what kind of risk-mitigation constructs should be incorporated in our reengineering models to obtain robust SCN designs? To answer these questions we need to look more closely at the notions of SCN *responsiveness* and *resilience*. At the operational level, short-term mitigation actions are required to deal with the variability of low-impact as well as high-impact business events: these are the domain of responsiveness policies. However, to deal with network threat situations, mitigation postures related to the SCN structure, but going beyond the standard reengineering decisions discussed previously, are required: these are the domain of resilience strategies. General discussions of enterprise resilience are found in [47] and on the Web site of the Center for Resilience (http://www.resilience.osu.edu) which defines resilience as "the capacity of a system to survive, adapt, and grow in the face of unforeseen changes, even catastrophic incidents". [6, 31, 37] conclude from empirical studies that business is in need of resilience strategies to deal effectively with unexpected disruptions. Several definitions of resilience and its relation with supply chain capabilities are reported in [29].

The aim of responsiveness policies is to provide an adequate response to short-term variations in supply, capacity, and demand. They provide a hedge against randomness and hazards to increase the SCN expected value. For a given network structure, they shape the means that can be used to satisfy demands with available internal resources and with preselected external resource providers. Responsiveness policies are typically associated to resource flexibility mechanism, such as capacity buffers [5, 33], production shifting [10], overtime, and subcontracting [2]; safety stock pooling and placement strategies [10]; flexible sourcing contracts [13, 22, 35, 37, 45]; and shortage response actions, such as product substitution, lateral transfers, drawing products from insurance inventories, buying products from competitors, rerouting shipments, or delaying shipments [12, 40, 45].

The aim of resilience strategies is to obtain a SCN structure reducing risks and providing capabilities for the efficient implementation of the responsiveness policies previously discussed. This can be done by avoiding or transferring risks [26], and/or by investing in flexible and redundant network structures [31, 38]. Avoidance strategies are used when the risk associated to potential product-markets, suppliers or facility locations is considered unacceptable due to, for example, the instability of the associated geographical area. This may involve closing some network facilities, delaying an implementation, or simply not selecting an opportunity. Another way to avoid risks may be through vertical integration, i.e. the internalization of activities. This may reduce risk through an improved control, but it converts variable costs into fixed costs. This is an incitation to produce internally for low-risk product-markets and to outsource production for higher-risk product-markets, thus transferring risks to suppliers. These are important trade offs that must be captured in SCN design models.

Responsiveness capabilities development may be flexibility or redundancy based. Flexibility-based capabilities are developed by investing in SCN structures and resources before they are needed. Examples of reengineering decisions providing such capabilities include selecting production/warehousing systems that can support several product types and real-time changes, choosing suppliers who are partially interchangeable, and locating distribution centers to ensure that all customers can be supplied by a back-up center with a reasonable service level if their primary supplier fails. Redundancy-based capabilities involve a duplication of network resources in order to continue serving customers while rebuilding after a disruption. An important distinction between flexibility and redundancy-based capabilities is that the latter may not be used [31]. Examples of redundancy-based capabilities include insurance capacity, that is maintaining production systems in excess of business-as-usual requirements, and insurance inventory dedicated to serve as buffers in critical situations [37]. The consideration of such responsiveness capabilities complicates SCN design models considerably. The main challenge is to elaborate resilience strategies providing an adequate protection from disruptions without reducing the SCN effectiveness in business-as-usual situations.

In summary, to achieve sustainable value creation in a disrupted world, one must formulate models seeking to maximize the discounted sum of the residual cash flows generated over a multi-period planning horizon, considering the revenues and costs of the operational and contingency actions required to satisfy customers demands, and taking the three types of uncertainties identified into account through a set of plausible future scenarios. The reengineering methodology presented in the next section is based on such models.

9.2.5 The Location-Transportation Problem

In order to illustrate some aspects of the reengineering methodology proposed, in what follows, we use a simple but common SCN design problem: the location-transportation problem (LTP) under uncertainty. The LTP is a hierarchical decision problem due to the temporal hierarchy between the location decisions and the transportation decisions, and it is described in detail in [20]. Briefly, the company considered purchases a family of similar products from a number of supply sources. This product is sold to customers located in a large geographical area and hence it must be shipped to a large number of ship-to-points. In order to serve its customers, the

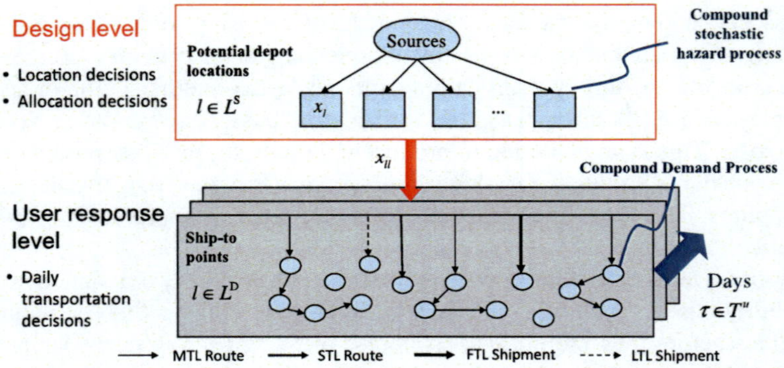

Fig. 9.4 The LTP structure under uncertainty

company must implement a number of capacitated depots with similar processes and technology. For a given day, the capacity of a depot reflects its maximum throughput in terms of a standard shipping unit (ex: pallets). In addition to its regular capacity level, we assume that under normal business conditions, the depot can provide supplementary daily capacity using overtime. Customer orders follow a stationary stochastic process, and the company wants to provide next-day delivery from a single source using common or contract carriers. To this end, several transportation options are available, namely: single customer full truckloads (FTL), single customer partial truckloads (STL), multi-drop truckloads (MTL) or less than truckload (LTL) transportation. For a given day, when all the orders are in, the company plans its transportation for the next day and it requests from its carriers the trucks required to deliver products to ship-to-points. However, the networks depots are under the threat of disruptions and, consequently, their capacity to respond adequately to ship-to-points orders can be perturbed. Therefore, in order to complete the orders received for a given day, the company relies first on the regular capacity of the depot assigned to customers, second on overtime, and third on order transfers between depots. If this is not sufficient, external resources can be used to satisfy all orders. Figure 9.4 illustrates the structure of the LTP under uncertainty. This problem is studied in depth in [18].

9.3 SCN Reengineering Methodology

Supply chain networks are very complex systems and they are rarely engineered on a green field basis. In most practical cases, when a SCN design project is initiated, the objective is rather to improve a portion of an existing supply chain. In fact, SCN reengineering can be seen as a cyclical improvement process where parts of the network are periodically restructured. The SCN reengineering methodology proposed in this section is illustrated in Fig. 9.5. The figure presents the main activities to be

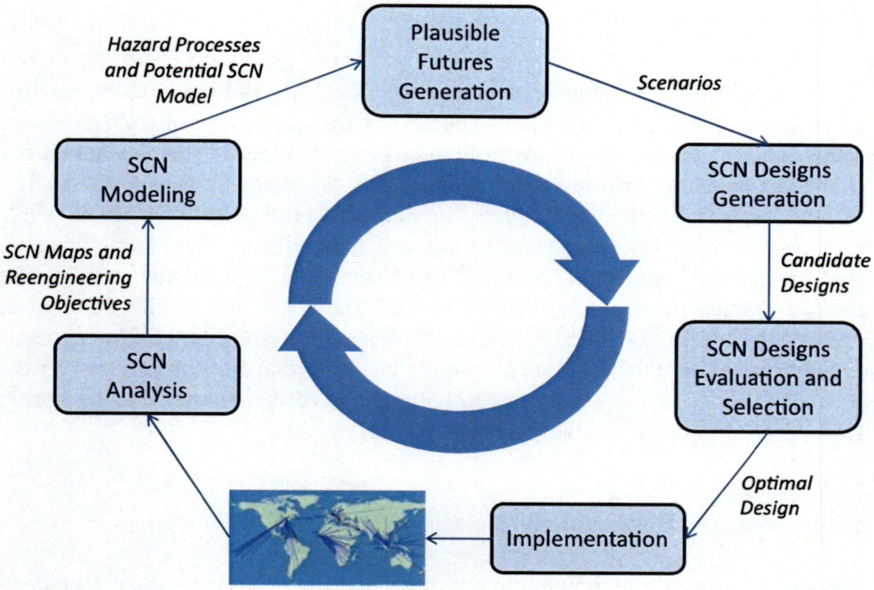

Fig. 9.5 SCN reengineering cycle

performed during a cycle and their respective deliverables (on the arcs). A reengineering cycle starts with a detailed study of the current SCN and of its business environment. This activity involves the collection and analysis of a lot of historical data, the prospective evaluation of evolutionary trends to identify future opportunities and threats, and strategic decisions on how the company would like to position its supply chain in the future. Based on these analyses, a SCN modeling activity is then undertaken. This involves mainly the elaboration of descriptive hazard models, and the formulation of normative design and user response models. The next activity involves the generation of plausible futures. These are mostly obtained using Monte-Carlo methods based on the hazard models and stochastic processes previously defined. Imaginative worst-case scenarios can also be specified. Samples of scenario are then used to generate instances of the normative design model previously formulated. When these models are solved, a set of candidate designs is obtained. In the following activity, candidate designs (including the status quo) are evaluated, using adequate performance measures with a large set of plausible scenarios, and the best SCN structure is selected. Finally, the optimal design is implemented and eventually a new reengineering cycle will be initiated. The following sections explain each one of these reengineering activities in detail.

9.3.1 SCN Analysis

The aim of this first activity is to understand the SCN and the business environment in which it is evolving, and to specify the reengineering cycle objectives and scope. The analysis must be done from three points of view: (1) the business is considered as an actor performing on the industrial scene, which leads to the analysis of markets, competitors, and industry structures; (2) the SCN is viewed as a complex socio-technical system to understand, which requires structural, functional, and performance analyses; and (3) the SCN is considered as a vulnerable system under threats, which gives rise to some risk analyses. These analyses provide the information required to perform a SCN diagnosis, thus identifying strengths, weaknesses, opportunities, and threats. Strategic supply chain development directions can then be elaborated to improve the competitive position of the company, and a specific reengineering project mandate can be specified.

9.3.1.1 Business Environment Analysis

Thorough business environment analyses are usually performed when a company prepares its strategic plans [7, 30, 44] and in the context of a SCN reengineering project, available studies may provide most of the information required on markets, competitors, and the industry. If the information is not readily available, however, one must at least make enquiries to identify emerging technologies (production, storage, transportation...), potential partners (suppliers, subcontractors, public warehouses, carriers, 3PLs...), and competitors (potential acquisitions, market shares...) that may contribute in improving the actual SCN. The nature of contracts with partners, capital and labor markets constraints, and environmental regulations must also be understood. The industry market structure and market development opportunities are particularly important inputs. The relationships between demand and value drivers such as product prices, response times and quality must be characterized, as well as long-term product-market expansion opportunities. Global economic trends and their impact on the company must also be examined. This leads to the elaboration of the set K of evolutionary paths to consider in the study.

9.3.1.2 SCN Processes, Structure and Performance Analysis

The structure and behavior of the SCN is usually studied using data extraction and exploration tools, geographical information systems (GIS), mapping formalizms, and statistical inference methods. These analyses require large amounts of historical data on products, customers, facilities, suppliers, subcontractors, carriers, sales, shipments, inventories, production quantities, etc., obtained mainly from the company databases and stored in a reengineering project database. At the structural level, the analyses performed provide a classification of products into product families, of

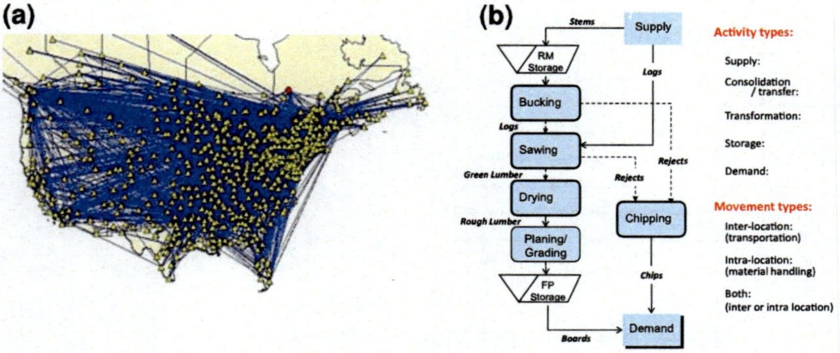

Fig. 9.6 **a** Current SCN flow map. **b** Activity graph for a Sawmill

ship-to-points into demand zones and product-markets, and of suppliers into supply zones. These classifications are often based on Pareto analysis, geospatial analysis, and cluster analysis methods. Snapshots of the resources associated to current facilities and current partners are also elaborated. Finally, maps of the current SCN flow patterns between supply zones, facilities, and demand zones are also produced. Such a map, created for a Pulp and Paper company using a GIS, is illustrated in Fig. 9.6a.

At the functional level, the analyses made yield process diagrams describing the logic of the company production-distribution activities and of its planning and control methods. Activity graphs, such as the one showed in Fig. 9.6b for a typical sawing company, are used to describe physical processes. SCN response policies and method, related for example to the assignment of orders to depots and to the elaboration of shipping routes, can be described using data flow diagrams, flow charts, or pseudo code. These, as we shall see, are required to formulate adequate design models and to evaluate alternative SCN designs. At this level, the supply, demand and resource consumption behaviors resulting from these processes, and the associated costs and revenues drivers, are also examined. This is done using data visualization methods such as histograms and resource usage profiles, as well as statistical methods such as time series analysis and regression analysis. Critical system parameters, probability distributions, and relationship functions are thus estimated. For example, Fig. 9.7a shows an inventory-throughput function obtained for the storage of utility poles in an electricity company, and Fig. 9.7b shows a plot of (cost/ton, distance) points for all the shipments made from a pulp and paper mill during a year for three modes of transportation (used to estimate transportation cost functions). It is at this stage that the business-as-usual random variables considered in the study are identified and characterized.

Another important dimension is the evaluation of SCN performances. Most companies follow their financial performances closely but they do not necessarily have a framework which directly relates supply chain activities to value creation. Also, in order to obtain sustainable value creation, one must not monitor performances only in term of expected value, but robustness measures and aversion to risk must be taken

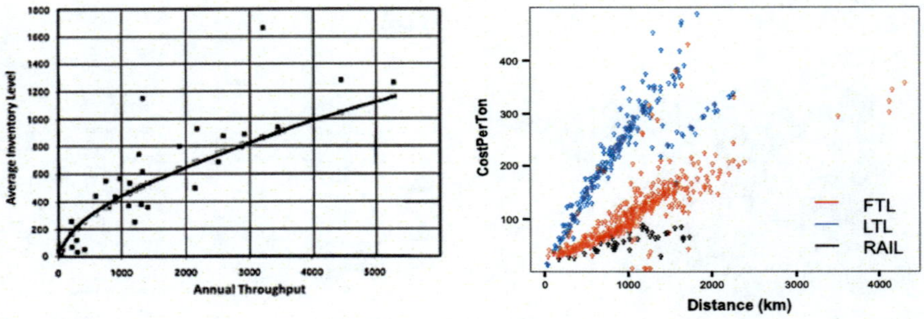

Fig. 9.7 **a** Inventory-throughput function. **b** Shipments cost-distance plot

into account. The issue here is not as much to complete a thorough evaluation of the current SCN, but rather to elaborate a performance evaluation framework that will enable the company to formulate adequate SCN design models, and subsequently to evaluate candidate designs in order to identify the best one.

9.3.1.3 SCN Risk Analysis

At this point in the project, risk analysis involves mainly the identification and classification of the SCN threats arising from hazardous and deeply uncertain events. More specifically, it involves the classification of SCN locations L into vulnerability sources $L_s,\ s \in S$, and of hazards into multihazards H, as well as the zonation of the territory into hazard zones Z and the classification of these zones into exposure levels $G_h,\ h \in H$. As illustrated in Fig. 9.8, when considering potential risks arising from natural, accidental and wilful hazards, a large set of vulnerability sources can be identified. However, the impact of hazards on these vulnerability sources can vary from catastrophic to low. At the strategic decision-making level, the number of vulnerability sources considered should be reduced to a manageable level. A filtering process based on a subjective evaluation of the vulnerability identified leads to the selection of the sources with potential strategic consequences to be considered in the reengineering project. The vulnerability sources S retained usually include the main internal production, distribution, and service resources influencing capacity (plants, warehouses, stores), the main product-markets or service-offers influencing demand, and the main vendors influencing supply (raw-material suppliers, energy suppliers). It is assumed here that all strategic vulnerabilities are related to SCN locations $l \in L$ and not to its arcs. The overriding criterion for the definition of a vulnerability source $s \in S$ is that all the locations $l \in L_s$ it covers must have a similar behavior in terms of impact intensity, time to recovery and recovery pattern when hit by a multihazard, so that they can all be described in terms of the same metrics. They must also be defined so that the sets $L_s \subset L,\ s \in S$, are mutually exclusive and collectively exhaustive.

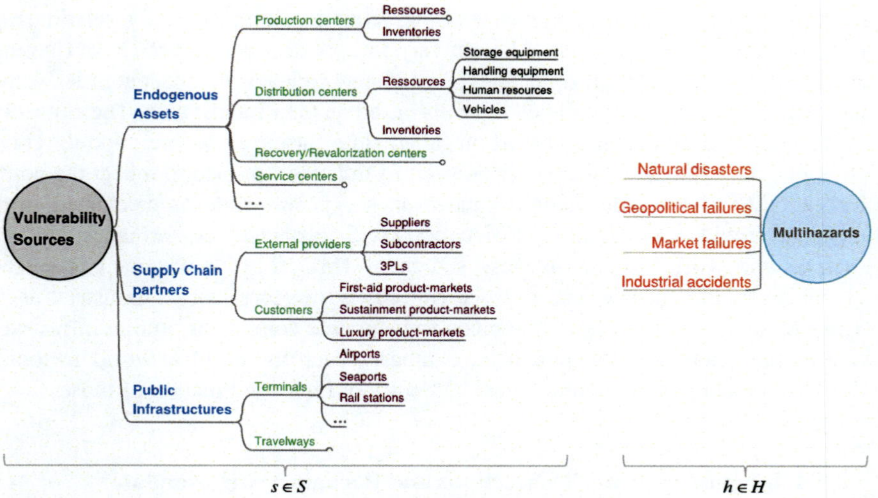

Fig. 9.8 Examples of vulnerability sources and multihazards

The specification of the multihazards set H to consider is also an important issue. Depending on the context of the study, some hazard types may not be relevant, and some vulnerability sources may be affected more by some hazards than by others. Also, the data related to some relevant hazards may not be available. The number of multihazards considered should be kept to a minimum while making sure that major (vulnerability source, multihazard) relationships are captured. This leads to the specification of vulnerability source threat domains that is, of subsets $H_s \subseteq H$ of multihazards which have an impact on a vulnerability source $s \in S$. Multihazards can be elaborated from the data provided by several public sources such as the Centre for Research on the Epidemiology of Disasters (http://www.cred.be), the Heidelberg Institute for International Conflict (http://www.hiik.de), the Federal Emergency Management Agency (http://www.fema.gov), and the U.S. Geological Survey (http://www.usgs.gov), and private sources such as Swiss Re (http://www.swissre.com) and Munich Re Group (http://www.munichre.com).

Zones Z delineating geographical areas with similar hazard characteristics must also be defined. These zones may correspond to 3-digit zip-codes, to counties, to states/provinces, to countries, or to a combination of those, depending on the level of precision desired and the data available. They must be constructed, however, to make sure that the SCN location aggregates defined (such as demand zones) fit uniquely in a hazard zone, and they must be large enough to consider the occurrence of extreme events in different zones as independent. They must also be defined so that the sets $L_z \subset L$ of locations in the zones $z \in Z$ are mutually exclusive and collectively exhaustive. Finally, exposure levels can be defined top-down or bottom-up, depending on the context. Exposure levels are sometimes associated to geographical regions, such as continents. The states in the continent then provide a relationship $g_h(z)$ between zones and levels. Alternatively, levels can be constructed by evaluating an exposure index for each zone, and then associating levels to adjacent index value intervals. Zones are then assigned to levels based on their index value. The exposure index used to do this can be based on failed state (http://www.foreignpolicy.com) and/or opacity (http://www.opacityindex.com) indexes designed to reflect the political stability of a region, natural catastrophes exposure indexes calculated from the data provided by CRED, FEMA, or USGS, economic performance indexes such as the world competitiveness scores of IMD (http://www.imd.ch) or the global competitiveness index of WEF (http://www.weforum.org), industrial accident indexes related to the claims made to insurance companies, public infrastructure quality indexes calculated from databases such as the CIA world factbook (http://www.cia.gov/cia/publications/factbook), or on a combination of those.

9.3.1.4 Diagnosis, Strategic Directions and Reengineering Mandate

The environmental, structural, functional, performance, and risk analyses discussed previously provide the information required to perform a SCN diagnosis. This is often done by identifying strengths, weaknesses, opportunities, and threats, and then by delineating critical success factors. Strategic supply chain development directions can then be elaborated to improve the competitive position of the company, and a specific reengineering project mandate can be specified. The former relates to product-market penetration targets, the specification of manufacturing stages and distribution echelons (by reengineering the activity graph), the identification of potential facility sites, suppliers, subcontractors and 3PLs, as well as of potential manufacturing, distribution and transportation capacity options. The later delimits the SCN boundaries for the reengineering project, and it specifies the planning horizon to cover, the random variables, vulnerabilities, multihazards and evolutionary trends to consider, the aggregation levels to use, the cost and inventory-throughput functions to apply, as well as the performance measures to utilize to evaluate alternative SCN designs.

To illustrate, for the location-transportation example introduced previously, the results of the analysis yields the following problem specifications. Since we are dealing with similar products having similar market and risk profiles, all products are aggregated in a single product family and aggregate demand quantities are measured in pallets. The company is considering a set L^S of sites where depots could be located, and it wants to provide next-day delivery to all the ship-to-points $L_l^D \subseteq L^D$ of its United States customers. For this reason, a depot $l \in L^S$ can supply only the ship-to-points L_l^D which are close enough to be reached in one day. Aggregate day-to-day demands from ship-to-points $l \in L^D$ are variable, and they follow a compound process with Exponential order inter-arrival times q_l and Log-Normal order size o_l. The cumulative distribution functions of inter-arrival times and order sizes are denoted, respectively, by $Exp_l(\cdot)$ and $LN_l(\cdot)$, $l \in L^D$. Also, to provide a better service, the company does not want to split customer orders. The fixed annual operating cost A_l incurred when a depot $l \in L^S$ is used was estimated, taking all relevant financial charges into account. The TL and LTL transportation means exploited by the company are classified into FTL, STL, MTL, and LTL transportation options, and cost formulas based on distance and load are derived for each option. Product sales prices u_l for ship-to-points $l \in L^D$ and purchase prices v_l for depots $l \in L^S$ were calculated using weighted averages. Vehicles capacity b^F in pallets/load and daily depots capacity $a_l,\ \ l \in L^S$, in pallets/day, were also calculated, as well as overtime/regular-time capacity proportions φ_l and overtime costs v_l^z. The response policy of the company is applied using specific procedures for order assignment, transportation option selection, and vehicle routing. Finally, when an order cannot be supplied by the primary or backup depot assigned to a ship-to-point, it can be shipped directly from the original supplier, or from a rival distributor, using a special expediting procedure. The average unit cost paid v^e when such recourse is used was also estimated.

Since the demand is stationary, a single engineering cycle constituted of one year of daily working periods needs to be considered, and evolutionary paths can be neglected. The main vulnerability sources to take into account are depots (S) and ship-to-points (D), i.e. $S = \{\text{S,D}\}$. Since the products distributed can be purchased from several alternative sources, and since their purchase price depends mainly on depots locations, supply sources are not considered as vulnerable, and they do not have to be modeled explicitly. The network is threatened mainly by natural disasters which are considered as a single multihazard, and it is sufficient to use US states as hazard zones. Exposure levels were specified bottom-up based on FEMA data, thus defining the sets of states $Z_g,\ \ g \in G$. These exposure levels are illustrated in Fig. 9.2.

9.3.2 SCN Modeling

Two related supply chain modeling activities are required at this point: the elaboration of descriptive hazard models which can be used to generate plausible future scenarios, and the formulation of a normative design model which can be used to generate candidate SCN designs. The former involves the definition of stochastic processes to describe how multihazards occur in space and in time, and to specify incidents intensity and duration. Also, recovery functions based on intensity and time to recovery variables must be specified to quantify the impact of a hit on network resources and demand. Potential SCN modeling, on the other end, can involve the formulation of a tailor-made optimization model, or the use of generic SCN design software. These two SCN modeling activities are discussed in detail in the following sections.

9.3.2.1 Hazard Processes Modeling

This activity is initiated by modeling the arrival, intensity, and duration of the multihazards $h \in H$ considered in the study. We assume that multihazards occur independently in hazard zones, and that the time between the occurrences of successive multihazards in a zone is characterized by a non-stationary stochastic arrival process depending on the evolutionary path considered. Note that these processes depend on the territory on which the SCN is deployed, but that they are independent of the SCN considered. Under evolutionary path $k \in K$, if an incident occurs in working period $\tau \in T^u$, then the time before the arrival of the next multihazard $h \in H$ in zone $z \in Z$ is a random variable $\lambda^h_{zk\tau}$ with cumulative distribution function $F^{\lambda^h}_{zk\tau}(\cdot)$. In practice, catastrophe models often use Poisson processes to determine the number of extreme events that can occur in a given period [1]. Accordingly, we consider that in most cases it is sufficient to assume that $F^{\lambda^h}_{zk\tau}(\cdot)$ is an exponential distribution $\text{Exp}(\mu^h_{zk\tau})$ with an expected time between multihazards $\mu^h_{zk\tau}$. Let $\phi^h_k(\mu^h_{z0}, \tau)$ be a function elaborated by experts, for evolutionary path k and multihazard h, to project the historical mean time between multihazards, μ^h_{z0}, estimated during the analysis activity, over the periods $\tau \in T^u$ of the planning horizon. Then, the required probability distributions are obtained simply by calculating $\mu^h_{zkt} = \phi^h_k(\mu^h_{z0}, \tau)$ for all h, z, k and τ.

The data provided by public or private sources such as FEMA, CRED, Swiss Re and Munich Re is often not sufficiently detailed to characterize the multihazards impact intensity and duration for each hazard zone. For this reason, the impact intensity and duration are usually modeled by exposure level. A hierarchical modeling approach based on conditional hazard zone hit probabilities, $p^h_{z|g}$, must then be used [19]. These conditional probabilities are estimated subjectively based on public or constructed indexes. For example, for geopolitical failures the *failed state index* can be used, and for natural disasters an incident frequency index calculated from CRED data can be used. We assume that when a multihazard $h \in H$ occurs in a zone $z \in Z$, its duration (in working periods) and its intensity (in a generic measure such as the loss level or the casualty level per period) are characterized by two correlated random variables related to the zone exposure level $g(z) \in G_h$, namely: the impact intensity β^h_g, with cumulative distribution function $F^{\beta^h}_g(\cdot)$ and the duration θ^h_g. The duration can be related to the intensity through incident *impact-duration functions* $\theta^h_g = f^h(\beta^h_g) + \varepsilon^h$, $h \in H$, estimated by regression, where ε^h is a Normally distributed error term. The impact distributions and the impact-duration functions are estimated from the incidents data.

The occurrence of an incident in a hazard zone does not necessarily result in a hit of all the SCN locations in that zone. When the hazard zones are large (countries or states), it is likely that only a part of the zone locations will be hit. Also, when considering the impact on product-markets, the SCN does not necessarily respond to all incidents. This leads to the estimation of *attenuation probabilities* α^h_l which are conditional probabilities that location $l \in L$ is hit when a multihazard $h \in H$ occurs in zone $z(l)$. It is clear that these probabilities are related to the hazard zones granularity. Large zones lead to small attenuation probabilities, and vice versa. Attenuation probabilities can be estimated by experts for each SCN location, based on experience and data available.

When a location $l \in L$ in zone $z(l)$ is hit by a multihazard $h \in H$, the severity of the incident for the SCN is characterized on two correlated dimensions: a SCN resource/market impact intensity and the time for recovery [37]. Clearly, these dimensions are related to the exogenous multihazard intensity and duration variables β_g^h and θ_g^h defined previously. However, the SCN impact severity must be expressed in units related to the capacity and demand of the vulnerability sources. It is assumed that the metrics used to characterize these two severity dimensions are the same for all the locations associated to a given vulnerability source, i.e. for all $l \in L_s$. Hence, for each vulnerability source $s \in S$, incident profiles must be specified for all locations $l \in L_s$, products $p \in P_s$ and multihazards $h \in H_s$. Damage on suppliers is typically assessed using an unfilled rate (% of material ordered during the incident not delivered) and the time required to restore supplies, whereas damage on production-distribution resources is usually assessed using a capacity loss rate and the time before production/distribution can resume. For vulnerability sources affecting demand, damage is usually assessed using an inflation or deflation rate expressing a demand surge or drop for a given period of time. Note that the evaluation of incidents severity may also be influenced by the state of the resources/partners associated to a vulnerability source.

Let ξ_l^h be a discrete random variable giving the time to recovery, in working periods, of location $l \in L$ when hit by a multihazard $h \in H_{s(l)}$. We assume that this time to recovery can be related to the multihazard duration $\theta_{g(l)}^h$ using an adequate translation function $\xi_l^h = q_{s(l)}^h\left(\theta_{g(l)}^h\right)$ specified for each vulnerability source $s \in S$ and multihazard $h \in H_s$. This function may be based on a proportion estimated from past instances or provided by experts. Consider a multihazard $h \in H$ hitting location $l \in L$ at the beginning of usage period $\tau' \in T^u$. Then, the impact of the hit lasts during periods $\tau = \tau', \ldots, \tau' + \xi_l^h - 1$. When a multihazard $h \in H$ hits a location l, its impact is not necessarily felt uniformly during the time to recovery ξ_l^h [37]. Several phases are usually observed, depending on the nature of the multihazard and of the vulnerability source. For example, when a manufacturing plant is hit by a natural disaster, production capacity drops quickly during a first phase, then there may be a stagnation period while recovery measures are organized, and during a third phase the capacity is gradually restored. Such phase-dependent impacts can be characterized by defining discrete recovery functions $\boldsymbol{\rho} = r_s^h(\beta, \xi, \boldsymbol{\rho})$, $h \in H$, $s \in S$, where $\boldsymbol{\rho} = [\rho_{\tau'}, \ldots, \rho_{\tau'+\xi-1}]$ is a vector of capacity/demand amplification percentages for the ξ working periods affected by the multihazard. The $\rho_{\tau'}, \ldots, \rho_{\tau'+\xi-1}$ values used as an argument in the function reflect amplification percentages before the hit and the function returns percentages after the hit. The use of these functions can be illustrated using the LTP case introduced previously. For this case, three types of recovery functions, associated respectively to customer inter-arrival times (q), to customer order sizes (o), and to depots capacity (a), must be defined:

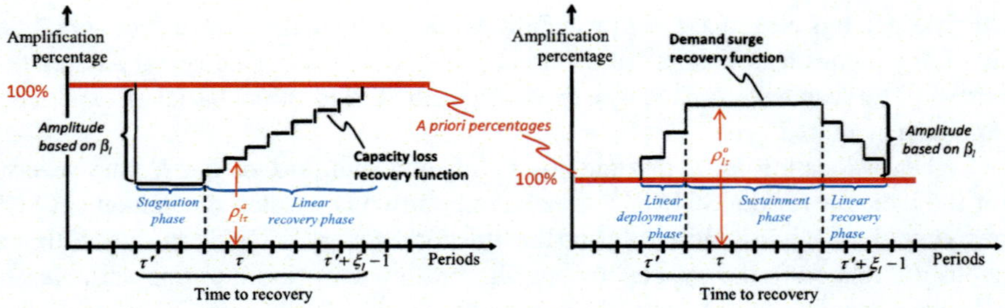

Fig. 9.9 Recovery function examples for depot $l \in L^S$ and ship-to-point $l \in L^D$

$\rho^x_{l\tau} = r^x(\beta_l, \xi_l, \rho^x_{l\tau}), \tau = \tau', \ldots, \tau' + \xi_l - 1, \; x = q, o, a$. As illustrated in Fig. 9.9, if the periods affected by the multihazard are not still recuperating from a previous incident, then the a priori percentages are $\rho^x_{l\tau} = 100\%, \; \forall x, \; l, \; \tau = \tau', \ldots, \tau' + \xi_l - 1$. The amplitude of the amplification depends on the multihazard impact intensity β_l. Using these recovery functions, the capacity and the demand can be calculated for specific periods and locations. For the order interarrival times and sizes, this gives rise to the perturbed random variables $q_{l\tau} = \rho^q_{l\tau} q_l$ and $o_{l\tau} = \rho^o_{l\tau} o_l, \; \tau = \tau', \ldots, \tau' + \xi_l - 1, \; l \in L^D$, and to their associated distributions functions $F^q_{l\tau}(\cdot)$ and $F^o_{l\tau}(\cdot)$. For the depots $l \in L^S$, this yield perturbed capacity levels $a_{l\tau} = a_l \rho^a_{l\tau}, \; \tau = \tau' \ldots, \tau' + \xi_l - 1$.

For the LTP case, two simplifying assumptions were made during the analysis: (1) there is no evolutionary trend, and consequently the index k can be dropped, (2) all hazards are grouped into a single multihazard, and thus the index h can be dropped. Under these assumptions, the hazard models can be simplified. The interarrival times for zone $z \in Z$ are characterized by the exponential random variable $\lambda_z \sim \text{Exp}(\cdot)$, and the impact intensity by the uniform random variables $\beta_z \sim U_{g(z)}(\cdot)$. We also assume that the duration translation functions $\xi_l = q_{s(l)}\left(\theta_{g(l)}\right)$ are based simply on proportions estimated from past instances so that the impact-duration functions $\theta_g = f\left(\beta_g\right) + \varepsilon$ can be substituted back in the former to get simplified time-to-recovery functions for each vulnerability source, i.e. to get the relations $\xi_l = f_{s(l)}\left(\beta_{z(l)}\right) + \varepsilon_{s(l)}, \; \varepsilon_{s(l)} \sim N\left(0, \sigma^{\xi}_{s(l)}\right)$. As we shall see, these hazard process models will be essential to generate plausible probabilistic scenarios $\omega \in \Omega^P$.

9.3.2.2 Potential SCN Modeling

The objective of this activity is to capture the essence of the SCN considered into an optimization model which can be used to generate good candidate designs. Since the future is uncertain, this model should ideally be a multi-stage stochastic program [3] with sufficient details to anticipate the operations of the network adequately. Moreover, as indicated previously, some risk-mitigation constructs should be incorporated in the model to obtain more resilient and robust SCN designs. Unfortunately, such models are extremely complex and difficult to solve. For this reason, complexity reduction measures are necessary. When design decisions are made on a rolling horizon basis, it is reasonable to use a multi-period two-stage stochastic program, which simplifies things considerably. Further complexity reduction is possible either by assuming that the future is known (to get a deterministic model), by considering resilience implicitly, and/or by using a crude approximation of the network operations to anticipate revenues and expenditures, which often leads to the use of location-allocation models [17, 21]. These simplified models may give good feasible designs which are, however, not necessarily optimal when evaluated with the performance metrics selected during the analysis activity. Two approaches can be utilized to model the network. One can exploit an existing generic model, such as the one proposed by [4] which is implemented in the SCN design software SCN-STUDIO,[1] or formulate a tailor-made model. The former may not provide a perfect fit for the company, but it requires no software development. The latter may provide a better fit but it involves major software development efforts. The two approaches are discussed below.

Most of the SCN modeling construct presented in this paper is supported by SCN-STUDIO. Therefore, in order to formulate a design model, a company using such a tool must simply exploit these modeling constructs to represent its business context. With the software interfaces, the company specifies the product families and product-markets to use, the alternative market policies (offers) to consider, the transportation means and options available and their capabilities, the activity graph to apply and the associated recipes, the potential supply, production-distribution and demand locations available, the relationship between inter-location lanes and transportation means, the current and potential platforms for production-distribution sites and the activity they can support, as well as all relevant cost, price, resource consumption, and capacity parameters. When some of these parameters are random variables, their probability distribution can also be defined. The user must then specify the planning horizon, the evolutionary trend functions, the scenario sample, and the objective function to employ. Functionalities to take hazard processes into account are also under development. When all this information has been entered, SCN-STUDIO is able to automatically generate the multi-scenario MIP model to solve to obtain candidate SCN designs, and eventually to solve it using CPLEX.

To illustrate the formulation of a tailor-made model, we use the LTP case discussed previously. Several stochastic programing models with different anticipation quality and resilience constructs, and therefore different complexity level, can be formulated for this problem [17]. We limit ourselves here to the presentation of a relatively simple stochastic location-allocation model with a multiple-sourcing strategy to increase the resilience of the designs obtained. More specifically, on a given day, we do not allow a customer order to be split but, on different days, orders do not have to be fulfilled from the same depot. Let $\underline{L}^S \subset L^S$ be the set of opened depots for a given SCN design. For a given scenario ω, on day τ, the depots capacity $a_{l\tau}(\omega)$, $l \in \underline{L}^S$, is known and the set of depots $\underline{L}^S$ can be partitioned into fully operational depots

[1] SCN-STUDIO was developed during a joint research project involving CIRRELT, at University Laval, Defence R&D Canada and Modellium Inc.

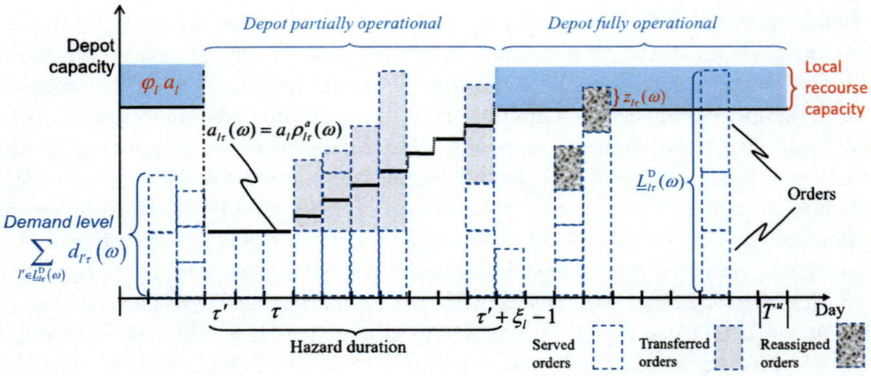

Fig. 9.10 Capacity-demand behavior at depot $l \in \underline{L}^S$ for a given scenario ω

$\underline{L}_\tau^S(\omega) = \{l|a_{l\tau}(\omega) = a_l\}$ and partially operational depots $\tilde{\underline{L}}_\tau^S(\omega) = \{l|a_{l\tau}(\omega) < a_l\}$. Similarly, customer demand $d_{l\tau}(\omega),\ l \in L^D$, is known and the set of ship-to-points can be specified. When a depot l is operational on day τ under scenario ω, additional capacity $\varphi_l a_l$, is available for use, where φ_l is a fixed proportion of regular daily capacity. As illustrated in Fig. 9.10, the stochastic demand level and depot capacity on each day dictate the kind of response decisions made. When depot l is operational it serves its primary ship-to-point orders and it can process reassigned orders from other depots. When it is partially operational, however, it serves only a subset of its primary ship-to-point orders and the remaining ones are transferred. Thus, based on depots primary missions, revised daily assignment decisions $\underline{L}_{l\tau}^D(\omega),\ l \in \underline{L}^S$, must be made to ensure that the depots capacity constraints $\sum_{l' \in \underline{L}_{l\tau}^D(\omega)} d_{l'\tau}(\omega) \le a_{l\tau}(\omega), l \in \underline{L}^S$ are respected. Following these reassignments, each depot ships its assigned orders using a combination of transportation means depending on order sizes. Also, as indicated previously, if internal capacity is not sufficient on a given day, then some customer orders must be fulfilled from an expensive external supply source. This stochastic capacity-demand behavior must be captured in the design model.

Although several TL and LTL transportation options are available on a daily basis to ship orders to customers, in order to simplify the model, they can be aggregated in weekly depots to ship-to-points flow variables with unit transportation costs $w_{ll'},\ l \in L^S,\ l' \in L^D$. These unit costs are estimated by regression using daily historical data [21]. This aggregation could be done by summing daily demands and capacities to get weekly demands and capacities. This would assume however that an order arriving any day of the week could be supplied from the depot any day of the week, which is not realistic. Since demand processes are stationary, a better approach is to use period sampling, i.e. to consider one randomly selected day per week. The planning horizon is then constituted of a subset $\hat{T} \subset T^u$ of daily periods. Under this complexity reduction mechanism, ship-to-point demands for scenario ω are denoted $d_{lt}(\omega),\ l \in L_t^D(\omega),\ t \in \hat{T}$, and depot capacities $a_{lt},\ l \in L^S,\ t \in \hat{T}$.

To formulate the model, the following decision variables are required:

x_l: Binary variable equal to 1 if depot l is opened, and 0 otherwise
$x_{ll'}$: Binary variable equal to 1 if ship-to-point l'can be supplied by depot l, and 0 other-wise
$F_{ll't}(\omega)$: Quantity of product supplied by depot l to ship-to-point l'on day t under scenario ω
$z_{lt}(\omega)$: Recourse capacity needed at depot l on day t
$F^e_{l't}(\omega)$: Quantity of products supplied to ship-to-point $l'$by the external supply source on day t under scenario ω

The variables x_l and $x_{ll'}$ are first stage design variables (denoted collectively by the vector **x**), and $F_{ll't}(\omega)$, $z_{lt}(\omega)$, $F^e_{l't}(\omega)$ are second stage recourse variables depending on the prevailing scenario $\omega \in \Omega^P$. This leads to the following two-stage stochastic program with fixed recourse:

$$R = \max_x \frac{|T^u|}{|\hat{T}|} \sum_{V=A,S} \pi^V \mathbf{E}_{\Omega^V|V} \left(R^u(\boldsymbol{x}, \omega) \right) - \sum_{l \in L^S} A_l x_l \tag{9.1}$$

subject to

$$x_{ll'} \leq x_l \quad l \in L^S,\ l' \in L^D_l \tag{9.2}$$

$$x_l, x_{ll'} \in \{0, 1\} \quad l \in L^S,\ l' \in L^D_l \tag{9.3}$$

where

$$R^u(\boldsymbol{x}, \omega) = \max \sum_{t \in \hat{T}} \left[\sum_{l \in L^S} \left(\sum_{l' \in L^D_t(\omega) \cap L^D_l} [u_{l'} - v_l - w_{ll'}] F_{ll't}(\omega) - v^z_l z_{lt}(\omega) \right) + \left(u_{l'} - v^e \right) \sum_{l' \in L^D_t(\omega)} F^e_{l't}(\omega) \right] \tag{9.4}$$

subject to

$$\sum_{l \in L^S} F_{ll't}(\omega) + F^e_{l't}(\omega) = d_{l't}(\omega) \qquad l' \in L^D_t(\omega),\ t \in \hat{T},\ \omega \in \Omega^P \tag{9.5}$$

$$F_{ll't}(\omega) \leq d_{l't}(\omega) x_{ll'} \qquad l \in L^S,\ l' \in L^D_t(\omega) \cap L^D_l,\ t \in \hat{T},\ \omega \in \Omega^P \tag{9.6}$$

$$\sum_{l' \in L^D_{lt}(\omega)} F_{ll't}(\omega) \leq a_{lt}(\omega) + z_{lt}(\omega) \qquad l \in L^S,\ t \in \hat{T},\ \omega \in \Omega^P \tag{9.7}$$

$$z_{lt}(\omega) \le \varphi_l a_l, l \in L_t^S(\omega);\ z_{lt}(\omega) \le 0, \quad l \in \widetilde{L}_t^S(\omega) \qquad t \in \hat{T},\ \omega \in \Omega^P \tag{9.8}$$

$$F_{ll't}(\omega) \ge 0,\ F_{l't}^e(\omega) \ge 0,\ z_{lt}(\omega) \ge 0 \qquad l \in L^S,\ l' \in L_t^D(\omega) \cap L_t^D,\ t \in \hat{T},\ \omega \in \Omega^P \tag{9.9}$$

In the objective function (9.1) of the first stage program, $\mathbf{E}_{\Omega^V|V}(\cdot)$ denotes the conditional expected value for acceptable-risk scenarios Ω^A or serious-risk scenarios Ω^S. The function $R^u(\boldsymbol{x}, \omega)$, gives the value of the second stage program (9.4–9.9) for a given design $\boldsymbol{x}$ and scenario ω. More specifically, it provides the net revenues generated during planning horizon $\hat{T}$. The expected net revenues are then multiplied in (9.1) by the horizon shrinking factor $|T^u|/|\hat{T}|$ to obtain an adequate approximation of the total expected profits. By using the probabilities π^A and π^S in the objective function (9.1), the total expected profit is obtained, which is adequate for a risk-neutral decision-maker. However, if the decision-maker is risk-averse, these probabilities need to be replaced by weights $\hat{\pi}_S > \pi^S$ and $\hat{\pi}_A = 1 - \hat{\pi}_S$ to give more importance to serious-risk scenarios [17].

As it stands, this model cannot be solved because the set Ω^P contains an infinite number of scenarios. As we shall see, it can be solved relatively easily, however, if we restrict ourselves to a sample of randomly selected scenarios. Several other models were also proposed for facility location problems under uncertainty [41, 42].

9.3.3 Plausible Futures Generation

It should be clear by now that plausible future scenarios are a central element of our SCN reengineering methodology under uncertainty. We have seen how random business-as-usual events, high-impact disruptions, and evolutionary paths can be modeled using stochastic processes. However, the joint impact of these events over the planning horizon must be represented in terms of a set of plausible future scenarios. In other words, the stochastic processes defined must be used to generate the sets of scenarios required to create and evaluate candidate designs. This can be done using standard Monte-Carlo techniques. To obtain a scenario $\omega \in \Omega^P$, we start by generating as many independent pseudo-random numbers, uniformly distributed in [0, 1], as we have random events in our stochastic processes. Then, interpreting these numbers as cumulative probabilities, we use the inverse probability distributions of the random events to generate a scenario instance. When doing this, however, the hierarchical structure of the stochastic process models developed must be respected. In our context, the resulting scenario generation procedure obtained has five main steps:

1. An evolutionary path is randomly selected.
2. A chronological list T_z of all the multihazards arrival periods is constructed for every hazard zone $z \in Z$.
3. The intensity and duration of the disruptions considered are generated and used to calculate amplification factors with the recovery functions.

4. The amplification factors are used to calculate the working periods capacity and demand. The numeric value of hazard-independent random variables is also computed.
5. The working period quantities obtained are aggregated into planning period quantities.

The last step is required because the SCN design generation activity needs scenarios expressed in terms of planning periods $t \in \hat{T}$. The design evaluation phase however usually uses scenarios expressed in terms of working periods $\tau \in T^u$.

To illustrate the approach, for the LTP case, the Monte-Carlo procedure required to generate depots capacity and ship-to-points demand values, for a scenario instance ω, is given in Sect. 9.3.3.1. In the procedure, u denotes a pseudorandom number, $\Phi^{-1}(\cdot)$ the inverse of the standardized Normal variate, and $T_t^u \subset T^u$ the set of days in planning period $t \in \hat{T}$. Note that Step 1 is skipped because, in the LTP case, evolutionary trends are not considered. This procedure can be used repeatedly to generate the scenario samples required in the next reengineering activities.

9.3.3.1 Monte-Carlo Procedure for the Generation of a Scenario

2. For all $z \in Z$, do:
   Using the distribution of λ_z, generate multihazard arrival moments $T_z \subseteq T^u$
3. Set $\rho_{l\tau}^x = 1, x = q,o,a, \forall l \in L_z, \tau \in T^u$
   For all $\tau' \in T_z$, do:
      Compute $\beta_z = U_{g(z)}^{-1}(u)$
         For all $l_z \in L \mid u \le \alpha_l$, do:
         Compute $\xi_l = f_{s(l)}\left(\beta_{z(l)}\right) + \sigma_{s(l)}^{\varepsilon} \Phi^{-1}(u)$
         Compute $\rho_{l\tau}^x = r^x\left(\beta_{z(l)}, \xi_l, \rho_{l\tau}^x\right), \tau = \tau', \ldots, \tau' + \xi_l - 1, x =, q, o, a$
         End For
   End For
   End For
4. For all $l \in L^S$ and $\tau \in T^u$: Compute the daily capacity $a_{l\tau}(\omega) = a_l \rho_{l\tau}^a$
   For all $l \in L^D$, do:
   $\eta = 0;\ d_{l\tau}(\omega) = 0,\ \tau \in T^u;\ \tau = 1; F_{l\tau}^q = Exp_l$
   While $\eta \le |T^u|$, do:
      Compute the next order arrival time $\eta = \eta + F_{l\tau}^{q^{-1}}(u)$ and $\tau = \lceil \eta \rceil$
      Derive $F_{l\tau}^q$ from Exp_l and $\rho_{l\tau}^q$; Derive $F_{l\tau}^o$ from LN_l and $\rho_{l\tau}^o$
      Compute the daily demand $d_{l\tau}(\omega) = d_{l\tau}(\omega) + F_{l\tau}^{o^{-1}}(u)$
   End While
   End For
5. Select a day τ in the weeks $T_t^u, ct \in \hat{T}$, to get $a_{lt}(\omega),\ l \in L^S$, and $d_{lt}(\omega), l \in L^D$

9.3.4 SCN Designs Generation

During this activity, the design models previously formulated are solved to obtain candidate SCN designs. They are usually solved with commercial solvers, such as CPLEX (either directly or indirectly as is done when using a SCN design tool such as SCN-STUDIO), or with a tailor-made heuristic method. When a stochastic programing model (such as 9.1–9.9) is formulated, if the set of probabilistic scenarios Ω^P is large, then the model cannot be solved directly. The best that can be done is to solve it for some scenarios samples generated using the Monte-Carlo procedure discussed in the previous section. The approach involves the generation of several scenario samples, and the solution of the resulting sample average approximation (SAA) models [36], to get a set of candidate SCN designs. Given our partition of probabilistic scenarios in two subsets Ω^A and Ω^S, the idea is first to generate a large independent sample of M equiprobable scenarios $\Omega^M \subset \Omega$, and to partition it into the subset Ω^{M_A} of M_A acceptable-risk scenarios and the subset Ω^{M_S} of M_S serious-risk scenarios (using the hazard tolerance level κ, as illustrated in Fig. 9.3). An estimate of the probabilities π^A and π^S is then given by $\overline{\pi}_A = M_A/M$ and $\overline{\pi}_S = 1 - \overline{\pi}_A$, and these estimates can be used to specify the value of the risk-aversion weights $\hat{\pi}_A$ and $\hat{\pi}_S$ to use in the SAA model. Second, a small sample Ω^{m_A} of m_A acceptable-risk scenarios is randomly selected in Ω^{M_A} and a small sample Ω^{m_S} of m_S serious-risk scenarios is randomly selected in Ω^{M_S} to get $\Omega^m = \Omega^{m_A} \cup \Omega^{m_S}$. This hierarchical sampling procedure provides equiprobable scenarios, with probability $1/m_A$ and $1/m_S$, respectively. These small scenario samples are then used to formulate a SAA model.

The approach can be illustrated with the LTP case discussed previously. The Monte-Carlo procedure in Sect. 9.3.3.1 is first used to generate several samples of M scenarios, Ω_i^M, $i = 1, \ldots, I$. Then, proceeding as indicated in the previous paragraph, small scenario samples $\Omega_i^{m_A}$, $\Omega_i^{m_S}$, $i = 1, \ldots, I$ of size m_A and m_S, respectively, are selected, and the weights $\hat{\pi}_A$ and $\hat{\pi}_S$ are specified. The SAA model to solve for a given scenario sample Ω_i^m is the following:

$$\max_x \frac{|T^u|}{|\hat{T}|} \sum_{V=A,S} \frac{\hat{\pi}_V}{m_V} \sum_{\omega \in \Omega_i^{m_V}} \times \left(\sum_{t \in T} \left[\sum_{l \in L^S} \left(\sum_{l' \in L_\tau^D(\omega) \cap L_l^D} [u_{l'} - v_l - w_{ll'}] F_{ll't}(\omega) - c_l^z z_{lt}(\omega) \right) + (u_{l'} - v^e) \sum_{l' \in L_t^D(\omega)} F_{l't}^e(\omega) \right] \right) - \sum_{l \in L^S} A_l x_l \tag{9.10}$$

subject to

$$x_{ll'} \le x_l \quad l \in L^S,\ l' \in L_l^D \tag{9.11}$$

$$\sum_{l \in L^S} F_{ll't}(\omega) + F^e_{l't}(\omega) = d_{l't}(\omega) \qquad l' \in L^D_t(\omega),\ t \in \hat{T},\ \omega \in \Omega^m_i \tag{9.12}$$

$$F_{ll't}(\omega) \le d_{l't}(\omega) x_{ll'} \qquad l \in L^S,\ l' \in L^D_t(\omega) \cap L^D_l,\ t \in \hat{T},\ \omega \in \Omega^m_i \tag{9.13}$$

$$\sum_{l' \in L^D_{lt}(\omega)} F_{ll't}(\omega) \le a_{lt}(\omega) + z_{lt}(\omega) \qquad l \in L^S,\ t \in \hat{T},\ \omega \in \Omega^m_i \tag{9.14}$$

$$z_{lt}(\omega) \le \varphi_l a_l,\ l \in L^S_t(\omega);\ z_{lt}(\omega) \le 0 \qquad l \in \tilde{L}^S_t(\omega),\ t \in \hat{T},\ \omega \in \Omega^m_i \tag{9.15}$$

$$F_{ll't}(\omega) \ge 0,\ F^e_{l't}(\omega) \ge 0,\ z_{lt}(\omega) \ge 0;\ l \in L^S,\ l' \in L^D_t(\omega) \cap L^D_l,\ t \in \hat{T},\ \omega \in \Omega^m_i \tag{9.16}$$

$$x_l, x_{ll'} \in \{0, 1\} \qquad l \in L^S,\ l' \in L^D_l \tag{9.17}$$

Model (9.10–9.17) is a mixed integer program and it can be solved relatively easily with recent commercial solvers, such as CPLEX-12. These solvers incorporate generic heuristics (such as the feasibility pump) to find good initial solutions and they are able to solve surprisingly large SCN design problems. Moreover, since our objective here is to generate good candidate designs, and since commercial solvers tend to take a lot of time to prove optimality after they found the optimal solution, larger optimality gap parameter values (say 0.01 instead of the default value of 0.0001) can be used to reduce computation times. Also, several heuristic methods were proposed in the literature to solve deterministic location-allocation problems (see [20] for a review) and some of them can be extended relatively easily to solve stochastic versions of the problem.

Several instances of Model (9.10–9.17) were solved in [18] for realistic cases involving as many as 15 potential depots and 706 ship-to-points located in Northeast and Midwest US states. A one-year planning horizon including 240 working days was considered, and the next-day delivery requirement was implemented through a 400 miles limit on the distance between depots and ship-to points. Exponential order inter-arrival times were utilized with log-normal order quantities. Diverse cost and demand structures were examined for networks with different ship-to-point density and potential depots size. The models were generated with OPL Studio 6.1 and solved to optimality with CPLEX-11 on a 64 bits server with a 2.5 GHz Intel XEON processor and 16 GB of RAM. Samples of 1,000 scenarios were used to estimate the probability $\overline{\pi}_A$ and $\overline{\pi}_S$ of acceptable and serious-risk scenarios (a hit histogram is presented in Fig. 9.3 for one of these samples). Samples of 30–50 scenarios were embedded in the SAA models, and four replications were solved for each case ($I = 4$). The largest models included 883,766 variables and 905,351 constraints, and their optimal solution was found in 86 minutes, on average, with a 60 min standard deviation.

The SCN designs obtained with the I scenario samples generated are not necessarily all different. Also, additional candidate designs can be obtained by modifying the risk-aversion weights $\hat{\pi}_A$ and $\hat{\pi}_S$. The model as it stands tends to select two supply depots for each ship-to-point, i.e. to provide designs such that $\sum_{l \in L^S} x_{ll'} = 2, l' \in L^D$. The depot with the largest flow can then be considered as the primary supply depot $l_1^S(l')$ for ship-to-point $l' \in L^D$, and the other one as a backup depot $l_2^S(l')$. Nothing guarantees, however, that this will happen for all samples $\Omega_i^m, i = 1, \ldots, I$, and constraints could be added in the model to impose this characteristic. Other model variants could also be used to generate alternative designs. Moreover, when inspecting the solutions obtained, the company may want to modify some designs to make them more attractive from their point of view. In any case, when this activity is completed, a number of candidate designs $\boldsymbol{x}^j, \ j = 1, \ldots, J$, are available and one of these designs must be selected.

9.3.5 Designs Evaluation and Selection

Since several complexity reduction mechanisms are typically used to generate candidate designs, the SCN designs $\boldsymbol{x}^j, \ j = 1, \ldots, J$, are not necessarily optimal when considering all problem facets and all relevant performance measures. Consequently, candidate designs must be evaluated and compared based on operational response processes and performance metrics which are as close as possible to those of the company. Moreover, before changing its current SCN design, $\boldsymbol{x}^0$, the company will want to ensure that the new design provides significant value added. The status quo design $\boldsymbol{x}^0$ must therefore always be included in the list of designs to evaluate. Ideally, the designs $\boldsymbol{x}^j, \ j = 0, 1, \ldots, J$, should be evaluated for all plausible futures $\omega \in \Omega$. However, as indicated previously, this is not possible because there may be an infinite number of plausible scenarios. Again, the best that can be done is to perform the evaluation for a sample of scenarios. This sample must however be independent of the samples generated to get candidate designs, it must be much larger, and it must include worst-case scenarios in Ω^U. The value of a given design-scenario pair $(\boldsymbol{x}^j, \omega)$ is obtained by solving one or several models representing the response and SCN adaptation decisions taken over the planning horizon when the design considered is implemented. The set of scenario values thus obtained for a given design are used to evaluate performance measures. An adequate SCN design evaluation must be based on expected value, deviation and robustness measures, and it must take the decision-makers risk attitude into account. Based on these measures, classical multicriteria filtering and selection techniques are finally used to select the design to implement.

This design evaluation and selection approach can be illustrated with our LTP case. The Monte-Carlo procedure in Sect. 9.3.3.1 is first used to generate a large sample Ω^N of N scenarios, with $N \gg M$, and the sample is then partitioned into acceptable-risk scenarios Ω^{N_A} and serious-risk scenarios Ω^{N_S}, based again on the hazard tolerance level κ. From these samples, two moderate size subsets of scenarios are randomly selected to perform the design evaluation: a subset $\Omega^{n_A} \subset \Omega^{N_A}$ of n_A acceptable-risk scenarios, and a subset $\Omega^{n_S} \subset \Omega^{N_S}$ of n_S serious-risk scenarios. In order to obtain worst-case scenarios, a subset of tail scenarios are also selected in the distribution of the number of hits (see Fig. 9.3). These scenarios are then taken as it is, or modified manually by adding imaginative elements, to get the required set of worst-case scenarios Ω^{nU}. Finally, an historical scenario ω^0 can also be defined from the customer orders and the disruptions which occurred during the last year. In practice, companies often want to know how candidate designs would have performed over the recent past.

The value of designs $\boldsymbol{x}^j$, $j = 0, 1, \ldots, J$, must then be obtained for scenarios $\Omega^n = \Omega^{n_A} \cup \Omega^{n_S} \cup \Omega^{nU} \cup \{\omega^0\}$. If we were applying the standard SAA approach [36], the design evaluation would be done by solving the second stage program (9.4–9.9), with scenarios $\Omega^{n_A} \cup \Omega^{n_S}$, and the design with the highest expected value would be selected. However, since the SAA model is based on several approximations, there is no reason to restrict ourselves to such a gross assessment. The evaluation of the designs should be based on a response optimization procedure as close as possible to the real-operational processes. Also, to obtain (9.10), we simplified the objective function, but when comparing the designs, there is no reason not to apply more adequate performance evaluation measures.

The assignment-transportation procedure described in Sect. 9.3.5.1, can be utilized to calculate the net revenues $\hat{R}^u(\boldsymbol{x}, \omega)$ generated over the planning horizon for a given SCN design $\boldsymbol{x}$ under a given scenarios ω. This procedure is based on the response policy of the company specified during the analysis activity. In the procedure, $R_{l\tau}(\boldsymbol{x}, \omega)$ denotes the daily revenues of opened depot $l \in \underline{L}^S$, $C_\tau^z(\boldsymbol{x}, \omega)$ the network loss on day τ for external recourses, and L is a priority list ranking ship-to-points $l \in L^D$ in decreasing order of their importance. On a given day τ, based on the set $L_\tau^D(\omega)$ of ordering customers, list L is used to assign the most important customers to their primary depot $l_1^S(l')$ and to transfer the remaining orders to the backup depot $l_2^S(l')$ or to an external supply source. Note that if the design $\boldsymbol{x}$ considered does not specify the backup depot explicitly, then $l_2^S(l')$ is specified using a distance or flow-based rule. In the second part of the procedure, a transportation problem must be solved for each day and for each depot $l \in \underline{L}^S$ which are opened in design $\boldsymbol{x}$. This problem is solved in two steps: (1) for orders that are larger than a truckload b^F, a decision is made to ship as much as possible in FTL; (2) a routing algorithm is used to construct delivery routes for the residual order quantities. The routing algorithm employed should be as close as possible to the procedure used in practice by the company. The *UserResponse* procedure yields the net SCN revenues associate to day-to-day operations during the planning horizon. In order to obtain the value added by the SCN the fixed depots costs must also be taken into account. The value of the designs for each scenario is thus calculated as follows:

$$\hat{R}(\boldsymbol{x}^j, \omega) = \hat{R}^u\left(\boldsymbol{x}^j, \omega\right) - \sum_{l \in \underline{L}^S} A_l\left(x_l^j\right), \quad j = 0, 1, \ldots, J, \quad \omega \in \Omega^n \tag{9.18}$$

The design values by scenario provided by (9.18) can finally be used to compute the selected performance measures. For the LTP case, the following four measures would be relevant. The expected value $\overline{R}(\boldsymbol{x}^j)$ of a design $\boldsymbol{x}^j$ is provided by:

$$\overline{R}(\boldsymbol{x}^j) = \sum_{V=A,S} \overline{\pi}_V \overline{R}_V(\boldsymbol{x}^j);\ \overline{R}_V(\boldsymbol{x}^j) = \frac{1}{n_V} \sum_{\omega \in \Omega^{n_V}} \hat{R}(\boldsymbol{x}^j, \omega),\ V = A, S \quad (9.19)$$

where $\overline{R}_A(\boldsymbol{x}^j)$ and $\overline{R}_S(\boldsymbol{x}^j)$ are conditional expected values for acceptable and serious-risk scenarios, respectively. Since downside deviations from mean returns are undesirable, an adequate variability measure to assess a design $\boldsymbol{x}^j$ is the mean-semideviation $\mathrm{MSD}(\boldsymbol{x}^j)$ given by:

$$\mathrm{MSD}(\boldsymbol{x}^j) = \sum_{V=A,S} \overline{\pi}_V \mathrm{MSD}_V(\boldsymbol{x}^j)$$

$$\mathrm{MSD}_V(\boldsymbol{x}^j) = \frac{1}{n_V} \sum_{\omega \in \Omega^{n_V}} \max\left[\left(\overline{R}_V(\boldsymbol{x}^j) - \hat{R}(\boldsymbol{x}^j, \omega)\right); 0\right], V = A, S \quad (9.20)$$

where $\mathrm{MSD}_A(\boldsymbol{x}^j)$ and $\mathrm{MSD}_S(\boldsymbol{x}^j)$ are conditional mean-semideviations for acceptable and serious-risk scenarios, respectively. Worst-case scenarios, can be used to evaluate an absolute robustness measure proposed in [23]. For design $\boldsymbol{x}^j$ this measures the minimum return $\overline{R}_U(\boldsymbol{x}^j)$ under all worst-case scenarios, calculated as follows:

$$\overline{R}_U(\boldsymbol{x}^j) = \min_{\omega \in \Omega^{n_U}} \left\{\hat{R}\left(\boldsymbol{x}^j, \omega\right)\right\} \quad (9.21)$$

Measures (9.19–9.21) provide the basis for a multi-criteria evaluation of the designs considered. The value of the designs $\hat{R}(\boldsymbol{x}^j, \omega^0)$ for the historical scenario ω^0 can also be taken into account. Any multi-criteria analysis method can then be used to eliminate dominated designs and to rank non-dominated designs. A compound utility measure is also provided by the following expression:

$$R(\boldsymbol{x}^j) = (1 - \psi) \sum_{V=A,S} \hat{\pi}_V \left(\overline{R}_V(\boldsymbol{x}^j) + \gamma_V \mathrm{MSD}_V(\boldsymbol{x}^j)\right) + \psi \overline{R}_U(\boldsymbol{x}^j) \quad (9.22)$$

where $\gamma_V \in [0, 1]$, $V = A, S$, are variability aversion weights for acceptable and serious-risk scenarios, and where $\psi \in [0, 1]$ is an extreme event aversion weight. Using this approach, the best design can be selected for implementation. Note that the computational efforts associated to this evaluation and selection activity is much smaller than for the design generation phase. For the LTP, when a standard SAA approach is applied, calculating the value of a design $\boldsymbol{x}^j$ with (9.18) involves the solution of the second-stage program (9.4–9.9) for one scenario $\omega \in \Omega^n$ at the time, and this program can be separated per day. The resulting linear programs are easily solved with CPLEX. When the approach suggested here is applied, the user response problem becomes even simpler because it decomposes per depot.

9.3.5.1 User Response Procedure

UserResponse $\left(\boldsymbol{x}, \left(l'(p), l \in L^D\right), \left(L^D_\tau(\omega), \tau \in T^u\right), \text{L}; \hat{R}^u(\boldsymbol{x}, \omega)\right)$
For all $\tau \in T^u$, do:
    Assign orders to depots
    For all $l' \in L^D_\tau(\omega)$, in order of the priority in L, do
        If capacity available at $l^S_1(l')$ then assign ship-to point l' to depot $l^S_1(l')$
        Else, If capacity available at $l^S_2(l')$ then assign ship-to point l' to depot $l^S_2(l')$

        Else use the external supply source to serve l'
    End do
    Compute depot revenues and expenditures
    For all $l \in \underline{L}^S$, do
        Solve the day τ *transportation problem* to get the transportation costs $\hat{C}^u_{l\tau}(\omega)$
        Compute depot l net revenues

$$\hat{R}_{l\tau}(\omega) = \sum_{l' \in \underline{L}^D_{l\tau}(\omega)} \left([(u_{l'} - v_l)\, d_{l'\tau}(\omega)] - v^z_l z_{l\tau}(\omega) \right) - \hat{C}^u_{l\tau}(\omega)$$

    End do
    Compute the transportation costs $\hat{C}^u_{0\tau}(\omega)$ for the external supply source
End do
Compute the SCN design net revenues $\hat{R}^u(\boldsymbol{x}, \omega) = \sum_{\tau \in T^u} \left(\sum_{l \in \underline{L}^S} \hat{R}_{l\tau}(\omega) - \hat{C}^u_{0\tau}(\omega) \right)$

9.4 Conclusion

This paper presents a relatively generic supply chain network reengineering methodology for businesses operating in an uncertain environment. A particular attention was given to the modeling of high-impact disruptive events. General guidelines were given and, to illustrate the approach, more detailed explanations were provided for the case of the location-transportation problem under uncertainty. Although the methodology is comprehensive, that is, it considers strategic location, capacity, sourcing, transportation, and marketing decisions in an integrated manner, these decisions are rarely reconsider all together in practice. Doing so, would give rise to huge projects with extremely difficult models to solve. At a given point in time, a company seeks rather to reoptimize only a subset of these decisions, which yield manageable projects and models. The methodology should therefore be perceived as a continuous improvement process.

The objective pursued is to provide sustainable SCN value creation by seeking designs which are both effective and robust. The methodology evaluates robustness through a high-quality anticipation of user decisions, for a sample of adequately selected plausible future scenarios. The risk attitude of the decision-maker is also considered. Complementary work performed to test the methodology [17, 21] indicates that it offers a judicious accuracy-solvability trade off. The approach is also the backbone of a commercial SCN design tool (SCN-STUDIO) which was developed to support businesses in their effort to improve their supply chains.

References

1. Banks E (2006) Catastrophic risk: analysis and management. Wiley Finance
2. Bertrand JWM (2003) Supply chain design: flexibility considerations. In: de Kok AG, Graves SC (eds) Supply chain management: design, coordination and operation. Handbooks in OR and MS, vol 11. Elsevier, Amsterdam
3. Birge JR, Louveaux F (1997) Introduction to stochastic programming. Springer, New York
4. Carle MA, Martel A, Zufferey N (2010) The CAT metaheuristic for the solution of multi-period activity-based supply chain network design problems. Research document CIRRELT-2010-52, CIRRELT, Université Laval, Canada
5. Chopra S, Sodhi MS (2004) Managing risk to avoid supply-chain breakdown. MIT Sloan Manag Rev 46:52–61
6. Christopher M, Peck H (2004) Building the resilient supply chain. Int J Logist Manag 15(2): 1–13
7. D'Amboise G, Martel A, Oral M et al (1995) Lélaboration de projets stratégiques. In: Martel A, Oral M (eds) Les défis de la compétitivité. Publi-Relais, Montreal
8. Dong M (2006) Development of supply chain network robustness index. Int J Serv Oper Inform 1(1/2):54–66
9. Gogu R, Trau J, Stern B et al (2005) Development of an integrated natural hazard assessment method. Geophys Res Abstr 7:03724
10. Graves SC, Tomlin B (2003) Process flexibility in supply chains. Manag Sci 49:907–919
11. Grossi P, Kunreuther H (2005) Catastrophe modeling: a new approach to managing risk. Springer, New York
12. Gunasekaran A, Patel C, McGaughey RE (2004) A framework for supply chain performance measurement. Int J Prod Econ 87:333–347
13. Gutierrez GJ, Kouvelis P, Kurawala A (1996) A robustness approach to uncapacitated network design problems. Eur J Oper Res 94:362–376
14. Haimes YY (2004) Catastrophe modeling: a new approach to managing risk. 2nd edn. Springer, New York
15. Hendricks KB, Singhal VR (2005) Association between supply chain glitches and operating performance. Manag Sci 51(5):695–711
16. Klibi W, Lasalle F, Martel A et al (2010) The stochastic multi-period location-transportation problem. Transp Sci 44(2):221–237
17. Klibi W, Martel A (2009) The design of effective and robust supply chain networks. Research document CIRRELT-2009-28, CIRRELT, Université Laval, Canada
18. Klibi W, Martel A (2010) Modeling approaches for the design of resilient supply networks under disruptions. Research document CIRRELT-2009-27,CIRRELT, Université Laval, Canada
19. Klibi W, Martel A (2011) Scenario-based supply chain network risk modeling. Research document CIRRELT-2011-30, CIRRELT, Université Laval, Canada
20. Klibi W, Martel A, Guitouni A (2010a) The design of robust value-creating supply chain networks: a critical review. Eur J Oper Res 203(2):283–293
21. Klibi W, Martel A, Guitouni A (2010a) The impact of operations anticipations on the quality of supply chain network design models. Research document CIRRELT-2010-45, CIRRELT, Université Laval, Canada
22. Kouvelis P (1998) Global sourcing strategies under exchange rate uncertainty. In: Tayur S, Magazine M, Ganeshan R (eds) Quantitative models in supply chain management. Kluwer, Norwell, pp 625–667
23. Kouvelis P, Yu G (1997) Robust discrete optimization and its applications. Kluwer Academic Publishers, Norwell
24. Lee H (2004) The triple-A supply chain. Harvard Bus Rev 82:102–112
25. Lempert RJ, Groves DG, Popper SW, Bankes SC (2006) A General, analytic method for generating robust strategies and narrative scenarios. Manag Sci 52(4):514–528
26. Manuj I, Mentzer J (2008) Global supply chain risk management. J Bus Logist 29(1):133–155

27. Martel A, Benmoussa A, Ezzedine I et al (2010) Military missions scenario generation for the design of logistics support networks, ILS 2010 Proc. Casablanca, Morocco
28. Mo Y, Harrison TP (2005) A conceptual framework for robust supply chain design under demand uncertainty, Chapter 8. In: Geunes J, Pardalos PM (eds) Supply Chain Optimization. Kluwer Academic Publishers, Dordrecht
29. Pettit TJ, Fiksel J, Croxton KL (2010) Ensuring supply chain resilience: development of a conceptual framework. J Bus Logist 31(1):1–21
30. Porter M (1980) Competitive strategy: techniques for analyzing industries and competitors. Free Press
31. Rice J, Caniato F (2003) Building a secure and resilient supply network. Supply Chain Manag Rev 7(5):22–30
32. Rosenblatt MJ, Lee HL (1987) A robustness approach to facilities design. Int J Prod Res 25(4):479–486
33. Sabri EH, Beamon BM (2000) A multi-objective approach to simultaneous strategic and operational planning in supply chain design. Omega 28:581–598
34. Scawthorn C, Shneider PJ, Shauer BA (2006) Natural hazards—The multihazard approach. Natural Hazards Rev 7(2):39
35. Semchi-Levi D, Snyder LV, Watson M (2002) Strategies for uncertain times. Supply Chain Manag Rev 6(1):11–12
36. Shapiro A (2003) Monte carlo sampling methods, chapter 6. In: Ruszczynski A, Shapiro A (eds) Handbooks in OR and MS, vol 10. Elsevier
37. Sheffi Y (2005) The resilient enterprise: overcoming vulnerability for competitive advantage. MIT Press Books
38. Sheffi Y (2007) Building a resilient organization. Bridge 37(1):30–36
39. Shell Global Scenarios to 2025 (2005) The future business environment: trends, trade-offs and choices. Shell International Limited (SIL)
40. Shen ZJM, Coullard C, Daskin MS (2003) A joint location-inventory model. Transp Sci 37(1):40–55
41. Snyder LV (2006) Facility location under uncertainty: a review. IIE Trans 38(7):537–554
42. Snyder LV, Daskin MS (2005) Reliability models for facility location: the expected failure cost case. IIE Trans 39(3):400–416
43. Snyder LV, Daskin MS (2006) Stochastic p-robust location problems. IIE Trans 38(11):971–985
44. Thompson A, Strickland A (1999) Strategic management. 11th edn. Irwin, New York
45. Tomlin B (2006) On the value of mitigation and contingency strategies for managing supply chain disruption risks. Manag Sci 52(5):639–657
46. Van der Heijden K (2005) Scenarios: the art of strategic conversation. 2nd edn. Wiley, New York
47. Van Opstal D (2007) The resilient economy: integrating competitiveness and security. Council of competitiveness
48. Yucesan E (2007) Competitive supply chains: a value-based management perspective. Palgrave Macmilan, New York

Chapter 10
Optimizing the Hunter Valley Coal Chain

Natashia L. Boland and Martin W. P. Savelsbergh

10.1 Introduction

Coal remains the most important energy source for power generation, providing 37% of the world's electricity. As the global population grows, and as living standards improve in the developing world, a international demand for energy is increasing at a rapid rate. Coal is still the most abundant, widely distributed, safe, and economical fossil fuel available to meet this escalating energy demand.

Australia's coal industry is a major contributor to the country's social and economic development. Black coal is Australia's largest export, worth close to A$50 billion and representing more than 20% of Australia's commodity exports in 2008–2009 [4]. Coal deposits are found in most Australian states, but are particularly abundant in New South Wales and Queensland, which account for around 97% of Australia's black coal production [3]. Black coal in New South Wales is mined near the eastern and western edges of the large Sydney–Gunnedah Basin. Mines in the Sydney and southern areas are typically underground mines, while mines in the Hunter Valley, and those near Gunnedah, (which is also served by the Hunter Valley rail and port system), are mainly open cut mines. Most black coal in Queensland comes from an area known as the Bowen Basin.

The export coal industry in Australia is serviced by coal loading terminals located in Queensland and New South Wales. As a result of expansion work in recent years,

N. L. Boland (✉)
School of Mathematical and Physical Sciences, University of Newcastle,
Callaghan, NSW 2308, Australia
e-mail: Natashia.Boland@newcastle.edu.au

M. W. P. Savelsbergh
CSIRO Mathematics Informatics and Statistics,
North Ryde, NSW 1670, Australia
e-mail: Martin.Savelsbergh@csiro.au

H. Gurnani et al. (eds.), *Supply Chain Disruptions: Theory and Practice of Managing Risk*, DOI: 10.1007/978-0-85729-778-5_10,

Table 10.1 Australian coal exporting ports during 2007–2008 [3]

Port	Annual capacity (million tonnes)	Export loadings (million tonnes)
Abbot Point	21.0	12.5
Brisbane	6.0	5.5
Dalrymple Bay	68.0	43.5
Gladstone	75.0	54.1
Hay Point	44.0	36.9
Newcastle	102.0	88.9
Port Kembla	16.0	12.7

the terminals currently have a total handling capacity in excess of 300 million tonnes of coal a year, with further expansion planned or in progress (see Table 10.1 for details on the capacity and loadings of the largest coal export ports).

An efficient, economic transport system is a key element in the viability of coal mining in Australia and coal mining is a major contributor to the viability of the state rail systems in New South Wales and Queensland. The oldest producing areas are located close to the coast where domestic transport is well established. However, as more remote deposits are being discovered and worked and as production volumes increase, it is crucial that the transport infrastructure is simultaneously developed and expanded.

The majority of coal is carried to its destination by rail. In the major producing states, coal is the single most valuable rail freight item and where state rail systems did not originally extend to coal fields, the installation of new track and rolling stock has been funded by the mining companies themselves. Trains transporting coal are among the longest in the world, in the Hunter Valley with as many as four locomotives and 90 wagons amounting to a length of more than 1.5 km. A train of that size can carry about 8,500 tonnes of coal.

As can be seen from Table 10.1, Newcastle is by some margin home to Australia's largest coal export operation by volume. It is also the world's largest: the Newcastle port throughput in 2008 was around 92 million tonnes, or more than 10% of the world's total trade in coal for that year. This coal was handled by Port Waratah coal services (PWCS), which operates two terminals at the Port of Newcastle. The terminals are a shared facility, serving around 30 mines owned by 14 different coal mining companies in the Hunter Valley.

The PWCS terminals include stockyards, where cargoes of (typically blended) coal product are assembled in stockpiles using stacking machines, and then reclaimed using bucket wheel reclaimers for transport via conveyor belts to shiploaders at the berths. Although in and of itself a challenging logistics operation, the port is just one end of the logistics chain for coal export, a chain which is critically dependent on a shared rail system to transport coal from load points at the mines, to the port stockyard. Whilst some track is owned by individual mining companies, the majority of track in the system, and its key shared sections, are leased and managed by the Australian Rail Track Corporation, with train operators such as Pacific National and QRNational operating the rolling stock used to transport coal. The collection of mining companies, rail operators, the track owner, and PWCS together constitute the

Hunter Valley coal chain (HVCC). Demand for coal has increased significantly in recent years, and is expected to increase still further in the years to come. To meet that demand, and to best capitalize on Australia's coal resources, PWCS and the HVCC are committed to significant expansion of the coal export operation. However, the ability to increase exports is limited by the capacity of the coal chain; the Hunter Valley can only export as much coal as it can get to and through the port.

It was recognized early that to achieve the desired expansion, coordinated planning was needed. In an Australian landmark for collaborative logistics, the HVCC founded the Hunter Valley coal chain logistics team (HVCCLT) in 2005, now incorporated as the Hunter Valley coal chain coordinator limited (HVCCC), to provide a single integrated planning and scheduling center for the HVCC. The HVCCC does modeling work to support major infrastructure expansion decision-making, and is tasked with all planning activities up to within 12 h of actual operation. HVCCC facilitates communication between planning and scheduling staff, and so enables greater plan coordination. With integrated planning comes greater complexity; even highly experienced human planners cannot consider the combined interplay of scheduling decisions across the whole chain. Thus optimization technology is necessary to support HVCCC's efforts to maximize the efficiency of the coal chain.

In this chapter, we give an overview of the Hunter Valley coal export operation, and present a range of models that either have been developed, or are under development, to support and automate various aspects of coal chain planning. These are for the most part discrete event simulation and optimization models, although currently one important planning activity is still supported largely by spreadsheet models. While some of these models are strategic and focus on capacity expansion decisions to accommodate future growth, others are operational and focus on train scheduling and stockyard management to achieve the necessary daily throughput. To handle the increased flow of coal through the system, system components are being optimized and interfaces between components are being streamlined.

To the best of our knowledge, there are only a few papers in the literature on the use of optimization models in bulk goods supply chains of this complexity. Everett has written a series of papers on approaches to supply chain management for iron-ore export (see for example [14]), but these largely focus on blending to meet customer demand specifications, and do not address integrated transport scheduling. Some models of coal stockyard management are available, such as [10], but again the focus is blending, in this case for a power station with uncertain demand; rail operations are simplified and there is no need for integration with vessel transport.

Of course, even the best-laid plans are subject to uncontrollable disruption. Daily disruptions due, for example, to unexpected rail delays, or mechanical failures, whilst usually minor, can still have an important cumulative effect on the supply chain efficiency. As a consequence, disruption management is the next frontier for modeling and decision-support technology. Therefore, we also discuss opportunities for optimized disruption handling in the Hunter Valley coal chain. (The coal export chain of Dalrymple Bay, the second largest coal exporting port, was investigated by Ernst et al. [13].)

The focus of this chapter is likely to be somewhat different from the focus of the other chapters. Whereas other chapters focus on how to prepare for and handle supply disruptions, this chapter focuses, in a sense, on how to prevent supply disruptions. The Hunter Valley coal chain represents one of the supply points in the larger global coal supply chain. Its biggest challenge is how to handle increased demand and the daily fluctuations in demand (it may be too strong to speak of demand disruptions). Erasmus said "Prevention is better than cure", and therefore we believe a chapter on supply-side efficiency is certainly appropriate.

Furthermore, the Hunter Valley coal export chain differs from typical supply chains studied and has several characteristics that make it interesting in its own right. It is a large, multifaceted, and dynamic system consisting of several interacting subsystems that are in themselves highly complex. Couple this with a large variety of products (coal is a blended product), a "build-to-order" organization, with a limited visibility of future demand, and it becomes enormously challenging to achieve the high level of throughput required to satisfy the growing demand. It is also a prime example of a logistics system that can only achieve its efficiency goals when the various stakeholders (and there are many) collaborate. Another interesting aspect, from an operations research perspective, is that it offers decision-making problems with varying timescales, e.g., strategic planning problems covering a 10-year period, tactical planning problems with horizons of a month, 3 months, and a year, operational scheduling problems that have to provide a detailed operational plan for the next day, and, finally, disruption handling problems that focus on schedule changes for the next minutes or hours. A major challenge in all decision-making problems is the level of detail that needs to be included in the models to ensure a sufficiently accurate representation of the underlying system. All in all, learning more about the HVCC, the various decision-making challenges, and the optimization models they lead to, should be of interest to practitioners and researchers alike.

The remainder of this chapter is organized as follows. In Sect. 10.2, we introduce HVCCC, the organization responsible for all planning and scheduling. In Sect. 10.3, we present a more thorough overview of the HVCC. In Sect. 10.4, we discuss a few key planning and scheduling problems. In Sect. 10.5, we delve into two optimization models that support decision-making at the HVCC. Finally, in Sect. 10.6, we conclude with some final remarks.

10.2 Collaborative Logistics: A Brief History of HVCCC

Up until 2003 there was little communication and coordination among the various entities responsible for the transport of coal through the HVCC. Not surprisingly, the result was an ineffective system operating mostly in a reactive mode as opposed to a proactive mode. The lack of coordinated maintenance activities, especially, led to significant problems. Trains expected to deliver the last coal for a stockpile scheduled to be reclaimed and loaded onto a vessel currently on its way to the berth would not arrive, as they were canceled due to track maintenance. This led to unnecessarily high demurrage costs. The impact of reduced track capacity due to maintenance was exacerbated when there was a lack of communication and coordination. Trains were being canceled without taking into consideration their urgency and their impact on downstream activities.

In early 2003, an Industry Review Team recommended the implementation of a centralized planning function, as it could potentially deliver enormous benefits to the HVCC. Following acceptance of this recommendation, the Hunter Valley Coal Chain Planning Group (HVCCPG), the forerunner of HVCCC, was established in June 2003. HVCCPG was initially established as a trial between PWCS and rail operator Pacific National. It quickly proved that centralized planning of coal chain activities could indeed release "latent" capacity to the benefit of the HVCC as a whole.

By 2005, all HVCC service providers had fully embraced the centralized planning model and took steps to formalise the HVCCPG. A Memorandum of Understanding was executed on July 5, 2005 and with this HVCCPG's name was changed to HVCCLT. Membership of HVCCLT then included all organizations responsible for the transport of coal from the Hunter Valley mines to the port and onto ships for export. Member organizations were PWCS as the operator of the cargo assembly and ship loading terminal, Pacific National and QRNational as the train operators, Australian Rail Track Corporation as the track owner, and Newcastle Port Corporation which manages all vessel movements in the Port of Newcastle.

Under the memorandum of understanding, HVCCLT represented a cooperative organization responsible for planning all coal exports for the Hunter Valley coal industry. HVCCLT was the first cooperative model of its kind in Australia implemented to maximize export opportunities through a coordinated approach to planning. Membership was open to any existing and future providers of transport and port infrastructure in the coal chain.

HVCCLT had two broad planning objectives, being:

- *Daytoday planning and scheduling*: coordinate vessel berthing, stockpile layouts, and train sequencing so as to fulfill customers' orders in the shortest possible time and
- *Long-term capacity planning*: assess the adequacy of the existing infrastructure and develop an integrated capital investment plan for a 10-year horizon so as to assist members with optimizing their investment decisions and to focus capital expenditure on the infrastructure essential to meet future coal export growth.

With a mix of federal, state, and privately owned organizations operating individual components of the coal chain, HVCCLT provided a single point of coordination for all planning decisions. It proved that by planning the coal chain as a single system, increased throughput and coordinated investment could be achieved.

In 2009, the HVCC went through a major restructure of the contractual arrangements for the movement of coal. The new arrangements provided greater certainty of long-term system capacity and contractual obligations. With the emergence of these new contractual obligations came the need to further evolve HVCCLT from a cooperative of service providers to a separate entity with legal status. This entity needed to be more representative of the coal industry. In particular it needed representation from coal producers as well as service providers. As a separate legal entity it would be better placed to more effectively meet its obligations in this new contractual environment.

On August 27, 2009 HVCCC was incorporated as a new independent legal entity and formally replaced HVCCLT. The membership of HVCCC had been expanded to include all current HVCC producers.

The HVCCC's mission is to plan and coordinate the cooperative daily operation and long-term capacity alignment of the HVCC. Its strategic objectives include:

- To plan and schedule the movement of coal through the HVCC in accordance with the agreed collective needs and contractual obligations of producers and service providers;
- To ensure minimum logistics cost and maximum throughput through the provision of appropriate analysis and advice on capacity constraints (whether physical, operational, or commercial) affecting the efficient operation of the HVCC; and
- To advocate positions to other stakeholders and governments on issues relevant to efficient operation, in order to maximize opportunities for improved coordination and/or further expansion of the coal chain.

Under its new structure HVCCC plays a pivotal role in determining the contractible system capacity, measuring the actual system performance and allocating system losses, and usage of contractual entitlement and administration of transfers of entitlement between parties.

HVCCC is a fine example of the value and benefits of logistics collaboration when done right. It has been recognized for its logistics excellence in the form of various awards:

- November 26, 2008. Australian Trader of the Year. The Logistics Magazine presented this award to HVCCLT for demonstrating highly efficient and innovative supply chain strategies for import and/or export. HVCCLT's innovative approach, as a cooperative planning organization, managing the export of close to 95 million tonnes annually, was rated by judges as an "industrial phenomenon". Recognized as key to HVCCLT's success was its focus on optimizing the entire supply chain, from coal mine to vessel, rather than tinkering with the individual components. This cooperative and coordinated approach was considered vital in enabling HVCCLT's objective of maximizing coal export volumes each and every day.
- June 12, 2009. Smart Awards for Supply Chain Excellence. The New South Wales Department of State and Regional Development awarded HVCCLT for Excellence in Supply Chain Innovation. The awards recognize organizations operating in New South Wales that have demonstrated supply chain innovation "at work". HVCCLT was chosen because it "excelled in identifying and optimizing innovative efficiency and investment improvements along the HVCC, the world's largest and most complex coal operation, responsible for up to 10 billion dollars in annual export revenue".

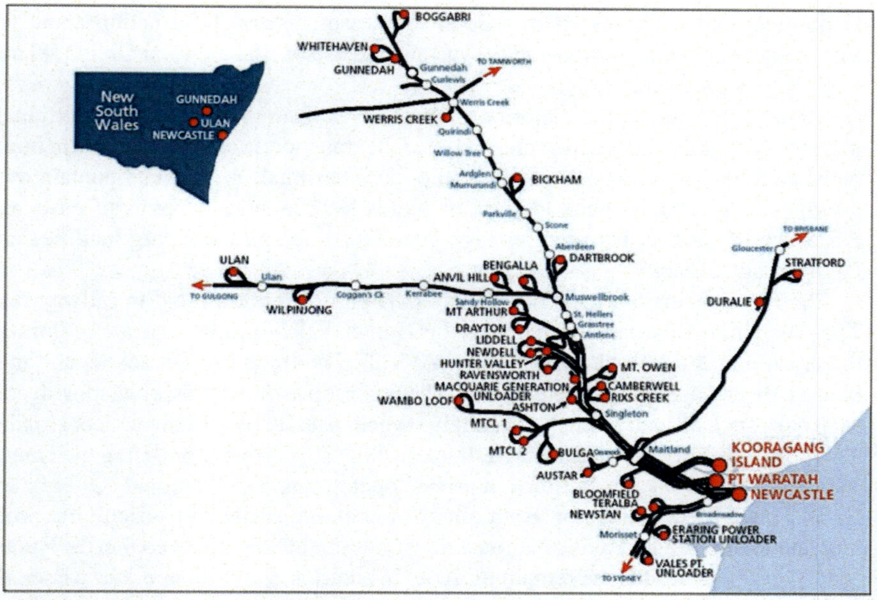

Fig. 10.1 Hunter Valley coal chain

Much of the material in this section, and further detail on some aspects, can be found at the HVCCC website http://www.hvccc.com.au.

10.3 The Hunter Valley Coal Export Chain: An Overview

The Hunter Valley Coal Chain, in physical terms, refers to the inland portion of the coal export chain in the Hunter Valley, NSW, Australia. It essentially follows the path of the Hunter River traveling south-east from the mining areas in the Hunter Valley to Newcastle.

Most of the coal mines in the Hunter Valley are open pit mines. The coal is mined and stored either at a railway siding located at the mine or at a coal loading facility used by several mines. The coal is then transported to one of the terminals at the Port of Newcastle, almost exclusively by rail; some coal is transported to the port by truck. (Some coal is also transported directly to power plants in the area, but this coal is not considered part of the export operation and so we do not discuss it here.) The coal is offloaded at a terminal onto stockpiles. Once the vessel for which the coal is meant arrives at a berth, the coal is loaded onto the vessel. The vessel then transports the coal to its destination. Figure 10.1 gives an overview of the HVCC.

In the remainder of this section, we discuss further details of the rail operations, the stockyard, and port operations, in turn. We also discuss sources of disruption in the coal export supply chain. (Good sources of general information about the technology of coal utilization and the sustainable use of coal are Coal Online [8] and IEA Clean Coal Centre [16]).

As indicated above, the majority of coal moves from the mines to the terminals in trains. The trains are used exclusively for the transportation of coal and run intact between a load point and the dump station at a terminal. A train can contain over 90 railcars. The use of such long trains yields benefits both in terms of costs and efficiency of service, but requires high-speed loading and unloading facilities and large storage capacity.

The railway corridor used in the HVCC is part of the Main North railway line. The Australian Government manages the Hunter Valley rail infrastructure through the Australian Rail Track Corporation (ARTC). The track is open access and may be used by any accredited rail operator. The other infrastructure associated with coal transport, such as load points, is privately owned, usually by a mine or a coal loader. There are currently two major above-track (rolling stock) operators using the Hunter Valley rail track: Pacific National, a private operator, and QRNational (Queensland Rail), a state-owned operator. Both transport coal, industrial and agricultural products, and other freight. Pacific National transports about 80% of the coal in the Hunter Valley and QRNational the remaining 20%. In addition, CityRail operates passenger services on parts of the track as the Hunter line.

A large section of track, referred to as the *main corridor* is, as mentioned above, shared with other freight and passenger rail operations. The trains are restricted to moving through the main corridor along *train paths* whose timing is predetermined and fixed by the track operators. A train path completely specifies the time and space path of a train along the main corridor. Outside of this main corridor, the movement of a train is not as constrained and a train may wait or travel slower than it otherwise could. Train paths are divided into up-paths, which travel from a load point to coal terminal, and down-paths, which travel in the opposite direction.

Once the coal arrives at a dump station at a terminal, it is unloaded on a belt conveyor that takes it to the stockyard where a stacker is used to create a stockpile. A stockpile has two main functions: it serves as temporary storage and it allows for blending of coals. Each train carries coal entirely of a single type. Trainloads of coal are built up on the stockpile as they arrive in the required proportions, to give the average composition of the desired blend when the pile is reclaimed. The trainloads destined for a particular stockpile may all be of the same coal type, or could be of several different coal types. In the first case, reclaiming homogenizes the coal and, in the latter case, combined blending and homogenization is achieved. A variety of stacking and reclaiming methods can be used to achieve blending and homogenization. For example, layers of coal could be built up to give a pile with a triangular cross-section. The coal is then reclaimed in a plane perpendicular to the layers. Ideally the reclaimed coal is thus a blend of the individual layers, reflecting the average composition of the stockpile. The reclaimed coal is collected on a belt conveyor and transported to a shiploader at the berth.

PWCS limited operates the current coal export facilities in Newcastle. Their coal export facilities consist of two coal loading terminals, located on either side of the

South Channel of the Hunter River. These are known as the Carrington Coal Terminal (CCT) in the suburb of Carrington and the Kooragang Coal Terminal (KCT) on Kooragang Island. Each of those terminals comprises equipment for the delivery and storage of coal to the terminal and for the loading of coal onto vessels. The stockyard at CCT has four pads, each 1 km long and 40 m wide for a total of about 0.6 million tonnes of "working" capacity. The stockyard at KCT also has four pads, two of them 2.5 km long and 56 m wide and two of them 1.3 km long and 56 m wide for a total of about 2.2 million tonnes "working" capacity. The terminals use different blending mechanisms. CCT uses the *windrow* stacking method with *slew-cut full face* reclaiming and KCT uses the *cone crescent* stacking method with *bench reclaiming*.

The coal export facilities operated by PWCS have a total capacity of 102 million tonnes per annum (Mtpa). CCT has a ship loading capacity of 25 Mtpa. It has berth space for two vessels and ship loading facilities that operate at 2,500 tonnes per h (tph) per berth. CCT is able to accept coal deliveries by either road or rail. KCT has a ship loading capacity of about 88 Mtpa. It has berth space for three vessels and ship loading facilities which can peak at 10,500 tph per loader. KCT is able to accept coal deliveries by rail only.

To accommodate future growth, a number of infrastructure initiatives are under way that have the potential to create capacity approaching the more optimistic forecasts from producers, suggesting in some cases, the need for up to 170 million tonnes of capacity by 2012. PWCS has announced a half-a-billion dollar upgrade to take port capacity to more than 110 million tonnes and the new Newcastle Coal Infrastructure Group (NCIG) terminal is targeting about a 30 million tonnes capacity by 2010. This latter terminal will be a private facility, shared by NCIG member companies. A fourth berth at the Kooragang terminal is also under development. More detail on the operation and future plans of PWCS can be found at http://www.pwcs.com.au, and on the NCIG at http://www.ncig.com.au; see also the presentation in [23].

Ships enter and leave the port under guidance from Port of Newcastle pilots. There are often restrictions on sailing times for vessels, with fully loaded larger vessels unable to sail except at high tide, and some vessels only able to navigate the channel in daylight.

HVCCC plans and schedules all movements of coal in the system, but is not (yet) responsible for the execution of the plans. It creates "inbound" plans, which cover the transportation of coal from the mines to the stockyard (i.e., load point activities, train schedules, dump station activities, and stacker activities), and "outbound" plans that cover the movement of coal from the stockyard onto the vessels (i.e., reclaimer activities and shiploader activities).

A preliminary inbound plan covering 36 h of operations is ready at midnight 24 h before it will go into effect. In the morning, the plan is updated using the latest information, e.g., regarding canceled trains, and then released at noon. An outbound plan also covering 36 h of operations is released at 5 pm 7 h before it will go into effect. The execution of the plans is the responsibility of the "Live Run" team at PWCS.

The coal chain experiences many forms of disruption, ranging from delays of only a few hours affecting a small portion of the operation, to major events that can stop all operations for several days.

Most major disruptions are weather-related. In bad weather, ships may not be able to enter or leave the port for periods of up to several days. Furthermore heavy rain can cause landslides, which can put sections of the rail track out of commission for anything from 1 to 10 days.

Disruptions due to issues with the ships themselves are rare. There have been incidents with the crew or cargo, for example, identification of a crew member with an infectious disease. Occasional delays in sailing can be caused by problems with the cargo paperwork, or unexpected issues arising in the ship loading, for example, due to errors in executing a loading plan leading to potential stability problems for the ship. However, the terminal operators have worked hard with ship operators in recent years to reduce incidents of these types, and delays from these causes are now very rare.

Occasional issues with the rail track can cause delays, for example, coal spillage on the track causes delays while the track is cleaned. Also track availability for coal movement can be affected by passenger and other freight rail problems. Any delay in the passenger rail system has the potential to impact coal train schedules, particularly as passenger trains are given first priority in the system.

Equipment breakdowns, or unplanned maintenance events are relatively frequent, but for the most part do not lead to major delays. Probably the most disruptive of these is a locomotive breakdown. In the case a locomotive breaks down, it may be able to be repaired in situ within a few hours. However if that is not possible, and the locomotive needs to be "rescued", recovery may take days. Two locomotives are required to effect the rescue: one to replace the affected locomotive in its train, and the other to transport it to a maintenance station. Incidents requiring locomotive rescue are relatively common, occurring about once per month.

Other equipment-related incidents may involve the shiploaders, conveyors, reclaimers, stackers, dumpsheds, rail wagons, and mine load point equipment. For example, wagons commonly experience problems with the doors not opening during dumping. Crippled wagons are moved to a siding, for later repair. A rotating pool of wagons is maintained on sidings at the port, so generally a fully operational wagon is available immediately to replace the crippled wagon. All other equipment experiences the usual "u-shaped" curve of minor breakdowns, and need for unplanned maintenance, with more frequent incidents when the equipment is new—being "bedded down"—and when the equipment is old.

All these delays can have flow-on effects. For example, delays at mine load points, which are generally at more remote locations, can cause problems with train crews, who cannot stay with the train indefinitely, but need to return to their home base within a reasonable window of time. The same is true with any train- or track-related incident, in which a train is delayed away from the port, or the crew's home base. Another example is the issue of train refueling. Recovery from rail-related delays can naturally lead to periods of higher rates of train arrivals at the port, which can cause queueing problems at the refueling stations. Both train refueling rates and actual track capacity for parking trains can limit recovery.

The impacts of the above types of incidents are generally felt as reductions in the throughput capacity of the system. However, direct costs are also incurred in several parts of the chain. Of course, delays in loading a ship can lead to demurrage costs. But train crew costs, in overtime, penalties, and the use of reserve crews, can also accumulate.

10.4 Planning and Scheduling in the Coal Chain

Now that we have an overview of what happens in the HVCC, we switch to how it happens and introduce a few of the key decision problems encountered. For further detail on some of these points, see the presentation in Vandervoort [22].

Infrastructure capacity planning. One of the most important and far-reaching decision problems faced by HVCCC is long-term capacity planning. The demand for coal is expected to almost double over the next 5 years. Even if the existing capacity could be used optimally, it would not be possible to accommodate the expected growth in demand. Thus capital needs to be invested in infrastructure upgrades and capacity expansion. Upgrading infrastructure and expanding capacity is enormously expensive, and thus it is crucial to carefully analyze the tradeoffs and ensure that money is invested in the "right" upgrades and expansions. Options considered include load point improvements, new train control technology, track capacity upgrades, additional tracks, new passing loops, new overpasses separating passenger and coal trains, new dump stations, additional stockyard space, new stackers, new reclaimers, new berths, entire new terminals, etc. Obviously, the cost and impact of the various options differ widely. Furthermore, the "right" thing to do may be to invest in a combination of upgrades and expansions. Optimization models to suggest one or more alternatives and simulation models to evaluate the impact of these alternatives are discussed in more detail in Sect. 10.5.

Coordinated maintenance planning. As discussed earlier, one of the benefits of centralized planning is that planned maintenance can be coordinated. Uncoordinated planned maintenance was identified as the most important reason for the effective system capacity to be substantially below the stand-alone capacity of the individual service providers in the coal chain. That is, none of the stand-alone capacities of the individual service provider caused a bottleneck in the system, but the actions of the individual services providers collectively resulted in an effective system capacity that was lower than the minimum of their stand-alone capacities. The goal of coordinated maintenance planning is to provide an annual coal chain capacity plan. On an ongoing basis, the goal is to align the monthly capacity plan with the daily targets.

Operational planning. The goal of operational planning is to provide rolling 2-week, optimized, and coordinated coal delivery and loading plans, where coal delivery refers to the transportation of coal from the mines to the stockyard at the terminal and loading refers to the movement of coal from the stockyard onto vessels at the berths. The first day of this plan provides the coal delivery and loading schedules to be executed. Operational planning is a continuous process of preparing and updating the plans to fulfill customer orders to load vessels. In a "cargo assembly" environment, i.e., a build-to-order environment, such as the HVCC, no coal moves unless there is a vessel scheduled to berth; coal is only assembled at the port in order to fulfill a vessel order.

The planning process at the moment is mostly manual and follows the following steps. It starts with an order being placed with PWCS for the loading of a vessel, which is either accepted or declined. Next, a plan is prepared for the accepted vessels that allocates vessels to berths and sets loading and sailing times. Subsequently, stockpiles at the port are planned to ensure that the cargoes are assembled and ready to be loaded onto the vessel at the vessel's appointed time. This cargo assembly plan is then turned into a rail schedule which specifies exactly what trains will run at which times to which load points and back. The end result is a set of plans for each day identifying exactly what assets are required to do which task at which times in order to maximize system throughput. The plans are continuously revised in response to changes and events that occur during the execution of today's plan and that have an impact on the next day's plan.

In Sect. 10.5, we discuss optimization technology that specifically focuses on the scheduling of the trains, which is one of the most difficult and time-consuming parts of the planning process at the moment.

Disruption management. So far, the emphasis of HVCCC has been on planning, i.e., creating a complete and detailed set of actions to achieve a set of predefined goals. However, the chance of the plan being executed in its entirety as prescribed is small due to unanticipated events that disrupt the system. In such situations the planned actions may need to be modified.

Of course, HVCCC has worked, and continues to work, at reducing the incidence of disruption events. Measures such as more aggressive preventative maintenance programs, and upgrade/replacement programs for the least reliable components of the system, are natural candidates for attention. However there is increasing appreciation for the need for more "surge capacity" in the system, to absorb the flow-on effects of disruption, and the need for accompanying infrastructure investment. For example, as a result of the modeling work by HVCCC, there are now plans to build new train provisioning (refueling) facilities inland, away from the port. The Port of Newcastle has also commissioned models to investigate the ability of its tug boat fleet to meet demand surges, for example as might be required in the wake of a weather-related port closure. These efforts are largely directed at ensuring that the coal chain is more robust to disruption.

Disruption management and short-term recovery planning is the responsibility of the PWCS "Live Run" team, who take over planning activities from HVCCC at 12 h out from operation. Of course, many disruptions spill well past the 12-h period, and hence recovery planning from both the disruption incident and the short-term recovery re-planning, is part of HVCCC's task. All such planning is currently carried out using sophisticated data management and visualization software, but without the assistance of optimization or other automated decision support tools.

10.5 Decision Technology

To be able to effectively and efficiently manage a multi-faceted, intricate system like the HVCC, with many interacting subsystems that are highly complex, decision support tools are indispensable. In this section, we discuss a few of the decision support tools that are in use or under development.

10.5.1 Forecasting

For strategic planning it is essential that forecasted mine productions are converted into a shipping stem, as the HVCC is a pull system, i.e., a build-to-order system, and the arrival of vessels at the port drives the HVCC. Generating a shipping stem that matches forecasted mine production and resembles a historic shipping stem is challenging in itself, but the process is complicated by the fact that new mines are brought on line, existing mines are (temporarily) shut down, new brands and new brand-recipes are introduced, and new terminals may start their operations.

Shipping stems are currently produced manually, which is extremely time-consuming (it may take up to 3 weeks to produce a single shipping stem). A shipping stem consists of a set of trips, where each trip represents the arrival of a vessel at the port. Each trip is characterized by its arrival time, the terminal at which the vessel has to be loaded, a cargo-profile, which specifies the various brands of coal that need to be loaded and their tonnage, the associated brand-recipes, which specify the various components, and thus the mines, that make up a brand and their tonnage.

Whilst good historical data—past shipping stems—is available, this data cannot simply be "scaled up" to match forecasted system production; more than scaling is required. The relative proportion of output from mines may change, in some cases quite dramatically, with some mines closing down and others opening. The relative proportion of brands may also be quite different from what it was in the past and brand-recipes may change. Furthermore, with the new NCIG terminal comes new requirements in terms of vessel classes, with agreements expected to restrict the proportion of vessels of a particular class berthing at the NCIG terminal and limiting what brands can, or must, pass through the NCIG terminal. Finally, HVCCC may want to understand the responsiveness of the coal chain to new practices, such as deliberate smoothing of demand over time. Therefore, a method for generating shipping stems that realistically reflect the nature of coal production must have some degree of randomization, so that possible future scenarios can be sampled and controlled for the types of variables just described.

An analysis of past shipping stems revealed that certain combinations of brands are much more likely to appear together as cargoes of a vessel, and are even more likely to appear as cargoes of a vessel of a particular class. Similarly, certain brands are much more likely to require coal from a particular mine, showing that specific brand-recipes were used repeatedly. Some demand smoothing was also apparent, with higher volume demands clearly spread more evenly across the year. As a consequence, simply sampling randomly from brands to put on vessels and sampling randomly from mines to put into brands is inappropriate. By doing so, we can potentially hide, or artificially create, coal chain bottlenecks.

In the end, we decided to base our approach on the single biggest factor likely to affect coal chain performance: vessel–brand combinations. Thus, we take the view that future vessel–brand combinations are likely to reflect history, with perhaps some trends towards or away from particular combinations. The result is a multi-phase approach for generating shipping stems that relies on integer linear programming and sampling. It allows for the generation of multiple different shipping stems for the same forecasted mine production. The different phases of the stem generation technology are briefly discussed below:

1. *Trip assessment*. This module establishes the likelihood of using a particular historical trip to satisfy forecasted mine productions. Since the difference between the mine production forecast and the historical mine production is likely to be different for different mines, some historic trips may be more useful than others when it comes to satisfying forecasted mine production. This difference is captured in the computed likelihoods.
2. *Shipping stem generation*
    a. *Trip sampling*. First, a cumulative distribution function (CDF) using the likelihoods determined in the trip assessment phase is constructed. Next, to sample a trip, a uniform random number is drawn in the range (0, 1) and the CDF is used to identify the corresponding trip. We repeatedly sample trips until the total tonnage of the sample exceeds the system production forecast.
    b. *Trip generation*. Trip generation resembles trip assessment, but is also distinctly different. During trip assessment, the historical trips are used as they were executed. In trip generation, the brand proportions in a cargo profile are allowed to change and so are the mine contributions to a brand, i.e., the fraction of coal sourced from a particular mine in a brand recipe is allowed to change. Thus, the historical trips are used to provide the likely brand combinations and brand-recipes, but the quantity of coal that is sourced from a particular mine may deviate, within bounds, from the quantity seen in the historical trip.
    c. *Trip timing*. This module determines the arrival times of the trips that will make up the shipping stem, so as to smooth vessel arrivals, demand for brands, mine production, and load-point activity over the planning period.

The *Trip assessment* phase has to be executed only once and the *Shipping stem generation* phase can be executed as often as desired.

Hierarchical optimization using quadratic and integer programming is used during trip assessment, trip generation, and trip timing. An acceptable shipping stem has to satisfy a large number of requirements. Some of these requirements must be satisfied, others take the form of desirable characteristics. For example, the total tonnage has to match the total forecasted mine production exactly, the fraction of the total tonnage handled by a terminal has to be within a certain range, and there cannot be too much variation in handled tonnage from month to month. To accommodate the various requirements a hierarchy of objectives is defined, ordered by priority, and a sequence of optimization problems is solved, each time optimizing a particular objective ensuring that the value of objectives higher in the hierarchy do not deteriorate. Furthermore, in case of a desirable characteristic, the objective function minimizes the deviation from some pre-specified target.

10.5.2 Simulation

An important decision support tool employed by HVCCC is a detailed simulation of the HVCC. The tool is used, among others, for the analysis of the impact of possible infrastructure expansions on throughput. Computer simulations have become very important in the analysis of large, complex, dynamic systems. Computer simulations provide an efficient and cost-effective way to study the behavior and performance of a system under various conditions.

A comprehensive, detailed discrete event simulation model of the HVCC has been developed. An early version of the model was described in [24]. The simulation model has two main components:

- *A master schedule generation module*. This module processes a shipping stem and prepares a master schedule to deliver coal from the mines to the terminals by the required date taking the various capacity considerations into account.
- *Dynamic network operation module*. This module analyzes the performance of the master schedule under dynamic circumstances. Coal trains are dispatched through the rail network according to the master schedule and are subjected to delays caused by other trains (passenger or other freight trains) and to queuing at load points and terminals.

The simulation requires demand data, load point data, track data, train data, terminal data, and maintenance data. Many of the data items have attributes that relate to their variability. For example, unloading rates at a dump station at a terminal may exhibit variability. Maintenance data is especially important. Due to the many, sometimes intricate and subtle, linkages between the various components of the system, it is difficult for planners to understand the full impact of planned maintenances, let alone the full impact of disruptions and unplanned maintenances. It is important to realize that some component of the coal chain being unavailable due to maintenance is not a rare event, it happens almost every day. Track maintenance probably occurs most frequently, but there is maintenance at dump stations, stackers, reclaimers, shiploaders, etc.

The simulation model generates key performance indicators, such as throughput, ship queues, ship turn-around times, ship loading times, demurrage costs, train cycle times, cargo build times, stacker utilization, reclaimer utilization, and shiploader utilization.

The model was calibrated and validated using historic data before running experiments with various anticipated operating scenarios. For validation purposes, the following model outputs were compared against the actual values: annualized export tonnage, delivery performance (in terms of late arrivals for each mine and overall late arrivals), average daily tonnage delivered, average number of train trips per day, and train utilization.

The key criteria examined when evaluating an actual scenario are the delivery performance and the annualized throughput. The simulation model is used to analyze strategic as well as operational decisions. At the strategic level, the simulation model is used to evaluate infrastructure capacity expansion options, e.g., converting single track sections into dual track sections or adding another dump station, but also the use of alternative materials handling/logistics options, e.g., the use of dedicated stockpiles in addition to cargo assembly stockpiles. At the operational level, the simulation model is used, among others, to analyze the impact of regular maintenance decisions, relaxed headway restrictions, point-to-point train travel times, and dump station rates.

10.5.3 Capacity Expansion

Strategic capacity planning is important in any industry and has therefore received a fair bit of attention in the literature; see for example [2, 5, 6, 15].

Strategic capacity planning is crucial for HVCCC. The amount of coal flowing through the HVCC is expected to grow substantially in the next decade (more than double). To accommodate that level of growth, infrastructure expansions will be necessary. To decide on the best (combination of) infrastructure expansions, both in terms of their cost as well as their effect on increasing throughput, an optimization model has been developed (see [20]), and is being refined.

The optimization model identifies a least cost set of expansions that leads to an infrastructure that can accommodate a shipping stem with little or no demurrage. To feel confident that the optimization model suggests the appropriate infrastructure expansions, the coal chain operations need to be modeled at an acceptable level of accuracy and for a time period long enough to capture typical variations in demand patterns. The choice was made to model decisions at the daily level and cover a year's worth of operations.

For presentational convenience, we discuss the optimization model assuming there is a single terminal; it is straightforward to extend the model to multiple terminals. Before discussing the model, we need to introduce some notation. Let V denote the set of vessels, $S(v)$ the set of stockpiles associated with vessel $v \in V$, and $S = \cup_{v \in V} S(v)$ the set of all stockpiles. (Implicitly, we are assuming here that each cargo is assembled on a single stockpile.) Let $J(s)$ denote the set of train jobs associated with stockpile $s \in S$ and $J = \cup_{s \in S} J(s)$ the set of all train jobs. Let L denote the set of load points and $J(l)$ the set of train jobs associated with load point $l \in L$. Let U denote the set of junctions and $J(u)$ the set of train jobs associated with junction $u \in U$. Finally, P is the set of all stockyard pads.

The following parameters are available: the first day s_v vessel $v \in V$ is available for loading, the last day t_v vessel $v \in V$ is available for loading, the amount of coal m_j of train job $j \in J$, the length l_j that will be occupied by that coal on a stockyard pad (in m), and the number of days d_s required to reclaim stockpile $s \in S$. [Implicitly, we are assuming here that once the reclaiming of a stockpile (and thus the loading onto the ship) commences, it will not be interrupted and will continue until the stockpile is reclaimed completely.] The length of stockpile $s \in S$, i.e., $\sum_{j \in J(s)} l_j$, together with the number of days d_s required to reclaim the stockpile defines the average length l_s of stockpile $s \in S$ that is reclaimed per day. Let L_p be the length (in m) of pad $p \in P$.

Let C_l^t be the loading capacity (in tonnes per day) of load point $l \in L$ on day t and let $\bar{C}_u^t$ be the junction capacity (in trains per day) for junction $u \in U$ on day t. Furthermore, let ρ_d be the dumping rate (in tonnes per hour), C_d^t be the dumping capacity (in tonnes per day) on day t, and $\hat{C}_d^t$ the dumping capacity (in hours per day) on day t. Let T_d be the set up time (in hours) for the dump station. Let ρ_s be the stacking rate (in tonnes per hour) and $\hat{C}_s^t$ be the stacking capacity (in hours per day) on day t. Let T_s be the "set up" time (in hours) for a stacker. Let ρ_r be the reclaiming rate (in tonnes per hour) and $\hat{C}_r^t$ be the reclaiming capacity (in hours per day) on day t. Let T_r be the set up time (in hours) of a reclaimer.

All capacities are relative to average handling rates. For example, if the length of a stockpile s is say 600 m and a reclaimer can reclaim at a rate of 400 m/day, then it is assumed that it takes $d_s = \lceil (600/400) \rceil = \lceil 1.5 \rceil = 2$ days to reclaim and l_s, the length of stockpile s that can be reclaimed per day averaged over all days required is $l_s = 600/2 = 300$.

The decision variables of the model are:

$$x_{jp}^t = \begin{cases} 1 & \text{if job } j \text{ is assigned to pad } p \text{ on day } t \\ 0 & \text{otherwise} \end{cases}$$

$$\forall p \in P \ \forall v \in V \ \forall s \in S(v) \ \forall j \in J(s) \ \forall t \le t_v - d_s + 1$$

$$z_{sp} = \begin{cases} 1 & \text{if stockpile } s \text{ is assigned to pad } p \\ 0 & \text{otherwise} \end{cases}$$

$$\forall s \in S \ \forall p \in P$$

$$y_{sp}^t = \begin{cases} 1 & \text{if loading of stockpile } s \text{ from pad } p \text{ starts on day } t \\ 0 & \text{otherwise} \end{cases}$$

$$\forall p \in P \ \forall v \in V \ \forall s \in S(v) \ \forall t \in [s_v, t_v - d_s + 1]$$

For convenience the following bookkeeping variable is used:

$$L_p^t = \text{the total length of the stockpiles on pad } p \text{ on day } t \quad \forall p \in P \ \forall t$$

The first sets of constraints force every train job to be performed and train jobs associated with the same stockpile to be assigned to the same pad in the stockyard:

$$\sum_{p \in P} \sum_t x_{jp}^t = 1 \quad \forall j \in J(s) \ \forall s \in S$$

$$\sum_{j \in J(s)} \sum_t x_{jp}^t = |J(s)| z_{sp} \quad \forall s \in S \ \forall p \in P$$

The second sets of constraints force every stockpile to be reclaimed from the pad to which it was assigned and only after all associated train jobs have been performed:

$$\sum_{p \in P} \sum_{t} y_{sp}^{t} = 1 \quad \forall s \in S$$

$$\sum_{t} y_{sp}^{t} = z_{sp} \quad \forall s \in S \; \forall p \in P$$

$$\sum_{j \in J(s)} \sum_{p \in P} \sum_{t' \leq t} x_{jp}^{t'} = |J(s)| y_{sp}^{t} \quad \forall s \in S \; \forall t$$

The next sets of constraints keep track of the coal that is on the pads in the stockyard each day and ensure that only the available pad space is used:

$$L_p^t = L_p^{t-1} + \sum_{j \in J} l_j x_{jp}^t - \sum_{s \in S} \sum_{t'=t-d_s+1}^{t} l_s y_{sp}^{t'} \quad \forall p \in P \; \forall t$$

$$0 \leq L_p^t \leq L_p \quad \forall p \in P \; \forall t$$

Note that we are assuming that stockyard space is not "reserved" for a stockpile. Stockyard space is only consumed when a train delivers coal and stockyard space becomes available dynamically when a stockpile is loaded onto a vessel. This is the most "aggressive" setting. It is possible to adjust the model to make it less aggressive, i.e., to assume that the space occupied by a stockpile is unavailable from the moment the first train delivers coal for that stockpile until the stockpile is completely loaded onto the vessel.

The final sets of constraints capture the various capacities that have to be respected, i.e., load point capacity, junction capacity, dumping capacity, stacking capacity, and reclaiming capacity:

$$\sum_{j \in J(l)} \sum_{p \in P} m_j x_{jp}^t \leq C_l^t \quad \forall l \in L \quad \forall t$$

$$\sum_{j \in J(u)} \sum_{p \in P} x_{jp}^t \leq \bar{C}_u^t \quad \forall u \in U \quad \forall t$$

$$\sum_{j \in J} \sum_{p \in P} m_j x_{jp}^t \leq C_d^t \quad \forall t$$

$$\sum_{j \in J} \sum_{p \in P} \left(T_d + \frac{m_j}{\rho_d} \right) x_{jp}^t \leq \hat{C}_d^t \quad \forall t$$

$$\sum_{j \in J} \sum_{p \in P} \left(T_s + \frac{m_j}{\rho_s} \right) x_{jp}^t \leq \hat{C}_s^t \quad \forall t$$

$$\sum_{s \in S} \sum_{p \in P} \sum_{t'=t-d_s+1}^{t} \left(T_r + \frac{\sum_{j \in J(s)} m_j}{d_s \rho_r} \right) y_{sp}^{t'} \leq \hat{C}_r^t \quad \forall t$$

More accurate versions of the stacking and reclaiming capacity constraints can be constructed by taking into account the fact that stackers and reclaimers can only serve one or two of the four pads in the stockyard.

The above capacity constraints illustrate the capacity constraints in terms of daily maximum capacity. However, the real model is more powerful and more flexible. It includes capacity expansion options and not only considers daily maximum capacities, but also long-term average capacity. We discuss this more elaborate structure in general terms. If C^t is the maximum capacity on day t, then a non-negative variable w_t is added to indicate any extra capacity needed in period t, i.e., the capacity limit on day t becomes $C^t + w_t$. A global capacity expansion variable w is introduced representing the actual addition of infrastructure. The objective function seeks to minimize cw, where c is the per unit cost of that type of capacity. To capture average capacity limits, we ask that the total capacity usage summed over all days is no greater than $T(C + w)$, where T is the number of days in the planning period, and C is the existing average capacity ($C < C^t$). Rather than requiring the capacity for every "peak demand" day to be accommodated by capacity expansion, we can ask, for example, that the capacity is enough to meet the demand on 90% of days, on average, via

$$\sum_t w_t \leq 0.9Tw.$$

This approach

1. Looks at an individual day t and recognizes that an artificial maximum capacity (C_t) for that day is insufficient;
2. Looks at the planning horizon and recognizes that the actual/average daily capacity (C) is insufficient and capacity expansion is needed; and
3. Asks that if the daily artificial maximum capacity gets violated too often (or by too much), we also need capacity expansion.

With this approach, if a few ships arrive early the daily artificial maximum capacity is insufficient, but we remain below the actual/average daily capacity and no expansion is suggested. This "damps" response to unpredictable surges in demand which can reasonably be accommodated operationally at the cost of demurrage, rather than by making under-utilized infrastructure investments.

A variation of the model discussed above has been in use at HVCCC and has helped generate infrastructure expansion suggestions that have been evaluated using the simulation model. This is not as easy as it sounds. The instances get extremely large and cannot simply be handed over to a commercial integer programming solver. Specialized integer programming heuristics, based on the "relax-and-fix" concept (see for example [17]), had to be developed. And even these integer programming heuristics can only handle a planning period of six months as opposed to the targeted planning period of one year.

A number of research questions arise as a result of these initial experiences. For example whether the capacity expansion decisions in an optimal solution to a model with 52 time periods, each corresponding to a week, will be different from the capacity expansion decisions in a, not necessarily optimal, solution to a model with 180 time periods, each corresponding to a day. And, if so, which solution is actually better. (The latter question may be answered by a simulation study.) Along the same lines, one can investigate whether the capacity expansion decisions in an optimal solution to a wrap-around model with 90 time periods, each corresponding to a day, will be different from the capacity expansion decisions in a, not necessarily optimal, solution to a model with 180 time periods.

Another relevant issue is uncertainty. The optimization model described above assumes that the future, in the form of a shipping stem, is known with certainty. Of course this is unrealistic. A two-stage stochastic optimization model with capacity expansion decision in the first stage and operational decision in the second stage, may be more appropriate, but may be computationally prohibitive.

These issues are currently under investigation and will hopefully result in even more useful optimization technology in the near future.

10.5.4 Rail Scheduling

At present, HVCCC receives relatively short notice of a vessel's arrival in the queue. In the near future, it is expected that the HVCCC will be able to reliably estimate vessel arrival times about 4 weeks in advance. This offers the opportunity to plan the transport of coal further in advance. An optimization model has been developed to assist in this task. The model is currently focused on rail and vessel scheduling; the stockyard is assumed to be uncapacitated.

The key model inputs are a list of vessels, together with details of their planned cargoes and arrival times, and a list of available train paths, together with knowledge of the available rolling stock. The key decisions made by the model are when to load each vessel, and when to bring each trainload of coal required for the vessel from the mine to the stockyard. These decisions are constrained by rail capacities, such as the maximum number of trains permitted at any one time, or the minimum headway between consecutive trains on a given piece of track. They are also constrained by the number of dump stations at the stockyard, the number of berths at the terminal, and, of course, by the number of trains in the system. The objectives considered by the model are to minimize demurrage, minimize dwell time of trains in the valley, and maximize throughput.

Although the mathematical modeling of ship scheduling problems found in container terminals has received increasing attention (see for example [21]), we are not aware of investigations using similar approaches for coal terminals (or bulk handling terminals in general), other than that of [1]. Rail scheduling has received much more attention in the literature, most of which involves creating train paths, i.e., determining the routing and frequency of trains subject to rail network constraints. Such investigations are often solved with heuristic approaches, e.g., [7, 18], or with discrete event simulation [12], as a large number of logic variables make the problem computationally intractable for exact methods (see [11]). Indeed [1] did address rail scheduling of this type, integrated with stockyard planning, for a coal export supply chain, using a combination of integer programming models and heuristic approaches. They found both rail scheduling without fixed train paths, and stockyard planning, to be challenging problems for optimization, and arrived at an approach that combined simulation with heuristics.

We use J to denote the index set of vessels (jobs to be performed), and r_j to denote the arrival time (release date) of vessel $j \in J$ in the port. From knowledge of the cargoes to be loaded and the loading rate at the terminal, the time required to load vessel j, denoted by p_j, is calculated. Note that p_j is required to model berth occupancy, hence depending on the type of vessel, p_j may include extra time needed for the vessel at berth waiting for a change in tide, or for daylight. Since the time of loading is to be decided by the model, this extra time is only an estimate. Each vessel $j \in J$ also has a due date, denoted by d_j, which indicates the time at which demurrage will start to be incurred if the vessel is not ready to sail. The demurrage cost for vessel $j \in J$ is given by c_j (dollars per unit time).

Each vessel $j \in J$ generates a set of train jobs, as follows. The cargoes to be loaded on the ship are sourced from a set of mines B_j. Each mine $b \in B_j$ has a preferred train type s_b, which determines how much coal can be carried on the train, and so from the known amount of coal to be loaded on the vessel, the number of trains n_{jb} needed to meet the demand of vessel j for coal from mine b, can be calculated. We define $B = \bigcup_{j \in J} B_j$ to be the set of all mines.

For each mine $b \in B$, there is a set of available *mine-to-port* train paths P_b, each of which specifies a path in time and space from the mine to the port. For each vessel j, n_{jb} of these paths must be selected to supply the coal from mine b needed for vessel j. In order to properly constrain train usage, for each mine-to-port path selected, a preceding port-to-mine path must also be selected, with enough intervening time at the mine for the train to be loaded; there is also a set of available *port-to-mine* train paths Q_b. Mine-to-port and port-to-mine paths can be operated independently, but there may be clashes between train paths within these two sets. Clearly, two train paths which occupy the same piece of track at the same time, or within too close a time, i.e., allowing insufficient *headway*, cannot both be selected. This allows us to define collections $\mathcal{P} \subseteq 2^P$, where $P \stackrel{def}{=} \bigcup_{b \in B} P_b$, and $\mathcal{Q} \subseteq 2^Q$, where $Q \stackrel{def}{=} \bigcup_{b \in B} Q_b$, of sets of mutually incompatible paths. It is also helpful to define $P_j \subseteq P$ to be the set of mine-to-port train paths that could be used to supply vessel $j \in J$, and $Q_i \subseteq Q$ to be the set of port-to-mine train paths that could be used by a train returning to the port on path $i \in P$.

The essential elements of the rail and vessel scheduling model can now be defined. For the rail aspect, we use binary variable y_{ij} to indicate that mine-to-port path $i \in P_j$ is selected to supply vessel $j \in J$, and binary variable z_{hi} to indicate that port-to-mine path $h \in Q_i$ is used by the train returning to the port on train path $i \in P$. The key constraints on the rail are:

$$\sum_{j \in J : i \in P_j} y_{ij} \leq 1, \quad \forall i \in P \tag{10.1}$$

$$\sum_{h\in Q_i} z_{hi} = \sum_{j\in J:i\in P_j} y_{ij}, \quad \forall i \in P \tag{10.2}$$

$$\sum_{i\in P_j\cap P_b} y_{ij} = n_{jb}, \forall j \in J, \quad \forall b \in B_j \tag{10.3}$$

$$\sum_{\substack{b\in B:\\ s_b=s}} \sum_{\substack{i\in P_b:\\ end(i)>u}} \sum_{\substack{h\in Q_i:\\ start(h)\le u}} z_{hi} \le a_s, \qquad \forall u \in U, \forall s \in S \tag{10.4}$$

$$\sum_{i\in R} \sum_{j\in J:i\in P_j} y_{ij} \le 1, \qquad \forall R \in \mathcal{P} \tag{10.5}$$

$$\sum_{h\in R} \sum_{i\in P:h\in Q_i} z_{hi} \le 1, \qquad \forall R \in \mathcal{Q} \tag{10.6}$$

where $start(i)$ and $end(i)$ denote the start and end times of train path i respectively, U is the set of start times of port-to-mine paths, a_s is the number of trains of type s in the fleet, and S is the set of all train types. The first two constraints are "train flow" constraints: (10.1) ensures that each mine-to-port path supplies at most one vessel, and (10.2) ensures that each mine-to-port path has a connecting port-to-mine path, if selected. The *demand* constraint (10.3) ensures that the demand for coal from each mine for each vessel is met. The *fleet size* constraint (10.4) counts the number of trains of each type in use at any one time, and ensures that this is no more than the number available (note that the end time of a mine-to-port path includes the time needed for the train to unload). The *headway* constraints are given by (10.5) and (10.6).

In addition to these constraints, there are similar constraints ensuring that at most one train occupies a mine's load point at any one time, that the number of trains to a mine in any 24-h period does not exceed a mine-dependent maximum, and that the number of trains unloading coal at the port does not exceed the number of dump stations.

The vessel scheduling aspect of the model has the form of a classic scheduling problem, in which each berth is viewed as a machine, for processing vessel jobs. If only one terminal is considered, it could be viewed as a job shop problem with parallel machines, release dates, due dates, and no pre-emption. We consider two alternative models, one based on a *positional date and assignment* formulation, and the other based on a *time indexed* formulation. A discussion of these types of models can be found in [19]. The former gives an accurate model of the port loading capacity and demurrage costs. The latter requires time to be discretized, which necessarily introduces approximation, particularly in regard to berth capacity. However, it appears to be by far the more computationally effective model in this context.

The rail and vessel aspects of the model are linked by two requirements: (a) vessel loading cannot begin until after all trains for that vessel have been unloaded, and (b) trains for a vessel cannot be scheduled earlier than a given length of time e_j prior to the scheduled loading start time of the vessel j. The latter requirement is a rough approximation to a stockyard capacity limit. It reflects a rule-of-thumb used in practice aimed at ensuring that the stockyard does not get too full by bringing down trains "just-in-time".

The two vessel scheduling models are defined below.

The *positional date and assignment vessel scheduling model*. This model is based on binary variables v_{jfk}, indicating that vessel $j \in J$ is the kth vessel loaded at berth f, for each $f \in F$, the set of berths, and for all $k = 1, \dots, K$, where K is an upper bound on the number of vessels that can be served by a single berth. The variable t_{jfk} records the time at which vessel j commences loading if $v_{jfk} = 1$; otherwise t_{jfk} is set to zero. Note that $\sum_{j \in J} t_{jfk}$ gives the loading start time for the kth job at berth f, and $\sum_{f \in F} \sum_{k=1}^{K} t_{jfk}$ gives the loading start time for vessel j. The constraints for the positional date and assignment (PDA) model are:

$$\sum_{f \in F} \sum_{k=1}^{K} v_{jfk} = 1, \quad \forall j \in J \tag{10.7}$$

$$\sum_{j \in J} v_{jf1} \leq 1, \quad \forall f \in F, \tag{10.8}$$

$$\sum_{j \in J} v_{jfk} \leq \sum_{j \in J} v_{jfk-1}, \quad \forall f \in F, \ \forall k = 2, \dots, K \tag{10.9}$$

$$\sum_{j \in J} t_{jfk} \geq \sum_{j \in J} (t_{jfk-1} + p_j v_{jfk-1}) - M(1 - \sum_{j \in J} v_{jfk}), \quad \forall f \in F, \ \forall k = 2, \dots, K \tag{10.10}$$

$$r_j v_{jfk} \leq t_{jfk} \leq M v_{jfk}, \qquad \forall j \in J, \ \forall f \in F, \ \forall k = 1, \dots, K, \tag{10.11}$$

where M is an upper bound on the latest start time of any job. The *job completion* constraints (10.7) ensure that each vessel is assigned a position at some berth. The *job ordering* constraints (10.8) and (10.9) together ensure that there is at most one vessel in any position at any berth, and that there are no "holes" in the assignment of vessels to berths, i.e., there is a vessel in position k only if there is also a vessel in position $k-1$. The *job timing* constraints (10.10) ask that kth the vessel at berth f not start loading until the (k-1)th vessel has finished processing (loading and sailing). The logic of the definition of the t variables is enforced by constraints (10.11).

The two requirements linking rail and vessel—that vessel loading cannot start until after train unloading, and that trains cannot be brought up to the port "too early"—are formulated in the PDA model via the constraints

$$\sum_{f\in F}\sum_{k=1}^{K} t_{jfk} \ge end(i)y_{ij}, \quad \forall j \in J, \forall i \in P_j, \text{ and} \tag{10.12}$$

$$\sum_{f\in F}\sum_{k=1}^{K} t_{jfk} \le (end(i) + e_j)y_{ij} + M(1 - y_{ij}), \quad \forall j \in J, \forall i \in P_j. \tag{10.13}$$

The *time indexed vessel scheduling model*. The time indexed (TI) model assumes a discretization of time, and is based on binary variables x_{jt} indicating that vessel $j \in J$ starts loading in period $t = \tilde{r}_j, \tilde{r}_j + 1, \ldots, T$. Here, T is the number of time periods in the discretization of the planning period $[0, M]$, (recall M is an upper bound on the latest start time of any job), and $\tilde{r}_j$ is calculated to be the release date in terms of number of periods. In Clement [9], a conservative approximation is used, so r_j is "rounded up" to the nearest period after the original release date, i.e., $\tilde{r}_j = \lceil r_j(T + 1)/M \rceil$. Similarly, the vessel processing time $\tilde{p}_j$ is recalculated in terms of periods, rounding up to the nearest number of periods. This ensures that any feasible solution for the *TI* model is feasible for the exact problem, but berth capacity will be under-approximated. The constraints for the *TI* model are:

$$\sum_{t=\tilde{r}_j}^{T} x_{jt} = 1, \quad \forall j \in J, \quad \text{and} \tag{10.14}$$

$$\sum_{j\in J}\sum_{t'=t-\tilde{p}_j+1}^{t} x_{jt'} \le |F|, \forall t, \tag{10.15}$$

where in summing over time periods, we ignore any indices outside the range $0, 1, \ldots, T$. The *job completion* constraints (10.14) ensure that each vessel starts loading in some period. The *berth capacity* constraints (10.15) ensure that at most $|F|$ berths are in use for loading in any one period.

The two requirements linking the vessel and rail aspects can be captured in one constraint in the *TI* model:

$$y_{ij} + \sum_{\substack{t=0 \\ start(t)<end(i)}}^{T} x_{jt} + \sum_{\substack{t=0 \\ start(t)>end(i)+e_j}}^{T} x_{jt} \le 1, \quad \forall j \in J, \forall i \in P_j,$$

where $start(t) = tM/T$ is the start time of period t.

Next, we discuss the objective functions for rail and vessel scheduling. Three objectives are of key interest to HVCCC in this planning exercise. The most important is to minimize demurrage. But when there is some flexibility in the rail, minimal demurrage solutions are free to "leave" trains at the mines unnecessarily long. Thus a secondary objective is to minimize "dwell", defined as the total time that trains spend at mine load points, over and above what they require to be loaded. These can be modeled as follows. If τ_j represents the tardiness of job j, and c_j the demurrage cost per unit time, then total demurrage is

$$Z_1 := \sum_{j \in J} c_j \tau_j,$$

while dwell is readily modeled as

$$Z_2 := \sum_{i \in P} \sum_{h \in Q_i} \theta_{hi} z_{hi},$$

where θ_{hi} is the time between *end*(h) and *start*(i) in excess of that required for loading a train of type s_b, where b is the mine such that $i \in P_b$. In the *PDA* model, the $\tau \geq 0$ variables are calculated via the constraint

$$\tau_j \geq \sum_{f \in F} \sum_{k=1}^{K} t_{jfk} + p_j - d_j, \qquad \forall j \in J.$$

In the *TI* model,

$$\tau_j = \sum_{t=0}^{T} \tilde{\tau}_{jt} x_{jt}, \qquad \forall j \in J,$$

where $\tilde{\tau}_{jt}$ is the tardiness of vessel j if it starts loading in period t. Since the model rounds up release dates, and since all capacity and train precedence relationships apply identically across the whole of a period, we can safely assume that each vessel j starts loading at the *start* of the period t for which $x_{jt} = 1$, and calculate $\tilde{\tau}_{jt}$ on that basis.

Finally, we note that both models presented assume a fixed planning horizon $[0, M]$, where M is chosen a priori. Clearly, if M is chosen too small, the problem will be infeasible. But if M is chosen too large, and there are no explicit incentives in the model to process vessels as soon as possible, (other than the demurrage cost), then the result will be reduced throughput. Thus a third objective is to maximize throughput. Currently this is not modeled explicitly; the issue is considered via a search on M.

An extensive computational study has been conducted with the above models using up to 5 weeks worth of HVCCC data, representing around 65 vessels to be served, with about 14 trains to be scheduled per vessel. Typically 8,100 mine-to-port train paths and 10,200 port-to-mine train paths are available each day. Time discretizations giving period lengths of 24, 12, 6, 4, and 1 h and planning horizons of up to 0, 2, and 4 days after the last release date were investigated. In all cases a "run-up" period of about 10 days was added at the beginning of the planning period to allow time for trains to be brought down prior to the start of the first vessel's loading jobs. The models were implemented using the Python interface to Gurobi v2.0, and solved on an Intel Xeon X5460 3.16 GHz dual quad core with 64 GB RAM running Red Hat Enterprise Linux 5.

Initial experiments confirmed that the demurrage and dwell objectives are closely aligned and that minimizing $Z_1 + 0.0001 Z_2$, where Z_1 represents demurrage and Z_2 represents dwell, always gives the minimum possible demurrage with a dwell that is only slightly higher than the minimum possible dwell. Thus in most experiments this weighted objective function is used.

The PDA model tended to be difficult to solve, with instances of only ten vessels needing more than an hour to solve to optimality; root node gaps and search trees were large. By contrast, the TI model solves surprisingly fast, with instances of up to 35 vessels solving to optimality within an hour; for a large fraction of the instances the solution to the linear programming relaxation was integer, and nearly all instances solved at the root node. Maybe even more surprising was the fact that the performance of the TI model was not severely affected by finer time discretizations, with instances based on the model with 4-h periods solving as quickly, and in some cases even faster than instances based on the model with 24-h periods. However, instances based on the model with 1-h periods did result in sharply increased solve times. More analysis is needed to fully understand this behavior. The tradeoff in solution quality versus discretization granularity was clearly apparent, with finer discretizations yielding substantially better solutions.

In an attempt to "recover" from the approximation in the TI model, without incurring the long run times of a complete PDA model solution, algorithms in which the TI and PDA models were solved in sequence, with some part of the solution from the TI model fixed in the PDA model, were tested. First, almost all the information from the solution to the TI model was transferred to the PDA model—the implied berth sequence and the selected train paths—so that the subsequent PDA model acts simply as a calculator that adjusts the loading start times of the vessels. Second, only part of the information from the solution to the TI model was transferred—only the implied berth sequence or only the train paths. With a coarse discretization for TI, the scheme takes substantially less time than stand-alone PDA and substantially improves the quality of the TI solution. However, for most instances, running TI with a finer discretization gives even better results in less time. More testing is still underway, and there do appear to be instances for which the combined TI-PDA approach performs better than either approach on its own.

10.6 Final Remarks

In this chapter, we have introduced the Hunter Valley export coal chain, an example of a complex bulk goods export chain. We have discussed both the necessity and the benefits of a collaborative approach to planning and scheduling the coal chain. The efficient and effective planning and scheduling of such a complex, multi-faceted, dynamic logistics system is beyond the capability of human planners and thus decision support tools are crucially important, especially given the expected future growth in demand for coal. We have reviewed some of the key decision problems and have presented an in-depth discussion of a few of the optimization models that have been developed.

In the end, it is all about answering two fundamental questions: "How to operate a given system configuration to maximize the coal that can be handled?" and "How to configure the system at minimal cost to handle a given amount of coal?" The initial steps towards answering these questions have been made, but there is still a long way to go. Fortunately, the ambition, the resources, the knowledge, and the talent is there to successfully answer these questions in the (near) future.

Acknowledgment We express our thanks to the other members of the research team involved in the development of decision technology for the HVCC for their support and valuable comments during the writing of this chapter, i.e., Bhaswar Choudhary, Tracey Giles, and Rob Oyston at HVCCC, Palitha Welgema at Rio Tinto, Andreas Ernst and Gaurav Singh at CSIRO, and Riley Clement, Faramrose Engineer, and Hamish Waterer at the University of Newcastle.

References

1. Abdekhodaee A, Dunstall S, Ernst AT et al (2004) Integration of stockyard and rail network: a scheduling case study. In: Proceedings of the 5th Asia-Pacific Industrial Engineering and Management Systems Conference, Gold Coast, pp 1–16
2. Ahmed S, Sahinidis NV (2003) An approximation scheme for stochastic integer programs arising in capacity expansion. Oper Res 51:461–471
3. Australian Coal Association (2010) http://www.australiancoal.com.au. Accessed 22 June 2010
4. Australian Department of Foreign Affairs and Trade (2010) Trade at a Glance 2009. http://www.dfat.gov.au/publications/taag/index.html. Accessed 22 June 2010
5. Bermon S, Hood SJ (1999) Capacity optimization planning systems (CAPS). Interfaces 29: 31–50
6. Birge J (2000) Option methods for incorporating risk into linear capacity planning models. Manuf Serv Oper Res 2: 19–31
7. Burdett RL, Kozan E (2009) A sequencing approach for creating new train timetables. OR Spectr 32(1):163–193
8. Coal Online. http://www.coalonline.org/site/coalonline/content/hom Accessed 21 June 2010
9. Clement R (2009) MIP models for scheduling the operations for a coal loading facility. In: Proceedings of the 44th Annual Conference of the ORSNZ, Christchurch, 3–4 December, pp 139–148
10. Conradie DG, Morison LE, Joubert JW (2008) Scheduling at coal handling facilities using simulated annealing. Math Methods Oper Res 68:277–293
11. Dessouky MM, Lu Q, Zhao J et al (2006) An exact solution procedure for determining the optimal dispatching times for complex rail networks. IIE Trans 38:141–152
12. Dorfman MJ, Medanic J (2004) Scheduling trains on a railway network using a discrete event model of railway traffic. Transp Res B 38(1):81–98
13. Ernst A, Krishanmoorthy M, Sier D et al (2008) Solving regional infrastructure bottlenecks: rail allocation policies for a coal terminal. Australas J Reg Stud 14(2):117–194
14. Everett JE (2007) Computer aids for production systems management in iron ore mining. Int J Prod Econ 110:213–223
15. Huh WT, Roundy RO, Cakanyildirim M (2006) A general strategic capacity planning model under demand uncertainty. Nav Res Logist 53:37–150

16. IEA Clean Coal Centre (2010) http://www.iea-coal.org/site/ieacoal/hom Accessed 21 June 2010
17. Pochet Y, Wolsey LA (2006) Production planning by mixed integer programming. Springer Science + Business Media, Inc. New York
18. Pudney P, Wardrop A (2008) Generating train plans with problem space search. Lect Notes Econ Math Syst 600:195–207
19. Queyranne M, Schulz AS (1994) Polyhedral approaches to machine scheduling, Technical report 408/1994, Department of Mathematics, Technische Universität Berlin, Berlin, Germany
20. Singh G, Sier D, Ernst AT et al (2009) Long term capacity planning for the Hunter Valley coal chain: models and algorithms. 20th National ASOR Conference, Gold Coast, September 2009, pp 27–29
21. Steenken D, Voß S, Stahlbock R et al (2004) Container terminal operation and operations research: a classification and literature review. OR Spect 26(1):3–49
22. Vandervoort J (2009) At the coal face—ensuring the performance of the Australian coal export supply chain. SMART Conference, Sydney, June 2009, pp 10–11. http://www.hvccc.com.au/HVCCCDocuments/Pages/MiscellaneousPresentations.aspx
23. Vandervoort J (2009) NSW coal supply chain and logistics. 5th Coaltrans Conference, Brisbane, August 2009, pp 17–18. http://www.hvccc.com.au/HVCCCDocuments/Pages/MiscellaneousPresentations.aspx
24. Welgama PS, Oyston R (2003) Study of options to increase the throughput of the Hunter Valley coal chain. In: Proceedings of MODSIM 2003, Townsville, July 2003, pp 1841–1846

Chapter 11
Risk Assessment of Supply Chain During New Product Development: Applications in Discrete and Process Manufacturing Industries

Atanu Chaudhuri and Kashi N. Singh

11.1 Introduction

Supply chains have become increasingly global and complex over the years with manufacturers sourcing raw materials and components from geographically-dispersed suppliers. Reduced costs, access to capacity, focus on core activities, etc are some of the advantages associated with outsourcing different aspects of product development, manufacturing and logistics. However, managing a complex network of global suppliers and sub-contractors to ensure cost-effective, high-quality and timely deliveries has become a daunting task for practicing managers. Moreover, customers have become increasingly demanding and there is pressure on companies across industries to develop and launch a wide variety of products in shorter time. To manage a complex supply network, it is an imperative that supply chain decisions are made integral part of the new product planning. Any glitch in the global supply chain not only will raise costs related to the remedial actions which are required after the occurrence of the event but also can delay the launch of products with serious financial losses to the participating companies in the supply chain.

Firms appreciate the critical role played by suppliers in product development. Kamath and Liker [12] and Ward et al. [21] defined the role of suppliers based on their capabilities and the responsibilities being taken up by them during the different stages of product development. Hoult [10] showed that large savings in product development can be generated by ensuring early supplier integration into the customer companys' integrated product development teams. This integration can be achieved by allowing suppliers to define the product architecture and by maintaining database commonality

A. Chaudhuri (✉) · K. N. Singh
Indian Institute of Management, Lucknow, India
e-mail: atchaudhuri@iiml.ac.in

K. N. Singh
e-mail: kns@iiml.ac.in

H. Gurnani et al. (eds.), *Supply Chain Disruptions: Theory and Practice of Managing Risk*, DOI: 10.1007/978-0-85729-778-5_11,

with suppliers. Thomke and Fujimoto [19] showed that development performance can be greatly enhanced by identifying and solving problems in the early stages of product development. Similarly, to create a well-functioning materials supply system, it is important to integrate the materials supply aspects early-on in product development projects [11].

However, research on modeling and analysis of supplier risk is limited. The current literature on supply chain risk management has analyzed different types of risks in supply chains and dealt with risk mitigating measures but has not adequately addressed the specific supply chain risks arising during product development stage. Similarly, the literature on suppliers involvement during product development has outlined appropriate mechanisms to improve buyer–supplier co-ordination but has not clearly established the linkage of the co-ordination mechanisms to the risks involved. In this paper, we develop a process for assessing supply chain risks during new product development and provide a framework for assessing vulnerability in supply chains. The framework also helps identify and prioritize control mechanisms to mitigate supply-related failures during new product development (NPD). We apply our framework to an aerospace supply chain and a pharmaceutical example.

11.2 Literature Review

Zsidisin et al. [23] analyzed different risk assessment techniques using case studies and showed how those can be used to verify suppliers activity in ensuring goal congruence and in reducing outcome uncertainty. Sinha et al. [18] described a generic methodology for mitigating risks and outlined five steps - 'identify risks', 'assess risks', 'plan and implement solutions', 'conduct failure and mode effect analysis' and 'continuously improve'. The authors explained their methodology through a case study in the aerospace industry. The methodology presented in this paper, however, did not identify the risks in the supply chain that might impact new product development. The paper neither provided any guidance for assessing the vulnerabilities of different links of the supply chain nor proposed specific action plans to mitigate the risks.

Wu et al. [22] developed a methodology for hierarchical classification of risk factors in inbound supply and used analytic hierarchy process (AHP) to rank risk factors for suppliers. Lee [13] used the concept of benefits, opportunities, cost and risk to arrive at a performance ranking of buyer–supplier forms by using a fuzzy AHP model. Lee's model provides a useful approach for selecting suppliers and the forms of relationships with them. We consider a situation where suppliers have been selected and the objective is to understand the overall vulnerability in the supply chain and perform a risk assessment to develop appropriate control plans during new product development. The above literature focuses on different applications of risk assessment methodologies but does not provide guidance on creating specific plans to mitigate those risks. It also does not discuss specific supply chain risks related to new product development.

There is another stream of literature dealing with suppliers involvement in new product development. Clark [5] found that suppliers involvement and strong supplier relationship accounted for about one-third of man-hour advantage and contributed to four to five months of lead time advantage for the Japanese automotive firms. A strong network of suppliers also allowed Japanese firms to have more unique parts in their designs, thus leading to improved product performance. Hartley [9] confirmed that on-time completion of suppliers' activities was significantly related to buyers overall project time performance. Moreover, working with technically competent suppliers reduced supplier-related product development delays.

The literature also identifies some best practices while collaborating with suppliers during different stages of product development. Ragatz [16] identified success factors for integrating suppliers into new product development and found supplier membership on the project team as the greatest differentiator between most and least successful integration efforts. The authors concluded that overcoming barriers to share proprietary information by having a formal trust development process and joint agreement on performance play important roles in integrating suppliers into new product development. "Austin [1] reported how companies in the personal computer industry were engaged in extensive collaboration efforts with suppliers to reduce the risk of suppliers not being able to ramp up fast enough in the product introduction phase. Gaudenzi [7] described how an aerospace, defense and security company closely cooperated with suppliers and external providers. Bozdogan et al. [3] showed through two case studies how architectural innovation can be generated by proactively integrating suppliers at concept exploration and definition stages of the product development [4]".

The review of the literature shows that there is a scope for research in developing a methodology for thorough risk assessment of supply chains during new product development and to develop coordination mechanisms with suppliers based on the identified risks. The existing literature does not specifically identify the supply chain risks during new product development. We try to bridge this gap by devising a step-by-step approach for vulnerability assessment of subsystems and suppliers during new product development using AHP, and develop a control plan to engage with suppliers using failure mode effect analysis (FMEA).

11.3 Understanding the Vulnerability of the Supply Chain During New Product Development

Identifying the weakest link in the supply chain during the early stages of product development can help firms meet the desired objective of launching the product on time while simultaneously meeting its performance and cost targets. Supply chain professionals can spend a lot of time in planning and assessing supply risk, if they do not prioritize their efforts by identifying the distinguishing characteristics [8]. This leads us to the question as to how can companies develop a better understanding of their supply chains and understand where the vulnerabilities lie during the early stages of product development.

"Project scope, i.e., the extent to which a new product is based on unique parts developed in-house has a significant impact on the engineering man-hours and lead time of product development projects [5]. In [5], a measure called 'new-in-house ratio', the ratio of total work done (man-hours) in-house by the project team to total engineering work (man-hours) is used. Uniqueness of the product and degree of suppliers' involvement can be major drivers of uncertainty towards the development of the product [16]. Novelty of the product can be due to its design, use of new materials for construction or due to its requirement for a new manufacturing process.

Degree of suppliers' involvement include prior experience of suppliers, degree of design customization carried out by the suppliers, impact of the suppliers' engineering metrics on product performance, new materials and new manufacturing processes used and also how a change in one component developed or manufactured by the suppliers impact others. Martin and Ishii [14] developed the coupling index-supply and coupling index-receiving to indicate the strength of coupling between components. These indices use a rating system for sensitivity of engineering metrics and measure how a change in one will affect the other. The stronger the coupling between components, the more likely a change in one will require a change in the other [4]".

For process industry, suppliers' involvement can be captured by parameters such as prior experience of suppliers, impact of suppliers process parameters on product quality and yield, etc.

As the complexity of the manufacturing process increases, failure to control such processes will have a negative impact on the product performance and cost, and may also result in a delayed product launch. A lengthy manufacturing process or dependence of the product performance on multiple process parameters and their interaction effects can contribute to that complexity. For process industry, complexity can be captured by loss of production due to poor quality of input. The later, the impact of poor quality input is observed, more difficult it is to ascertain the root cause. Also, in such cases the entire batch of material had to be scrapped which significantly increases costs, as well as, delivery time.

Logistical complexities also add to the vulnerability of a supply chain. Components with special logistical requirements create complexities in storage and transportation. Logistical delays in terms of packing, unpacking or transportation can lead to delays in the assembly schedule. For process industry, many raw materials have specific safety and handling requirements which have to be followed because of their hazardous nature. Dowlatshahi [6] showed the importance of involving logistics in the early phases of product design and development in a concurrent engineering environment.

Blackhurst et al. [2] observed that the lack of ability to measure capacity at different nodes in the supply chain (i.e., manufacturing facilities) and lack of capacity flexibility were areas of concern. Hence, lack of capacity flexibility can be considered to be a source of uncertainty in the supply chain. A supply chain with capacity that can be shared across suppliers is less vulnerable than the one in which there is a capacity dependence on a single supplier. Time to ramp up capacity is another variable which can determine the vulnerability of a supply chain.

In summary, the product and process characteristics that can be used for comparing different subsystems for assessing the vulnerability of different links in supply chains are degree of supplier involvement, process complexity, logistical complexity and manufacturing capacity.

11.4 Supply Chain Risk Assessment During NPD: Two Case Studies

We provide two case study examples one from the discrete manufacturing industry (an aircraft manufacturer) and another from the process manufacturing industry (a generic drugs manufacturer) to illustrate how thorough supply chain risk assessment during new product development can help companies to develop appropriate risk mitigating plans and hence avoid supply chain failures during or after the product launch.

Case Study A

A defense aircraft manufacturer faced delays in its combat aircraft programs. This company decided to develop a process of supply risk assessment. One of the authors who consulted the organization to create a process for analyzing vulnerability in its supply chain during product development to develop control plans and co-ordination mechanisms. Details of the aircraft manufacturer and its suppliers are masked for the sake of confidentiality. A part of the entire exercise is illustrated in this paper.

"Four subsystems of an aircraft- center fuselage, wing, empennage and forward fuselage were considered for supply chain risk assessment. For each of the subsystems, some parts of the design and manufacturing were outsourced to individual component manufacturers and subsystem manufacturers [4]".

11.4.1 Methodology Used for Risk Assessment

The vulnerability of each sub-system depends on four parameters - degree of supplier involvement, process complexity, logistical complexity and manufacturing capacity. Each of these parameters depends on multiple sub-parameters (Fig. 11.1). For example, degree of design customization by suppliers is one of the sub-parameters which determine the degree of supplier involvement. Similarly, degree of difference between prototype and finished product is one sub-parameter which impacts process complexity. 'New' suppliers are those which have designed or produced similar parts before but are contracting with the original equipment manufacturer (the aircraft manufacturer in our example) for the first time or those who have expertize in similar design or manufacturing process but have not produced or designed the parts, contracted by the producer. Using the above parameters, the different subsystems of the product were compared using AHP and a vulnerability score was calculated. Then for each subsystem, vulnerability of individual suppliers was determined using similar

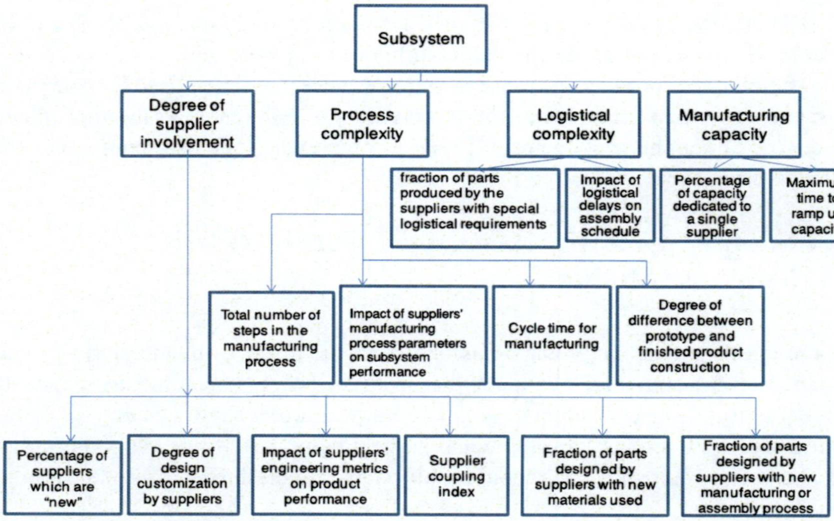

Fig. 11.1 Hierarchy of parameters for vulnerability assessment of the supply chain during new product development

metrics. A detailed failure mode effect analysis was then performed for the prioritized sub-system with a purpose to develop specific control plans and co-ordination mechanisms between suppliers and the original equipment manufacturers.

AHP introduced in 1971 by Saaty has become one of the most widely used methods for multi-criteria decision-making with a wide variety of applications across industries [20]. Figure 11.1 shows the hierarchy of parameters used for vulnerability assessment.

"The consistency ratio of AHP (CR) reflects the consistency of the pair-wise judgments. For example, judgments should be transitive in the sense that if A is considered more important than B, and B more important than C, then A should be more important than C. If, however, the user rates A as important as C, then the comparison is inconsistent and the user should revisit the assessment. Saaty [17] explains that CR is calculated by using the Eq. 11.1, where x stands for the maximum eigenvalue of the pair-wise matrix, and n is the size of the pair-wise matrix, RI is the random index value recommended by Saaty. AHP has some tolerance for inconsistency but comparisons made using a consistency ratio that exceeds 0.1 should be reconsidered.

$$\mathrm{CR} = \frac{x - n}{(n - 1)\mathrm{RI}} \tag{11.1}$$

We used a scale of 1, 3, 5, 7 and 9 for the pair-wise comparison of importance of degree of supplier involvement, process complexity, logistical complexity and manufacturing capacity. The parameters for which we use rating scales to compare the subsystems using AHP are degree of customization, impact of suppliers' engineer-

Table 11.1 Scale for degree of customization

Off the shelf design	1
Minor modification	3
Customization of an existing design	5
Customization of an existing design along with a new material of construction	7
Entirely new subsystem	9

ing metrics on product performance, impact of suppliers' manufacturing process parameters on subsystem performance, degree of difference between prototype and finished product construction and impact of logistical delays on assembly schedule. For degree of customization, the scale shown in Table 11.1 was used and for other parameters a 1, 3, 5, 7 and 9 scale with '1' being 'very low' and '9' being 'very high' was used. For all other parameters like cycle time for manufacturing, percentage of suppliers, which are 'new', actual raw data was used for comparison. [4]"

Our objective was to create a robust risk assessment process and create co-ordination mechanisms based on the prioritized risks. The importance weights for the parameters can vary during the course of the product development program. Hence, we generated four different scenarios for determining importance weights of degree of supplier involvement, process complexity, logistical complexity and manufacturing capacity. The scenarios were generated by varying the relative importance of the above parameters using the scale 1, 3, 5, 7 and 9. The scenarios represent probable development paths according to experts within the organization. For each scenario, a pair-wise comparison matrix was created. A team of experts from research and development, purchase and manufacturing functions of the organization along with the project leader and one of the author as facilitators completed the pair-wise comparison matrices for importance of each parameter using the suitable pre-defined scales and raw data, as applicable. The team first decided that their objective was to generate scenarios—one with equal weights for all the parameters, one in which degree of supplier involvement was most important, followed by process complexity, one in which manufacturing capacity was most important and the other in which degree of supplier improvement was most important, followed by logistical complexity. Accordingly, the team completed the pair-wise comparison matrix for each scenario.

Then using the Analytical Hierarchy Process (AHP), the eigenvector was solved for the matrix using the following steps:

Step 1: The pair-wise matrix was raised to powers that were squared each time;
Step 2: The row sums were then calculated and normalized;
Step 3: The computation in step 2 was continued until the difference between the sums in two consecutive calculations is smaller than a prescribed value.

This procedure was repeated with different pair-wise comparison matrix for each scenario. This procedure finally resulted in different weights of parameters for each scenario as shown in Table 11.2.

Table 11.2 Importance weights of parameters under different scenarios

Parameters	Weights			
	Scenario 1	Scenario 2	Scenario 3	Scenario 4
Degree of supplier involvement	0.57	0.25	0.60	0.25
Process complexity	0.24	0.09	0.16	0.25
Logistical complexity	0.07	0.04	0.18	0.25
Manufacturing capacity	0.13	0.61	0.06	0.25

Table 11.3 Scores of the subsystems under different scenarios

Subsystems	Scores			
	Scenario 1	Scenario 2	Scenario 3	Scenario 4
Center fuselage	1.489	0.983	1.456	1.128
Wing	1.391	0.843	1.402	1.015
Empennage	1.11	0.739	1.117	0.767
Forward fuselage	0.758	0.63	0.754	0.589

We then used AHP to rank the different subsystems under the different scenarios. The same team which conducted the pair-wise comparison of each parameter also conducted the pair-wise comparison of the subsystems for each parameter. Since, the team knew the parameters on which the subsystems were to be compared, each team member brought necessary supporting documents, available with them. For example, the R&D members provided supporting data and reasoning to justify the pair-wise comparison ratings for each sub-system on impact of suppliers' design parameters on product performance. The team member from production provided data on cycle time for manufacturing, number of steps in manufacturing process and provided justification about the degree of difference between prototype and finished production. The purchase and R&D members together calculated the supplier-coupling index based on their knowledge of specification flows between suppliers. Following this process, center fuselage and wing emerged as the two most vulnerable subsystem for all scenarios. Though the vulnerability ranks of the different subsystems did not change with the scenarios, the scenarios did make a difference in determining the vulnerability of the suppliers. The scores of the four subsystems under the scenarios are shown in Table 11.3.

Consistency ratios for the four scenarios are calculated as 0.066, 0.077, 0.021 and 0, respectively. As all the ratios are less than 0.1, we can assume that the results are consistent. Once the vulnerability of the sub-systems was assessed and subsystems were ranked, a list of critical suppliers were compiled for each subsystem. The individual suppliers for a subsystem were compared and ranked using similar parameters to identify the suppliers which may be the potential sources of risk. We used the same parameters to compare suppliers as that used for the subsystems except that we did not use the percentage of suppliers which are 'new'. We replaced the above parameter with a new one termed as experience of supplier. We used the scale shown in Table 11.4 for that purpose.

The company decided to rank the suppliers for each of the subsystems but since center fuselage was identified as the most critical subsystem under all scenarios,

Table 11.4 Scale for experience of supplier

Existing supplier with experience of design and or manufacturing for same or similar components	1
New supplier with experience of design and or manufacturing of same or similar components	3
Existing supplier with responsibility for design and or manufacturing of components it has not worked on before	5
New supplier with experience of same or similar manufacturing process but not for contracted components or assembly	7
New supplier new to the business but with requisite investments	9

Table 11.5 Rank of the suppliers under different scenarios for center fuselage

Subsystems	Scores			
	Scenario 1	Scenario 2	Scenario 3	Scenario 4
Supplier A	1.037	0.749	1.018	0.725
Supplier B	1.207	0.944	1.167	0.986
Supplier C	1.327	0.792	1.337	0.975
Supplier D	1.179	0.713	1.209	0.813

we illustrate our methodology only for the center fuselage. The company ranked the four suppliers for the center fuselage subsystem under four different scenarios using the AHP methodology, as outlined above. The results are shown in Table 11.5. The results show that supplier C was the most vulnerable supplier in scenarios 1 and 3 while supplier B was the most vulnerable supplier in scenarios 2 and 4. Thus, when degree of supplier involvement was the dominant parameter as in scenarios 1 and 3, supplier C turned out to be the most vulnerable one but when manufacturing capacity became more important in scenario 2, supplier B emerged as the most vulnerable one. When weights of all the parameters were equal, supplier B was the most critical closely followed by supplier C. This shows how vulnerability of suppliers depended on the scenarios. This information helped in creating specific control plans for each supplier using FMEA.

FMEA was conducted for all the suppliers for center fuselage and wing. As an illustration we show the results for suppliers B and C for center fuselage and identify the high risk priority number (RPN) areas considering the different scenarios. We use 1, 3, 5, 7 and 9 scales for Severity, Occurrence and Detection required to conduct the FMEA and to determine RPN which is product of the above three. The scales are shown in Tables 11.6, 11.7 and 11.8 respectively. For suppliers B and C the failure modes with the top five RPNs are shown in Tables 11.9 and 11.10, respectively.

The above FMEA shows that for supplier B, lack of understanding of other suppliers design parameters, interaction between design parameters and materials of

Table 11.6 Severity scale for FMEA

Very low	No or negligible effect	1
Low	Minor effect on product performance, project cost or time	3
Moderate	Moderate effect on product performance, project cost and time which will require control actions	5
High	High impact on product performance, project cost and time	7
Very high	Severe adverse impact on product performance, project cost and time	9

Table 11.7 Occurrence scale for FMEA

Remote	Failure is unlikely	1
Low	Low chances of failure	3
Moderate	Chances of occasional failure	5
High	Chances of frequent failures	7
Very high	Very high chances of failure	9

Table 11.8 Detection scale for FMEA

Very high	Likelihood that current controls, processes will surely detect and prevent the potential cause of failure	1
High	High chances of the current controls detecting and preventing the failure from taking place	3
Moderate	Current controls can only detect the failure but cannot prevent it from happening	5
Low	Current controls can detect the failure only before the final stages in the project	7
Very low	Current controls can detect the failure only at an advanced stage in the project	9

Table 11.9 Failure modes with high RPN for supplier B of center fuselage

Failure Mode	Severity	Ocurrence	Detection	RPN
Interaction between design parameters and materials of construction not properly understood	9	9	7	567
Impact of other suppliers' design parameters not properly understood	9	7	9	567
Materials of construction not proven under actual flight conditions and/or commercial manufacture	9	9	5	405
Specification limits/tolerances not appropriately set	9	5	7	315
Suppliers not able to ramp up capacity	5	9	7	315

construction, testing of materials of construction under actual operating conditions and commercial manufacturing are some of the key sources of risk. For supplier C, the critical failure modes are shortage of material for supplier, interaction between

Table 11.10 Failure modes with high RPN for supplier C of center fuselage

Failure Mode	Severity	Ocurrence	Detection	RPN
Shortage of materials for suppliers	7	9	7	441
Interaction between design parameters and materials of construction not properly understood	9	7	7	441
Inability to replicate prototype manufacturing process for final product	5	9	9	405
Suppliers not able to ramp up capacity	9	5	7	315
Manufacturing process technology not proven	7	9	5	315
Impact of other suppliers' design parameters not properly understood	9	5	7	315
Materials of construction not proven under actual flight conditions and/or commercial manufacture	9	7	5	315

design parameters and materials of construction not properly understood, and inability to replicate prototype manufacturing process for final product.

Based on the FMEA, the team at the defense company took some key actions, mentioned below, as part of the control plan for minimizing supply chain failures related to design, manufacturing process, capacity and sourcing during product development.

Step 1: A core team of designers and subject matter experts from the customer and the key suppliers was formed.

Step 2: The team decided to meet regularly at some given time intervals, and also when any early signal for design-related issues came up.

Step 3: The engineering metrics of each supplier and their potential impact on other suppliers were assessed upfront during the kick-off meeting of the core team. The team created a chart showing the specification flows across components and develops component-level coupling indices to understand the integration issues. A layout of the area surrounding the supplier's component system was also shared with them to develop a better understanding of how their parts fit with the adjoining parts similar to what Toyota practices with its suppliers [12].

Step 4: Subject matter experts from the customer and the concerned suppliers conducted necessary tests and simulations of the 'new' materials of construction on a prioritized basis. The team followed robust design principles to allow for variations in metal properties.

Step 5: Specific members of the team were assigned to identify deviations in process parameters and their root causes, conduct design of experiments and simulations to understand interaction effects of process parameters and improve the process to optimize performance. Wherever possible, the members worked on a virtual environment but if required, members from the suppliers' team were stationed at the customers facilities for some period of time.

A detailed project plan was created and a team of senior members of company A and its suppliers' groups conducted regular reviews at each milestone. Once the design validation was done, inputs from the design core team were used to proactively identify and document differences which were likely to occur in the commercial production. This ensured that teams of technology or process transfer were created by the suppliers for smooth transition and problem solving between prototype and final production. At the same time, the defence aircraft manufacturer analyzed the suppliers' financial health and worked with suppliers to plan for capital outlay. A team of manufacturing process engineers from the suppliers' and customer's sides along with some external experts provided technical guidance in ramping up production. All the above activities were included as part of the project plan and discussed during the milestone reviews.

The above discussion shows how a thorough understanding of the potential vulnerabilities in the supply chain followed by the risk assessment using FMEA can help create specific control plans and reduce the potential supply-related risks during product development.

Case Study B

A pharmaceutical company (company name concealed for confidentiality) developed a generic version of a blockbuster drug to reduce hypertension. It developed alternate routes of synthesis and successfully filed a Drug Master File with US Food and Drug Administration to obtain rights to sell the generic version of the drug. The product was successfully transitioned from a laboratory-scale production to mass-scale production in the plant. Within a few months after the mass scale production started, the company started facing variation in yield in the finished product. This indicated that less amount of finished product was obtained from the specified input quantities as per the route of synthesis developed by the company. As a result of this yield loss, not only did the cost of manufacturing increased, on some instances, the company was not able to ship the required quantity of finished product on time.

To understand the root causes of the variation of yield, the company formed a cross-functional team along with one of the authors as a consultant. Based on the understanding of the chemical process using a SIPOC model (Supplier, Input, Process, Output, Customer), and a root cause analysis, the team zeroed on a few critical steps in the process which can contribute to lowering of yield. A particular step of the process used a raw material from two suppliers. Both these suppliers had supplied their material during laboratory testing, as well as, during the scale-up in the manufacturing plant. The material supplied by both these suppliers met all quality criteria and hence were selected as "regular" supplies. But the particular raw material has an important effect in ensuring quality and yield of an intermediate stage of the product, which in turn affects the yield of the final product. Hence, the team decided to study the yield variation of the product with different batches of raw material supplied by the suppliers. Analysis revealed that for each supplier, yield improved when purity levels of the raw material supplied increased. Yields were low if the purity levels are closer to the lower specification limit. It was also observed that yield with raw material supplied by one supplier 'X' exceeded that with raw material supplied by 'Y' even though purity levels of 'X' were lower than that of 'Y'.

Similar results were obtained while conducting tests in the laboratory, as well as, from production batches over a longer period of time. This led to the conclusion that though purity levels were important, there was something else in the process which resulted in lower yield. Study of the route of synthesis of raw material supplied by 'Y' revealed likely formation of two other compounds 'R' and 'S'. 'S' turned out to be an optical isomer of the desired raw material which potentially impacts the yield of the product. As testing methods and facilities were not available with the customer company as well with supplier 'Y' for compound 'S,' its impact on yield was not well understood before, and the problem remained unnoticed until the yield variations were observed.

This analysis shows how failure to understand the impact of supplied raw material during product development stage can lead to serious consequences later. One typical approach which is used is to make the design robust so that variation in quality due to variation in input quality can be avoided, but, for chemical reactions it may not always be possible. In such cases it is important to analyze the impact of each step in the process on the yield and quality of the final product and analyze all possible impurities or additional compounds which can be developed during the process. During laboratory development, when the companies try to maximize yield they spend considerable effort in removing impurities and characterizing them with suitable testing methods. But potential impurities are usually not analyzed. For a raw material supplied by an external supplier, it is important to understand its route of synthesis and manufacturing process. On most instances, if the supplier is able to meet quality requirements as specified and is able to deliver adequate quantities as desired, detail analysis of supplier's route of synthesis is not done. Also companies fail to inspect suppliers' plants and processes and rely only on incoming supply inspection. Mayer et al. [15] showed through an application in the biotechnology industry how supply and plant inspections can be complementary when the internal spillover costs (i.e., costs which are incurred when the buyers' production process have to shut down due to failures or incur costs due to switching capacity to another product or would render work-in-progress valueless due to poor quality of input) and inventory costs are sufficiently high. In industries like chemicals and pharmaceuticals, such costs are typically high. For such industries, external spillover costs will also be high with severe reputation loss with customers, as well as, with regulatory agencies. The case study also underscored the fact that plant and supply inspections should be complementary for process industries.

If risk assessment has to be done during product development stage, the following approach can be followed:

1. Create a SIPOC chart and identify the critical to quality characteristics of each stage of the process
2. Identify the most vulnerable stage of the process based on the above analysis; AHP can be used for this purpose
3. Identify the most critical raw material for the vulnerable stages
4. If a critical raw material is outsourced to a supplier, analyze the route of synthesis followed by the suppliers for potential impurities

5. Conduct inspection of supplier's plants and acceptance sampling of suppliers' batches.

11.5 Managerial Implications and Scope for Future Work

Complex product development projects face several risks which get magnified due to involvement of multiple suppliers during different stages of development. Many organizations may fail to identify the potential vulnerabilities and to conduct thorough assessment of supply-related risks during new product development. Our approach shows how organizations can assess the vulnerabilities in supply chains by first identifying the subsystems and then suppliers who are more prone to risks. By generating multiple scenarios with different weights of parameters, we allow the practicing managers the flexibility to test the vulnerability of the subsystems and the suppliers under different conditions. This also allows organizations to understand which suppliers might need closer monitoring if, for instance, capacity flexibility becomes more important than the degree of supplier involvement. For a large number of parameters, subsystems and suppliers, a methodology using fuzzy numbers may be less time consuming and hence the preferred approach. Once the vulnerability analysis is conducted, FMEA for the suppliers of individual subsystems can be conducted. This will enable organizations to prioritize the failure modes and to develop specific control plans for them. We have suggested how suppliers can be involved to ensure that the control plans are adhered to by incorporating them in the overall product development plan.

For products like chemicals and pharmaceuticals, the most critical stages of reaction and vulnerable suppliers can be identified by AHP. Then specific control plans can be created based on FMEA for the suppliers for outsourced materials or for internal processes if the critical stages of manufacturing are conducted in-house.

One limitation of our approach is that it requires significant involvement of the customers and suppliers. Companies may like to outsource design and manufacturing activities to suppliers to reduce their involvement, bring down the costs and to take advantage of suppliers' capabilities. However, many complex product development projects will require high involvement of the customer and the suppliers. To ensure that risks are mitigated and the project objectives are met, a process needs to be created to achieve a balance between involvement and delegation of responsibilities to the suppliers. Creating vulnerability indices and empirically determining the vulnerability of supply chains during new product development for different industries can be another potential extension of this work.

Acknowledgments An earlier version of this paper was published in the proceedings of the PDMA India IV Annual International Conference, NPDC 09 held at Indian Institute of Technology, Madras between December 17 and 19, 2009. The authors thank the conference participants and reviewers for their valuable comments.

References

1. Austin AT (1998) Unlocking the supply chain's hidden value: a lesson from the personal computer industry. Supply Chain Manag Rev 2:24
2. Blackhurst J, Craighead CW, Elkins D et al (2005) An empirically derived agenda for critical research issues for managing supply-chain disruptions. Int J Prod Res 43(19):4067–4081
3. Bozdogan K, Deyst J, Hoult D et al (1998) Architectural innovation in product development through early supplier integration. R&D Manag 28(3):163–173
4. Chaudhuri A, Singh KN (2009) Vulnerability analysis and risk assessment during new product development: an integrated approach using AHP and FMEA. In: Proceedings of PDMA India IV international conference, NPDC, new product development: challenges in meltdown times, IIT Madras, India, 17–19 December
5. Clark KB (1989) Project scope and project performance: the effect of parts strategy and supplier involvement on product development. Manag Sci 35(10):1247–1263
6. Dowlatshahi S (1999) A modeling approach to logistics in concurrent engineering. Eur J Oper Res 115(1):59–76
7. Gaudenzi B (2008) Assessing risks in projects and processes. In: Zsidisin GA, Ritchie B (eds) Supply chain risk—a handbook of assessment, management and performance. Springer, New York, pp 67–81
8. Giunipero LC, Eltantawy RA (2004) Securing the upstream supply chain: a risk management approach. Int J Phys Distribution Logist Manag 34(9):698–713
9. Hartley JL, Zirger BJ, Kamath RR (1997) Managing the buyer–supplier interface for on-time performance in product development. J Oper Manag 15:57–70
10. Hoult DP (1997) Product development with suppliers—the role of database commonality. Society of automotive engineers (SAE) technical paper series, #970776, sp-1266, 7
11. Johansson E, Bellgran M, Johansson MI (2006) Evaluation of materials supply systems during product development projects. Int J Prod Res 44(5):903–917
12. Kamath R, Liker JK (1994) A second look at Japanese product development. Harv Bus Rev 72(6): 154–160 (November–December)
13. Lee AH (2009) A fuzzy AHP evaluation model for buyer–supplier relationships with the consideration of benefits, opportunities, costs and risks. Int J Prod Res 47(15):4255–4280
14. Martin MV, Ishii K (2000) Design for variety: a methodology for developing product platform architectures. ASME design engineering technical conferences, Baltimore, ASMA
15. Mayer KJ, Nickerson JA, Owan H (2004) Are supply and plant inspections complements or substitutes? a strategic and operational assessment of inspection practices in biotechnology. Manag Sci 50(8):1064–1081
16. Ragatz GL, Handfield RB, Scannell TB (1997) Success factors for integrating suppliers into new product development. J Prod Innov Manag 14:190–202
17. Saaty TL (1980) The analytic hierarchy process. McGraw-Hill, New York
18. Sinha PR, Whitman LE, Malzahn D (2004) Methodology to mitigate supplier risk in an aerospace supply chain. Suppl Chain Manag: Int J 9(2):154–168
19. Thomke S, Fujimoto T (2000) The effect of "front-loading" problem-solving on product development performance. J Prod Innov Manag 17:128–142
20. Vaidya OS, Kumar S (2006) Analytic hierarchy process: an overview of applications. Eur J Oper Res 169(1):1–29
21. Ward A, Liker JK, Cristiano JJ et al (1995) The second Toyota paradox: how delaying decisions can make cars faster. Sloan Manag Rev 36(3):43–61
22. Wu T, Blackhurst J, Chidambaram V (2006) A model for inbound supply risk analysis. Comput Industry 57:350–365
23. Zsidisin GA, Ellram LM, Carter JR et al (2004) An analysis of supply risk assessment techniques. Int J Phys Distribution Logist Manag 34(5):397–413

Chapter 12
Supply Chain Risk Management

Gary S. Lynch

The next time you have a bucket of water, try to release a single drop of water into the bucket without generating a ripple. You will notice how ripples immediately oscillate back and forth for quite some time. The ripples reach all the way to the sides of the bucket and bounce back, resulting in an infinite number of waves. The bucket represents the world of global trade; the water an infinite number of supply networks that support the movement of materials, products, cash, and information. An event the "drop" at a vulnerable point of the supply chain such as a catastrophic failure at the ports in Singapore (through which more than 50% of the world's goods is shipped), could initiate an economic tidal wave. Customers rarely understand, nor are they interested in, how their products are produced or get to the market. What is most important is that the product is available when and how they want it, will not harm them, and that its value lives up to the expectations promised. For customers, ignorance is bliss—for organizations, ignorance can be devastating.

A single drop of water represents not only a change to the environment but an opportunity to severely disrupt the norm. Any change, regardless of its scale, carries with it the potential for disruptive and/or systemic risk. Whether the change is anticipated: the migration to a new Enterprise Resource Planning platform; or unanticipated: a pandemic or volcano, the organization typically has the ability and experience to manage the risk brought about by change. However, fragile supply chains have evolved as they become more geographically disbursed, technologically advanced, and shared by a larger number of stakeholders with differing expectations and competing interests. Add the economic pressure of tight margins i.e. keeping inventory lean and consolidating suppliers/distribution centers, and you have the potential for single points of failure (SPOF) and systemic collapse.

Take for example the anticipated change in May 2009, when Atomic Energy of Canada Ltd. shut down of the NRU reactor in Chalk River, Ontario. The company

G. S. Lynch (✉)
Risk Intelligence and Supply Chain Resiliency,
Marsh, New York, NY, USA
e-mail: gary.lynch@marsh.com

H. Gurnani et al. (eds.), *Supply Chain Disruptions: Theory and Practice of Managing Risk*, DOI: 10.1007/978-0-85729-778-5_12,

scheduled a reactor shutdown for maintenance that was only supposed to take five days. The medical diagnostic isotopes produced at this facility represent approximately 50 % of global production and are a critical element in the North American healthcare supply chain. It is the *only* source of base isotope for technetium-99.[1] They are used for diagnosing and treating heart conditions and certain types of cancer, and are injected into patients in the United States 20 million times a year. Atomic Energy of Canada provides North America with it. What was supposed to be a routine shutdown resulted in an unforeseen 60-day delay. The extra 55 days deferred delivery of technetium-99 sending ripples throughout hospital, imaging services, and healthcare organizations throughout North America. Technetium-99 was a SPOF.

A SPOF is invariably singular in physical (one reactor) or virtual location but when realized, can trigger systemic effects across today's globally interdependent, integrated, and highly-synchronized supply ecosystem. The repercussions extend far beyond the individual organization, in many instances to unsuspecting stakeholders along the chain such as an organization within that industry or the economic and social well-being of an entire country.

More importantly, these SPOFs and their subsequent potential for systemic and catastrophic failure are not limited to a single function, process, resource, location, company, market, and/or geography.

For example, in July of 2007 the event was a 6.8 magnitude earthquake in central Japan. The SPOF was a Riken Corp.'s manufacturing plant in Niigata Prefecture, where the machinery had been displaced by the quake. The industry's SPOF was the inability to procure a single discrete low-value part, a $1.50 piston ring used by nearly half of Japan's automobile supply chain. Lean supply chains and just-in-time manufacturing all but eliminated the inventory that used to be held as a buffer. Although the factory was brought back on-line in a week replenishment took much longer and, the impact was significant: delayed production of more that 55,000 of Toyota Motor Corporation's vehicles and a slowdown or shutdown of approximately 70% of Japan's auto production assembly lines at Honda, Nissan, Mitsubishi, Mazda, Suzuki, and Fuji Heavy industries. Fortunately the outage was seven days but what if it were seven weeks?

The primary goal of supply chain risk management is to uncover, prioritize, measure, treat (mitigate and finance), and monitor the risk of these SPOFs; and to diminish the impact of an event through comprehensive and efficient resiliency practices. This chapter will focus on how to recognize and manage threats, identify and mitigate vulnerabilities, and establish resiliency throughout the global supply chains.

12.1 Kidney Failure and Supply Chain Risk Management: A SPOF Case Study

A drop in the bucket creates global waves? Nearly 300,000 people, mostly children, sick from acute kidney failure—the result of chemicals added to raw milk and animal

[1] International Atomic Energy Agency http://IAEA.org

feed (the "drop"). At first glance, this appeared to be a senseless and malicious incident. But upon further review, the evident cause and intended or unintended consequences were deeper and more complex. The Magnitude and duration of the ripple caused by the drop was something that the food industry had not previously experienced. It represents the daily challenge that global business managers face when trying to manage risk to their complex supply chains. It also represents how one drop or event can lead to a systemic "meltdown" can occur as a result of a SPOF.

This happened in the second half of 2008, when kidney-related illness skyrocketed amongst children in china. Investigations quickly determined that the most likely cause was the addition of melamine, an industrial chemical, to milk supplies in China. When it was over 21 companies in the supply chain were implicated and found guilty of involvement in this tragic series of events. The ripple effect to globally interdependent supply chains and the businesses they supported was immediate. Many countries banned all imports of Chinese dairy products, and Chinese authorities seized 2,176 tons of milk powder in a warehouse owned by San Lu, the primary culprit. More than 9,000 additional tons were recalled. Two million Chinese farmers were unable to sell their dairy products as demand ceased. Fonterra, one of the leading dairy companies in the world and a 43% owner of the San Lu group joint venture, wrote off its 2005 initial investment, estimated at $107 million. Fonterra's later impairment charges of $139 million included product recall costs, liability claims, and impairment of the San Lu brand. On September 27, 2008, San Lu declared bankruptcy.

The ramifications of the unaddressed risk included replacement by suppliers and distributors of Mengniu-Arla, a joint venture of two cooperatives (one each in Denmark and Sweden), which halted production as soon as the poisoning became known. Mengniu was a major milk supplier to Starbucks, which replaced the company with Viatsoy. Also affected were Kentucky Fried Chicken, the Hong Kong-based Lotte Group, Unilever (which recalled its milk tea powder used in Lipton products), Wellcome, Park'n Shop, Heinz baby cereals, and poultry feed stocks in France (where 1,200 tons of poultry feed had to be destroyed). Vietnam recalled and returned 26 dairy products imported from China; and Cadbury withdrew its 11 chocolate products manufactured in Beijing factories. More than 3,600 tons of tainted eggs and animal feed products were found in Hong Kong in November, 2008 and had to be destroyed. The systemic effects did not stop there. Further investigation revealed trace amounts of melamine in other food products including cheesecake, cookies, coffee products, and sweets/candy.

The San Lu case clearly illustrates the impetus for preemptive and proactive risk management initiatives to minimize SPOFs and prevent systemic failure. Most catastrophic supply chains events can be traced back to a SPOF and their impact minimized, or eliminated, if better understood. The exposure that laydormant in the SPOF in the melamine case was triggered by "change"; the decision by Fonterra (the New Zealand based global dairy giant) to enter into a business venture with the somewhat "unknown" Chinese-based Sanlu Corporation. Fonterra entered into business with Sanlu and assumed that its upstream suppliers (farmers, farm, and their cows) were exercising the same level of care that Fonterra demanded of all its suppliers. However, a lack of good hygiene, health, and nutritional practices led to poor product quality. The protein content testing consistently fell below western government standards. As a result, two million farmers and their families were at risk from losing their only source of income. They were motivated to resolve the issue and through the influence of greater "opportunists", melamine was added to the dairy product and the systemic effects were felt around the world.

Change brought about the opportunity to expand into new and emerging markets, but also introduced new uncertainty and the unknown. A drop in the bucket (business change), ripples appear (operations and technical changes needed to support business change) supply chains deployed or altered, and risk is unleashed.

The tragedy within the melamine story goes beyond the cost in lives and shareholder value. It is not a unique problem. SPOFs like this unfold every day whether caused by newsworthy headlines such as a volcano in Iceland, contaminated shipping pallets that taint consumer products, pirates in Somalia hijacking cargo vessels, nationalization of companies in Venezuela, an earthquake in Chile, or political unrest in Thailand. In virtually all SPOF instances, the problem is caused by a variety of different disastrous decisions, events, flaws, incidents, and outcomes from many causes. The failure may also be caused by more unpredictable and discrete events that are obvious at first analysis, such as faulty software that controls a mechanical device (e.g., acceleration and braking systems in an automobile), a political change that can ban a product from being distributed or a small components supplier going bankrupt. It is partly tragic because proper risk-related systems could detect and possibly prevent negative consequences caused by these and other events. Over my thirty-year career I have learned that in nearly all cases, the SPOF was identified in advance and there was at least one near or actual incident that preceded the systemic and/or catastrophic failure.

At a minimum, a greater risk consciousness at all levels to the warning signs and leveraging of corporate memory (to prior events and effects) could minimize the impact of adverse event.

12.2 Where to Begin?

Let's start by reviewing the basic risk terminology and elements of supply chain risk measurement and management. They are: threats, vulnerabilities, likelihood, impact, investment, and risk mitigation and risk financing.

- Threats are events that have the potential to cause the organization harm.
- Vulnerabilities represent weaknesses or those points in the supply chain network where the organization might be exposed and eventually exploited.
- Likelihood, or if it can be quantified—probability, is the possibility that an event will occur (i.e. a threat being realized). Likelihood is typically a qualitative and subjective judgment in risk management whereas probability is a calculation of what might happen based on historical evidence and statistical modeling. Neither method is a perfect science but the former, likelihood, is extremely suspect and of

little value in the world of supply chain risk management. It requires "organized guessing" and is typically heavily influenced by what one can imagine or is willing to manage. The "failure to imagine" was highlighted in the 9/11 commission report as a serious flaw in the United States' ability to anticipate and manage risk. Nassim Nicholas Taleb also highlighted this shortcoming in his best selling book, "The Black Swan".[2] An exception; it is possible to calculate with some certainty the probability of a hazard event (hurricanes/cyclones, earthquakes). Far from perfect, it should be noted that this is an accepted industry practice and the lifeline of the insurance business where there is significant loss of data and armies of actuaries and modelers. Now back to our terminology.
- Impact is the direct and/or indirect effect of an adverse event occurring. It can be measured in loss of revenue, margin, cash flow, asset. It also can be measured in loss of brand (i.e. customer confidence, market position), strategic value, inability to comply, and/or loss of life.
- The (risk) *investment* represents the allocation of time, management attention/focus, people, and capital the organization is willing to allocate to the final two terms: *risk-mitigation* and *risk-finance* programs.

The allocation process in supply chain risk management is critical, and the details are often overlooked. This economic process is critical to ensure the efficient and effective deployment of risk investment (time, management attention, capital, and resource) against value. Straightforward on the surface but rarely implemented in the world of catastrophic risk management and SPOFs.

12.3 Beginning with Change: A Successful Practice Case Study

In one situation of collective management taking responsibility, a cost optimization initiative quickly led to a cost-plus risk optimization solution for the organizations most critical product families. Management was initially concerned about the risks in its central distribution center. The company had acquired a half-dozen subsidiary organizations, each with its own small distribution center. The project had started out with the goal of consolidating and optimizing by moving inventory into the lowest cost-per-unit facility—at first glance, a sensible idea. However, executive management recognized that this created a different, potentially more serious aggregation risk, with ramifications to its customers in the North American market in the event of a catastrophic event. They decided that they needed a better understanding of the risk versus benefit of its optimization strategy. They decided to quantify the impact of a catastrophic event to the fill rates of its product and they did this by first rationalizing and prioritizing the 20,000 product SKUs (stock keeping units) into a dozen product families. Next, they rated and ranked the product families based on value (revenue, cash flow, strategic importance, brand visibility). Finally, they ran the analytics for

[2] "The Black Swan", Nassim Nicholas Taleb, Random House Trade Paperbacks; 2nd edition (May 11, 2010).

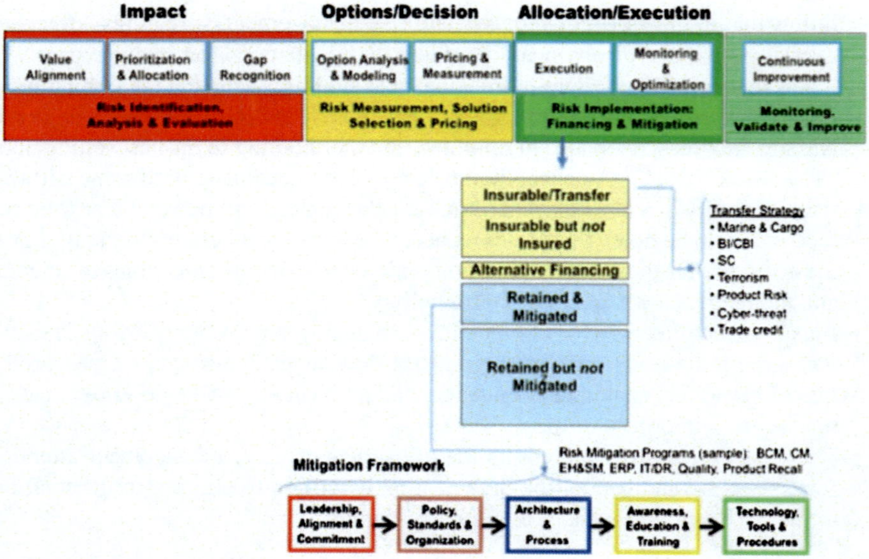

Fig. 12.1 Supply chain risk management *system*

each product family, and then the aggregate, of a one-versus-two distribution center model to understand the net effect of the investments versus risks mitigated (i.e. trade recovery time). In the end the analysis demonstrated that the benefit achieved through inventory diversification far exceeded the carrying costs of the second distribution center and additional inventory. Also revealed was the optimal resiliency placement of inventory for each of the product families. The simple solution—diversifying the risk—turned out to be not only the most cost-effective, but also the safest way to proceed (Fig. 12.1).

12.4 The Process

Value alignment (Fig. 12.2) is the first step in identifying, prioritizing, and determining the impact of supply chain risk. As previously mentioned, supply chain risk management is an economic exercise. Risk resources (time, management attention, capital, and labor) are extremely limited and the need to allocate these resources efficiently against an infinite universe of vulnerabilities is essential. A value-based approach establishes a target in which the allocation can be executed against. Value can be by product/service family/line, a specific product (e.g., blockbuster drug), a group of SKUs, a market, geography, and/or customer. The point here is that for the risk investment to be justified, we must know the *value at risk* and the *impact of failure*. Long gone are the days of trying to sell the risk program based on just qualitative judgments. In the prior case study, the organization segmented value by a

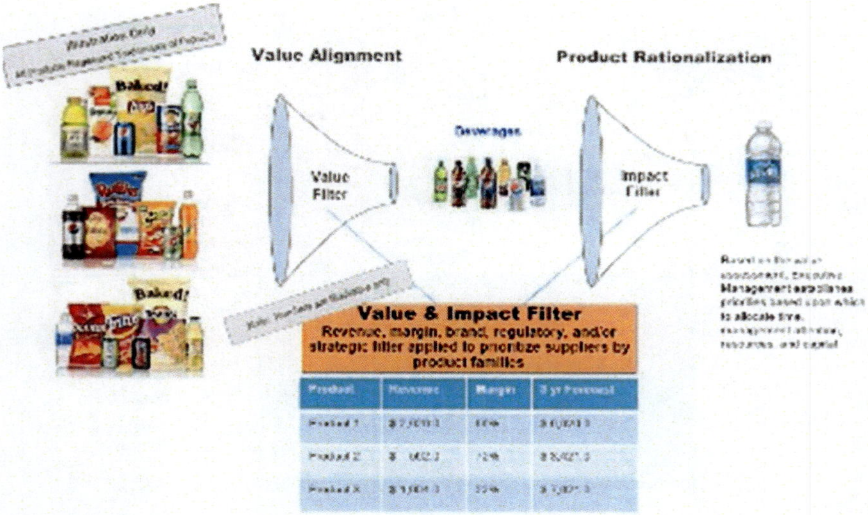

Fig. 12.2 Value alignment (products above are all ® registered trademarks of PepsiCo)

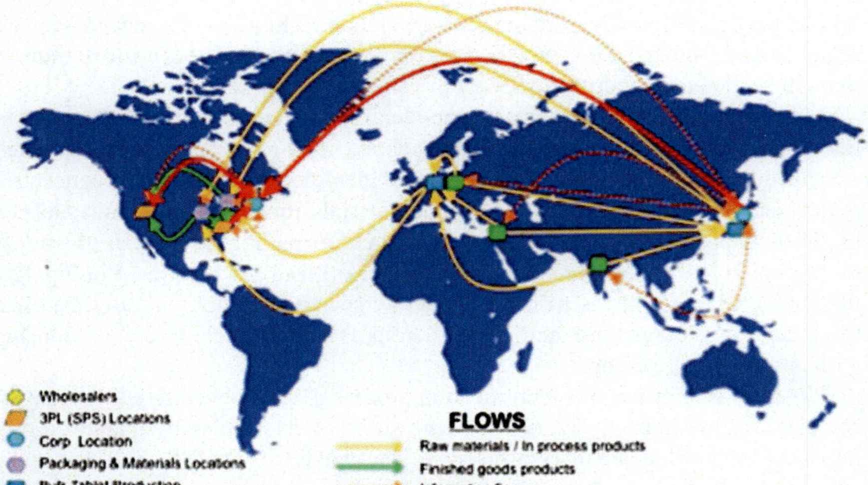

Fig. 12.3 Product, material, cash, and information flows

dozen key product families. Revenue, cash flow and margin data were provided and then used to prioritize the families. This analysis provided a target in which impact could be determined and investment could be allocated.

The next step is to map the flow of materials and products, cash and information (Fig. 12.3). Once the flows are understood, a more detailed resource mapping

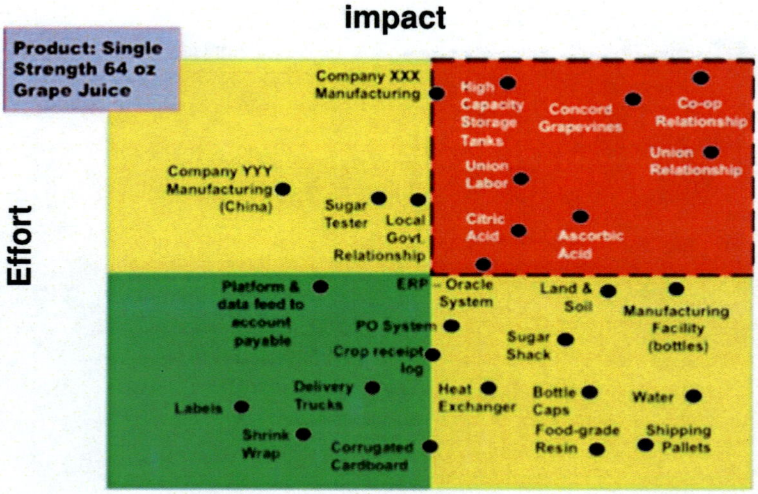

Fig. 12.4 Resource prioritization (Source: "Single point of failure", Gary Lynch, Wiley, 2009)

will be needed to identify critical processes, labor, technology, physical assets, and relationships. The step is essential in order to create the master list of critical resources that are needed to support the delivery of value to the market.

Now that critical resources have been identified, the next step is to calculate the maximum failure for each resource in the context of *value at risk*. For example, let's assume that a consumer product company has identified 64-ounce grape concentrate as its most critical product line. The flow of materials, product, information, and cash has been mapped and critical resources identified. As part of the analysis, the impact of maximum loss is calculated and resources prioritized. As illustrated in Fig. 12.4, the most critical resources from an impact perspective are not just physical assets (high capacity storage tanks and Concord grapes) but also include key relationships with unions and the co-op.

The third step of the risk identification process is to establish risk tolerance and calculate the *net* impact. The risk tolerance level is determined by quantifying the impact to value of the loss of a key resource. The goal is to identify the inflection point where the failure of a key resource becomes unmanageable and material. Figure 12.5 below illustrates this point. The *net* or *risk adjusted value* takes into account already implemented risk mitigation and financing practices.

12.5 Iceberg, Right Ahead!

Let's take a closer look at whats driving increased supply chain risk. In dealing with the reality of our interconnected world, business and operations managers must

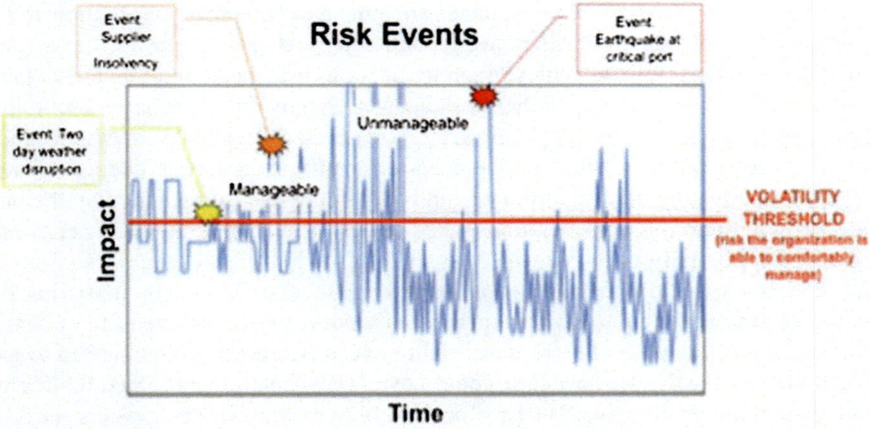

Fig. 12.5 Manageable and unmanageable supply chain risk

contend with the drama created from the latest supply chain risk event. Let's face it; you don't hear many stories on the six o'clock news about how lean, efficient, and diligent global supply chains are you tend to hear public alerts about tainted cough medicine or fickle gas pedals. What problems could these external events create? To name a few: transportation delays, manufacturing malfunctions, or labor strikes due to unfit working conditions. This is a small chunk of the risks globally stratified supply chains encounter. Stakeholders will continue to become more aware of the risks and exposures that could personally impact their investment. Here are a few examples:

- Increasing client concern about the effectiveness of their critical business partners' risk practices.
- A "raising of the bar" or minimum standard for risk management practices including the deployment and monitoring of governance, ethics, and compliance standards and the implementation and monitoring of CSR (corporate social responsibility) practices.
- Increasing concern and influence by industry watchdogs, central governments, and NGOs (non-government organizations) as well as heightened regulatory and compliance pressure. An example is the Congo Amendment (adopted from the Conflict Minerals Trade Act [H.R. 4128)] that was included in the 2010 US financial reform bill. The amendment requires companies to trace the origin of the minerals and file an annual disclosure with Securities and Exchange Commission detailing whether the materials (tin, tungsten, tantalum, gold) originated in the Congo or nine adjacent countries.
- Geopolitical pressures including those used to protect an economy via taxes, tariffs and other trade-adverse policies.
- Emerging risks created by the deployment of new technology (e.g., nano-technology).

These and other stakeholder demands are motivating organizations to proactively manage risk to their business networks and supply chains. However, the actions that organizations took to achieve such tremendous efficiencies and financial gains often conflict with the supply chain risk agenda. Supply chain managers (sourcing, production, logistics, demand planning, etc.) are now faced with two issues: educating business leaders and partners on the need to identify, measure, and better manage risk when pursuing new opportunities and addressing exposure created by the past decision to forgo risk considerations when achieving such enormous efficiency and revenue opportunity gains.

The risks we tend to manage in our supply chains today are often limited to just the ones we can see and control. Like it or not, risk tends to be event-driven and a reactive topic. That is today's reality, as many senior executives have communicated to me, "we do not make that we have little control over". We can mitigate vulnerabilities but we cannot mitigate threats. We have no control over whether most events (threats) will or will not happen but we can minimize our exposure (vulnerability). Like an iceberg, what is visible is only a small portion of what can cause harm. The risks that lie beneath the surface, and often turn up in our assessments, are the ones that our supply chain risk management "system" needs to address. Let's face it, the world has changed (and will continue to change—guaranteed) and so has the fundamental operating model of many organizations. For most, more than 70% of their business depends on others to deliver value to the market. However, the scope of their risk activities typically begins and ends at their organizational perimeter. The good news is that this practice is changing rapidly. Organizations are now beginning to consider all public and private sector stakeholders in their massive supply and distribution networks. As you might imagine, this is a daunting task when you are a high-tech or chemical company and the upstream supplier base is in excess of 45,000.

12.5.1 Conflicting or Converging Agendas?

At the root of the conflict are the business goals and risk return that managers are facing when providing value in global markets. Increased competitive pressures such as escalating customer services demands, rising complexity, and accelerating costs are a few examples of the risk/reward conflicts. From the operations perspective, the *value* imperative by which business and supply chain managers are measured consists of four key objectives that often conflict with the risk agenda:

1. *Increase the velocity of cash*: a financial goal of any business is to constantly ensure that the cash moves into the supply chain from customers, faster than it is paid out to suppliers (or as close to that as possible). This means the corporation wants to maximize both inflow and outflow of cash at the highest possible volume, to create a robust movement of supplies, materials, final products, and ultimately, profits. Dell Computer Corporation proved the benefit of this in the early 1990s with its configure-to- order strategy and lean manufacturing techniques. It is likely that an

incentive to streamline internal controls in the interest of more rapid movement of goods will also invite problems.

2. *Constantly improved profit margins*: performance at all levels is ultimately judged by how well margins are squeezed. This has led to reducing (leaning) inventory, outsourcing manufacturing (to reduce labor, capital, taxes), and procuring materials from lowest-cost sources.
3. *Compliance*: management at all levels of the supply chain is expected to ensure compliance with all regulatory, statutory, and contractual rules. This is notably complex and cumbersome for all international movement of goods based on border-related issues, maritime laws, and safety regulations (which merely scratch the surface of compliance). The pressures of compliance may also have an unintended consequence. For example, in the San Lu case, contractual requirements for specific protein levels indirectly led to the use of dangerous and toxic additives, leading to the poisoning of over a quarter million people.
4. *Supply availability to anticipate and meet demand*: supply chain managers know all too well that stock-outs are either a lost opportunity or a clear sign of supply chain failure. Products have to be available when customers require delivery. Ironically, this requirement might create supply chain risks and subsequent losses. Companies over-stocking to avoid stock-outs pay higher insurance premiums, have greater losses from obsolescence and theft, smaller margins, and may create unintended inefficiencies such as poor use of limited warehouse space.

Accompanying the importance of operational efficiency and maintenance of margins is the overriding need for corporate governance and responsibility. Management is charged with the job of growing the organization and its profits, however this should not occur at the cost of lower safety standards, product quality, or relationships between suppliers and management. A natural rift exists between the goal of improved margins and the basic corporate responsibility to its stakeholders (communities, customers, investors, business partners, regulators, and governments). Most supply chain risks are created by this conflict; therefore the answer to how supply chain risk is reduced has to reside in the methods employed by management to reconcile these conflicting interests.

12.6 Justifying the Investment

The next piece of the risk puzzle is risk measurement and investment. Here, risk savvy investors determine where their scare resources will have the greatest impact and what the optimum level of financial commitment will be necessary. The process begins with option analysis and modeling.

In the prior phase we focused on identifying, quantifying, and then prioritizing risk impact and exposure. Most organizations will be tempted to jump right into creating and deploying risk mitigation (e.g., business continuity, crisis management, supplier viability) and/or financing (e.g., insurance, hedging) solutions. Although it may be a generally accepted practice, this approach will eventually yield less than optimal results in the world of supply chain management, where performance and every investment decision is measured and then measured again. Modeling and analytics is what drove businesses to efficient supply chain network design. A qualitative judgment or worse, the use of fear, uncertainty, and doubt simply will not justify investments in CFO or supply chain management circles.

It is for this reason that detailed option analysis and modeling are important. The process begins by identifying risk mitigation choices, or as it is sometimes referred to, *levers*. Here are examples of some of those levers, if the analysis involves inventory resiliency:

- Direct ship from plants to customers/maintain inventories at plants
- Source product from other, global distribution centers
- Increase inventory investment within the organization's channel
- Substitute product
- Develop segmented inventory allocation/prioritization scheme by key customers, products, markets, geographies, etc.
- Salvage/repackage product
- Develop decision model for implementing strategies/levers

The detailed analysis is used to determine what effect the investment choices have on reducing impact. Once the optimal mix has been identified then the mitigation and financing programs can be funded and executed. Remember, you must include the cost of maintaining the program that you secured the funding against. Be very careful here, the run rate for the maintenance and monitoring of your risk mitigation program can be 10–15 times higher than the deployment cost. The investment will track back to an overall allocation against the value, fleshed out in the first step of our process. The case below is a succinct example of creative methods to optimize value while minimizing risk.

12.7 Aligning Risk Investments Against Supplier Risk: A Successful Practice Case Study

A global high-tech organization was considering how to manage the risk to 4,500 suppliers in light of having already identified, measured, and invested in risk mitigation programs—the use of on-site audits, surveys, policies, and legal contracts that were not enough. The company believed that although these risk mitigation techniques were important, they measured too much of what happened rather than indicating what might happen. They decided to augment their overall supply chain risk management strategy by applying analytics (e.g., rating revenue exposure, event probability analysis, scenario planning) and real-time monitoring technology to manage the overall risk of their key business partners. However, with more than 200 product families and 8,500 products, the deployment of a deeper level of anticipatory tools and real-time monitoring systems would be expensive. They decided to conduct an analysis of their top revenue and value drivers and concluded that 25 product families covering about 100 products made up more than 50% of their total revenue. Accordingly, they implemented a tier-based monitoring system for this small but critical handful of product families. The company mapped its supply chain from the sourcing of raw materials and commodities, through contract manufacturing and onto distribution. Their risk mitigation strategies identified solutions at the manufacturing level (e.g., capacity, pre-qualified alternate site usage, starting and finished goods inventory buffers), inventory level (i.e. semi, finished goods, raw materials, subs), and for test equipment (e.g., looked at lead times/standardization/asset visibility, component buffers/second source, alternate site qualification). Once these key components had been identified, supply chain management identified external threat monitoring scenario to focus on failures to key components, plants, ports, and foundries. The company identified its speed in response/maturing a crisis management program, activation of business continuity practices, and resiliency strategies such as product substitutes, deploying strategic inventory, supply chain diversification, and preventive measures on many levels. The threat scenarios analyzed both isolated and regional types of extended disruptions.

12.8 Executing, Validating, and Monitoring the Program

Congratulations! You've been able to successfully measure your exposure and secure investment for risk mitigation and financing programs. I will not have enough room in this chapter to go into all of the details of comprehensive risk mitigation and financing programs so let me highlight two key points: program management and program content. Like any good program, there are basic elements to executing, monitoring, and improving the supply chain risk management program. Figure 12.6 illustrates these elements.

A comprehensive supply chain risk management program consists of five critical pieces: product risk, supply chain resiliency, sustainability, crisis management, and business intelligence and analytics. Each of these program elements consists of a set of comprehensive sub-programs. At a macro-level, Fig. 12.7 illustrates these programs. More specifically, a comprehensive supplier risk program may require dozens of "sub-programs" (Fig. 12.8).

Besides risk mitigation, risk financing is an important part of the portfolio of solutions for managing risk. The two most frequently used risk-financing strategies are hedging and insurance.

A supply chain program to hedge risk will typically consist of agreeing upon price today of critical commodities and/or supplies, in anticipation that the price will go up in the future (or the commodity will become scarce). A hedging strategy would have limited the downside risk in the following example:

Fig. 12.6 SCRM—program management elements

Execution
(Integrated) Program Management
Decision support execution
Intelligence management
Monitoring & Optimization
Transfer optimization analysis & modeling
Mitigation optimization analysis & modeling
Risk portfolio management analysis & modeling
Continuous Improvement
Validation
Root Cause Analysis
Program Improvement
Integration Planning

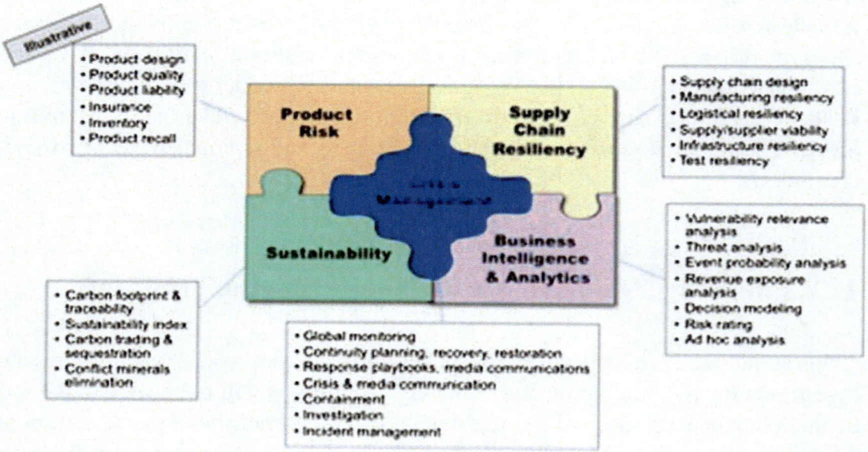

Fig. 12.7 SCRM—program content

> Pulp prices rapidly increased by over 50% after the two leading producers suffered significant disruptions; an earthquake in Chile and a dock strike in Finland. The packaging market which relies on pulp is more than a $400 billion dollar global industry and accounts for nearly 15% of total retail food and beverage prices. So what did global purchasing managers do? They rapidly shifted their orders to Canada, Germany, and other sources but had to pay a premium for the paperboard (pulp). The risk-financing strategy would hedge against a long-term outage in the case of the Chilean sawmills and pulp factories being out for an extended duration.

In addition to hedging, the organization has the opportunity to transfer "measurable" risk to third parties. More appropriately referred to as *insurance*, risk transfer protects against extreme financial loss caused by named perils such as rising water or wind damage. This option can be costly, depending on the value of goods, exposure to known perils (threats), amount and type of coverage (e.g., property, marine and cargo, product liability), deductibles, and a number of other considerations. In Chap. 10 of my book, "Single Point of Failure: The Ten Essential Laws of Supply Chain Risk

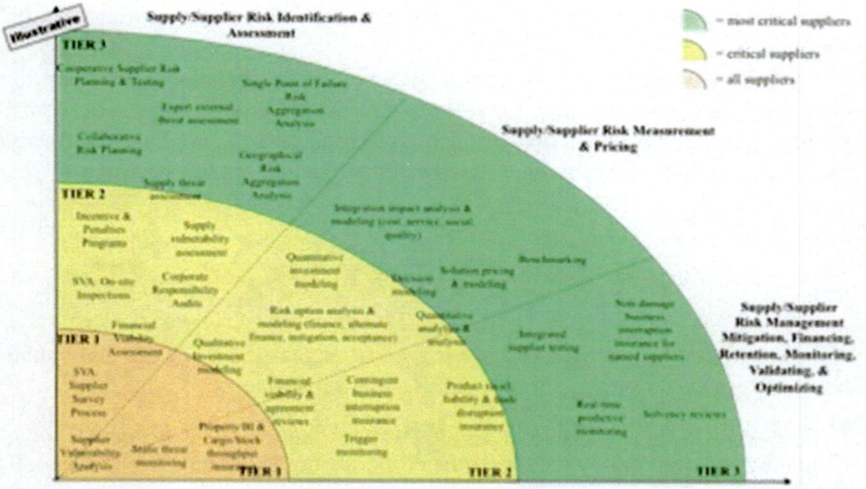

Fig. 12.8 Supplier resiliency programs

Risk	GSS/SCI	Trade Disruption Insurance	Cyber Risk	Product Recall	CBI/BI
Loss of Supply	✓				✓
Inability to Provide or Obtain Service	✓				✓
Political Upheaval	✓	✓			
Pandemic (Zurich at supplier site/infectious disease)	✓				
Terrorism Disrupting Transportation (GSS)	✓	✓			
Natural Disaster	✓	✓			✓
Labor Stoppages/Strikes	✓	✓			
Improper Placement or Cargo Holding		✓			
Theft	✓				✓
CAT events	✓	✓			✓
Unanticipated Supply/Supplier Constraints	✓				
Pollution & Contamination	✓				
Delivery Delays	✓	✓			
Cyber Attack/Theft, IT Loss, Virus/Hacker (Zurich)	✓		✓		
Product Recall & Resolution				✓	

Fig. 12.9 Supply chain risk transfer (insurance) options

Management", Wiley, 2009, I go into great detail about the type and limitations of risk-financing options. It is a complex topic that requires a thorough understanding of value and exposures. An insurance broker usually handles the placement for this reason (as well as gaining access to the best pricing in a larger market). Figure 12.9 illustrates the types of supply chain risk coverage that are available in the market, as of the time of this writing. The column labeled GSS/SCI represents emerging risk transfer products that provide non-property-related damage insurance for failed suppliers.

The message here is that in order to survive, the organization must adapt to the changing risk paradigm and to the rules in effect in the risk universe. This demands supply chain flexibility, alternatives, diversified sourcing of suppliers and manufacturers, and management strategies designed not to react, but to anticipate.

12.9 Lessons Learned: Avoiding the Icebergs

A pattern of *common obstacles* has emerged from years of fieldwork and countless hours of research. Here are three of those common obstacles:

1. *Management may suffer from a blurred or obstructed view*: it often is the case that a manager, even at the top of the corporate organization chart, does not have visibility upstream or downstream (especially beyond the first tier). This is but one variation of the assumption that SPOF also means single *cause* of failure. This is a typical challenge for pharmaceutical, food and beverage, or component manufacturers that deal primarily with brokers, contractors, forwarders, and wholesalers on the upstream/supply side and distributors on the downstream/distribution side of the chain. If you look at the typical organization, the reasons are obvious. The *assumed* trust between an organization and its third parties is usually characterized in convoluted contract language, and does not include more than the first party relationship. This means that, even if you are responsible for managing risk on a broad scope, your frame of reference is limited in a serious manner.
2. *Management may lack ownership of supply chain and supply chain risk, and may be operating on unsubstantiated assumptions*: this is an enormous problem. Managers may assume that their responsibility is limited to their function (functionally designed organizations drive functional/silo behaviors). For example, as a procurement manager my job is to manage suppliers, as a plant manager my job is to oversee production or to maintain the appropriate inventory, and as a warehouse manager I am responsible for throughput levels in the warehouse. The Human Resource manager handles labor risks (including the recent pandemic scares), the Environmental, Health and Safety Manager a mixture of environmental and compliance risks, the Security Manager for facility, people and sometimes product risk and the list continues. This is a recipe for massive fragmented risk initiatives, conflicting agendas, and significant inefficiencies. All employees are responsible for risk within their own domain but what about concern over unaddressed vulnerabilities arising as a result of interdependent processes, relationships, skills, technology, or assets somewhere along the supply chain? Perhaps keeping up appearances is going to fail if and when a risk becomes reality, so it makes sense for management to take full responsibility for the full risk universe. All of the managers along the supply chain function as a series of dominoes within the SCRM culture, and if one refuses to acknowledge the interaction between all,

then the entire supply chain is at risk. Unsubstantiated assumptions include the common belief that "it is someone else's job" to prevent losses or to address supply chain risks. For example, if your organization relies on a large overseas manufacturing and supply provider, whose job is it to ensure that products meet safety standards? If you assume it is the job of the supplier, you might be wrong. Remember, if one of your customers is injured because *your* product is defective, the blame is going to be pointed right at your organization. No one cares if the organization relied on a supplier in China or a manufacturer in India. In the case of Mattel's problem with lead paint used on imported toys, it was widely viewed as a Mattel problem and not as a problem caused by anyone else. Mattel probably assumed that the manufacturer was exerting quality control standards, but did not take steps to verify this. In fact it was both a design and production flaw. They, as well as the manufacturers, relied instead on unsubstantiated (and as it turns out, unjustified) assumptions.

3. *The common default policy is to rely on passive monitoring and management*: the de facto policy for too many organizations is to wait for a risk to materialize and then decide how to react or to deploy passive validation techniques (e.g., supplier audits once a year). This grows from the incorrect assumption that singular flaws or incidents are to be blamed for failures. For small, unexpected risks, this may be the most appropriate and cost-effective method even when not entirely realistic. However, passive monitoring contains no preemptive measures, and prevention is the best way to reduce losses and protect the company and its stakeholders from the unexpected and unintended consequences of big losses. Passive monitoring is typically deployed during the second phase of a supplier viability program. These programs typically begin by sending out surveys and self-assessments. They tend to mature with predictive intelligence and real- time monitoring.

12.10 Closing

To summarize, to deploy a supply chain risk management *system* the organization needs to: identify, measure, mitigate and finance, validate, and monitor risks.

Of course, this so-called *system* must be aligned with the business priorities and value, tied into the enterprise risk management framework, and then continuously improved from critical learning. Effective risk management also requires direct ownership and accountability not only for the resources of labor, technology and processing, physical assets, and relationships, but also for the processes, and flows of cash, information, product, and materials.

The solution to this range of challenges is to be able to flesh out priorities and build the business case to justify the needed investment. Risk management does not come cheap, and because potential losses are often intangible (some would even say unlikely to materialize), a lot of resistance to investment is going to be met in the boardroom.

By being aware of the differences between the SPOF and the more complex series of root causes of failure, management is able to develop a truly effective and meaningful version of supply chain risk management. Lacking this observed reality, it would never be possible to change either, the organizational culture or the big picture view of management, to overcome the tendency for loss to prevail. The realistic approach is to abandon inaccurate assumptions, most notably of the idea that SPOF means not only location, but root cause as well; and to revisit the entire realm of risk-related assumptions. This not only improves the organizational experience level of loss from these failures; it further improves the efficiency of the supply chain itself.

Good supply chain risk management practices are grounded in detail and acknowledge that no organization can exist in isolation.